H.M. King George V.
Colonel-in-Chief of the Royal Regiment of Artillery.

THE HISTORY
OF
THE ROYAL ARTILLERY

FROM THE INDIAN MUTINY TO THE GREAT WAR

DEDICATED BY
HIS MAJESTY'S GRACIOUS PERMISSION TO
KING GEORGE V.
COLONEL-IN-CHIEF OF THE ROYAL ARTILLERY

VOLUME III—CAMPAIGNS
(1860–1914)

MAJOR–GENERAL SIR JOHN HEADLAM, K.B.E., C.B., D.S.O.
LATE COLONEL-COMMANDANT, R.A.

The Naval & Military Press Ltd

published in association with

FIREPOWER
The Royal Artillery Museum
Woolwich

Published by
The Naval & Military Press Ltd
Unit 10 Ridgewood Industrial Park,
Uckfield, East Sussex,
TN22 5QE England
Tel: +44 (0) 1825 749494
Fax: +44 (0) 1825 765701
www.naval-military-press.com

in association with

FIREPOWER
The Royal Artillery Museum, Woolwich
www.firepower.org.uk

*In reprinting in facsimile from the original, any imperfections are inevitably reproduced
and the quality may fall short of modern type and cartographic standards.*

PREFACE.

In the two preceding volumes of this History an account was given of the principal events in the life of the Regiment during the period between the Indian Mutiny and the Great War, and an attempt was made to recapture something of the spirit as well as to recall the events of those bygone years. But nothing was said of active service, although between 1860 and 1914 the British Army was engaged in many wars in which the Royal Artillery bore an honourable part. The reason for this notable omission was that none of these campaigns—with the possible exception of that in South Africa—affected its development to a degree at all comparable with the influence exerted by the European wars of the period, or by the march of science. To have interspersed accounts of the British campaigns throughout the narrative would therefore have interrupted the story of tactical and technical progress without throwing light on its causes. It was accordingly decided, when the work was planned, that the campaigns should be treated separately, and this volume is the result.

The War Services of the Royal Regiment of Artillery require different treatment from those of Cavalry and Infantry Regiments. "UBIQUE" and "QUO FAS ET GLORIA DUCUNT" have a wider significance than the longest roll of Battle Honours. A record of the war services of every battery, or of the exploits of every regimental hero, would involve volumes of portentous length, overburdened with names of little general interest now. A sense of proportion must be observed. The place for such details is in Battery rather than Regimental History.

This volume is primarily concerned with those campaigns that have marked some real epoch in the life of the Regiment, or exerted some distinct influence on its development.

During the period covered by it artillerymen stood to their guns against the rushes of fanatics armed with spears and dahs, swords and scimitars : they served them under the fire of slings and arrows, match-locks and jezails, Tower-muskets and Dane-guns, as well as under that of modern rifles. But it was only rarely that they had to face bursting shell. The wars to be studied with care are obviously those in which the guns were not all on our side, and in considering these the action of the artillery will be described in detail, with such comments as appear called for in order to elucidate its lessons.

This is not to imply, however, that nothing is to be learnt from campaigns where this condition was absent. Expeditions against the most primitive of people, contemptible as regards armament and military science, have often called forth the highest qualities in artillery officers. Over snow passes, across waterless deserts, through equatorial swamps, they have brought their guns with the aid of coolie or camel, horse or mule, ox or elephant. And they have fought them with gunners drawn from every race within the Empire, trained and inspired by their example. In many a hazardous enterprise the reputation of the Regiment has been safe in the keeping of a subaltern with nothing to depend upon but his own determination and powers of improvisation. Some campaigns which exemplify such qualities and methods of warfare have therefore been chosen for detailed examination, and opportunity has been taken to note in passing how often such minor operations offered the first chance of distinction to those who subsequently rose to fame.

The object in view throughout has been to concentrate attention on the part played by the artillery. Any discussion of the causes—political or military—which led to the outbreak of hostilities, or which dictated the terms of peace, would be alien to the subject, and has been omitted. Similarly the description of the country and of the military operations has been confined to the provision of such information as is required for an assessment of the action of the artillery. The naming of persons and places has been reduced to a minimum, and it is hoped that most places mentioned will be found on the maps. Care has been taken to avoid the use of "native" words as much as possible, but some of these, especially those descriptive of local features, cannot be avoided without inconvenience. They are, for the most part, familiar to all, but a glossary will be found on page xii.

In conclusion I must express my gratitude for all the ungrudging assistance which I have received. Without the help of those who placed documents at my disposal, or were good enough to read my accounts of events in which they took part, it would have been impossible for me to reproduce the outlook of the moment. They are too numerous to admit of individual mention here, as are the authors of the many books and articles which I have consulted. I hope that they will accept this acknowledgment, and pardon any plagiarisms of which I may have been guilty.

JOHN HEADLAM.

CONTENTS.

INTRODUCTION.

PART I—INDIA.

PART II—CHINA.

PART III—NEW ZEALAND & CANADA.

PART IV—NORTH & EAST AFRICA.

PART V—WEST & SOUTH AFRICA.

PART VII—THE SOUTH AFRICAN WAR.

APPENDICES.

PORTRAITS.

INTRODUCTION.

DURING the long period covered by this volume there were naturally many changes in the organization, armament, and training of the artillery. These have been fully dealt with in the previous volumes of this history—Vol. I, 1860—1899, Vol. II, 1899—1914. But, to avoid the necessity for frequent references, it has been thought advisable to recapitulate here the salient features of these changes.

Organization.

In the first year with which we are concerned—1860—the "Brigade System" was introduced in the Royal Artillery, and batteries of all branches were designated by a combination of their battery and brigade letter or number. In the horse artillery both were lettered; in the field artillery batteries were lettered and brigades numbered; in the garrison artillery both were numbered. Thus "A/A" denoted a horse battery, "A/1" a field battery, and "1/1" a garrison battery.

In the re-organization of 1882 a still further complication was introduced by the addition of territorial divisions in the case of the garrison artillery. Thus we get such cumbrous titles for batteries as "1/1 South Irish".[1] But in 1889 the brigade system was abolished in the horse and field artillery, and the batteries lettered or numbered consecutively. The garrison artillery followed suit in 1901, and with the separation of the mounted and dismounted branches the designations became simple—A/R.H.A., 1/R.F.A., and 1/R.G.A.—to remain so until the post-War period.

---

[1] In 1889 these divisions were reduced to three—Eastern, Southern and Western.

The brigades referred to above were purely administrative groups, entirely unconcerned with tactical command. It was not until the South African War had forced the tactical importance of the lieut.-colonel's command that it was made a permanent formation, under the title of "brigade-division"—to give place in 1903 to "brigade."

Batteries were organized as at present in sections—then called "divisions"—of two guns under a subaltern. Horse, Field, and Mountain had six guns, Heavy and Siege four. Batteries were commanded by "1st captains" until the restoration of the rank of Major in 1871.

It must be remembered, however, that, except in the Horse Artillery, these batteries were really companies of Foot Artillery, which "went into battery" (as the term was) for a strictly limited period. It was not until the 70's that field batteries established their position as mounted units. Mountain batteries became a separate branch in 1889, Heavy and Siege not until the next century. Meanwhile there was nothing in their title to show whether, for the nonce, they were serving as field, mountain, heavy, or siege, and this has therefore usually been indicated in the following pages.

Armament.

In 1859 the Royal Artillery made the momentous decision to adopt rifled guns, and these began to reach batteries in the first year of the period covered by this volume, although smooth-bores lingered for years in the service. And since the rifled guns were to undergo many a metamorphosis, we shall come across a very large number of different equipments in the campaigns about to be chronicled. To assist readers these may be broadly classified as follows :—

Previous to 1860 —S.B. = Smooth-Bores.

    In the   60's—R.B.L. = Rifled Breech-Loaders— the "Armstrongs".

   ,, ,,   70's—R.M.L. = Rifled Muzzle-Loaders.

In the 80's & 90's—B.L. = Breech-Loaders.

,, ,, 1900's —Q.F. = Quick-Firers.

It may also be noted that in the smooth-bore days all batteries contained both guns and howitzers. With the adoption of rifled guns, however, howitzers disappeared from the field artillery not to re-appear until nearly the end of XIXth Century, and then in separate batteries and brigades.

The table on the next page, showing the principal natures in each catagory, may be found useful for reference.

Conclusion.

Finally a word must be said regarding the inclusion in this account of units not forming part of the Royal Regiment of Artillery. In the first place the Indian mountain batteries did not do so, strictly speaking, during the period covered by this History. They were, however, regular batteries of the Indian Army, and an account of the campaigns in which the Royal Artillery has been engaged without mention of the Indian batteries would be absurd. In another category come the Artilleries of the Dominions; in still another the Local Corps which were frequently raised in Colonies or Protectorates, and exhibited every variety of constitution, although they were in most cases officered from the Regiment. All these have also been included, since they served under the British Flag. But in this latter respect the Artillery of the Egyptian Army was on a different footing. Here again, however, a large proportion of the officers were drawn from the Royal Artillery, and when batteries of both served together they were under the same command. No objection can, it is thought, be taken to the inclusion of Egyptian batteries when they fought side by side with those of the Royal Artillery.

DECADE	CATEGORY	HORSE	FIELD	MOUNTAIN	HEAVY
The 50's	S.B.	6-pr. gun 12-pr. how.	9-pr. gun 24-pr. how.	3-pr. gun 12-pr. how.	18-pr. gun 8" mortar
The 60's	R.B.L.	9-pr. gun	12-pr. gun	6-pr. gun	40-pr. gun 8" mortar
The 70's	R.M.L.	9-pr. gun	9 & 16-pr. guns	7-pr. gun	40-pr. gun 6·3" how.
The 80's	B.L.	13-pr. R.M.L.¹ 12-pr. gun (7-cwt.)	13-pr. R.M.L.¹ 12-pr. gun (7-cwt.)	2·5" (R.M.L.)	do.
The 90's	do.	12-pr. gun (6-cwt.)	15-pr. gun 5" how.	do.	do.
The 1900's	Q.F.	13-pr. gun	18-pr. gun 4·5" how.	10-pr. (B.L.) 2·75" (B.L.)	30-pr. gun (B.L.) 5·4" how. (B.L.) 60-pr. gun (B.L.)²

¹ The 13-pr. was an effort to stave off a reversion to breech-loading, but re-armament with it had not been nearly completed before it was superseded by the 12-pr. B.L.

² The 60-pr. was introduced when heavy batteries were formed at home. All the previous entries in this column apply only to batteries in India.

MAPS & SKETCHES.

NOTE.

During the period from 1860 to 1914 there were many changes in the nomenclature employed to denote units of artillery. These changes have been indicated above, but to have followed them throughout the narrative generally would have been confusing to the reader. The term "brigade " has therefore been used for the lieut.-colonel's command generally, and "company" for the major's command in the garrison artillery. These were the terms in use at the beginning and end of the period.

For the same reason the expression "heavy artillery" is used in the sense it bore up till the Great War, that is as meaning "heavy field", or "position" artillery, corresponding to what is now termed "medium" artillery.

Abbreviations have been avoided as a rule, but to save space the words "Royal Artillery" have been omitted when it appeared that this could be done without risk of misunderstanding, as, for instance, when speaking of "the Mess"—"the Institution"—etc.; and the various publications of the latter have been referred to throughout simply as "the Proceedings". Nor has it been thought necessary to repeat "of the Mounted and Dismounted Branches" whenever mentioning the "Separation".

The term 'gun' has been used generally to include all types of ordnance, except where it has been necessary to differentiate.

The extracts from Regulations and Orders are published with the kind permission of the Controller of H.M. Stationery Office.

GLOSSARY.

Berg	—	Mountain.
Burg	—	Town.
Boh	—	Officer or Leader.
Dah	—	Sword.
Donga	—	Water-course.
Dorp	—	Village.
Drift	—	Ford.
Dacoit	—	Robber, Bandit.
Dervish Fakir	—	Member of Mahomedan religious fraternity.
Ghazi	—	Mahomedan Fanatic.
Impi	—	Armed Force.
Inspan	—	Harness up.
Jebel	—	Hill.
Jezail	—	Long muzzle-loading musket.
Jingal	—	Light swivel-gun.
Jong	—	District Headquarters, usually a Fort on a hill.
Kloof	—	Ravine.
Kop, Kopje	—	Hill.
Kotal	—	Pass in hill range.
Kraal	—	Village.
Krantz	—	Cliff.
La	—	Pass in hill range.
Laager	—	Camp, Convoy.
Lashkar	—	Armed Force.
Mealies	—	Maize, Indian Corn.
Nek	—	Pass in hill range.
Nullah	—	Ravine, water-course.
Pa	—	Stockade.
Pan	—	Pool.
Pont	—	Ferry.
Poort	—	Pass in hill range.
Rand	—	Range of hills.
Sangar	—	Stone breastwork.
Schanz	—	do. do.
Spruit	—	Water-course.
Tangi	—	Defile or gorge.
Tsawbwa	—	Chief.
Veld	—	Country, as opposed to town.
Vlei	—	Stream, Pool.
Yak	—	Cattle used for transport.
Zareba	—	Enclosure, usually of thorn bushes.
Zarp	—	Transvaal Police.
Zwart	—	Black.

PART I.

INDIA.

CHAPTER I.

The North-West Frontier prior to the Second Afghan War.

The Border Hills—The Border Tribes—Punitive Expeditions—Baluchistan—Waziristan—The Afridis—The Jowaki Expedition—The
Mohmands—The Hindustani Fanatics—Ambela—The Black Mountain.

Comments.

The "Mountain Trains"—The Horse and Field Batteries—List of Units
—Medal & Clasps.

Map 1.

BEFORE commencing an account of the operations on the
North-West Frontier of India, it is necessary to say something of the nature of the country, the characteristics
of its inhabitants and their relations with the Government.
It can only be a bird's eye view, but those who desire
further guidance will find many books on the subject, the
work of soldiers and civilians who can speak with the
authority of intimate personal knowledge, among whom
can be numbered names well-known in the Regiment.

Running approximately parallel to the broad bed of the
Indus, and at an average distance of some fifty miles
beyond it, the Border Hills form a mountain barrier
between India and Afghanistan from the sea to the
Kabul River—approximately eight hundred miles. Then,
turning eastward, they circle round the Peshawar plain
until they cross the Indus and reach the frontier of
Kashmir.

They form, for the most part, a tangled web of ridges
and ravines, in which the water-courses serve as the only
roads. The few passes which traverse this formidable
barrier have been the avenues of commerce and of conquest
from time immemorial. Commencing from the coast, the
first are those leading from the Sind desert to the uplands

of Baluchistan—the Bolan and Harnai, with Quetta at their head. Next come the Gomal and Tochi through Waziristan to Ghazni in Afghanistan. Further north the Kurram Valley leads from Kohat to Kabul. Finally we reach the Khyber, familiar to all, with Peshawar standing ward at its mouth.

Any attempt to explain the ramifications of the tribes that inhabit this mountain belt would greatly complicate the account of the expeditions, without elucidating their military lessons. For the purpose of this history it is sufficient to say that the whole border-land of which we have been treating is divided between two nations, the Baluch and the Pathan. Both are Mahomedan, and both are accustomed to eke out the meagre resources of their barren hills by looting the caravans traversing their passes and raiding their neighbours in the plains. For centuries they have defied the rulers of India, boasting that the plains, within a night's run of the hills, are their hunting-ground.

How to deal with such neighbours is no simple problem, and it must be confessed that British policy has vacillated sadly. At the close of the Indian Mutiny the position was much the same as that taken over on the annexation of Sind and the Punjab. The Sikhs had left the hill-men severely alone, and the East India Company had followed their example except when its hand was forced by some outrageous foray. Fortunately we are not concerned with the arguments for or against the "open" or "closed" frontier, the "forward" policy or the reverse. It suffices that for the period covered by this chapter the limit of British authority was the foot of the hills—all beyond that was in sober truth a *terra incognita*. To inflict punishment on its inhabitants for crimes committed upon British territory there were but three courses open to us. The simplest was the infliction of a fine, with the payment of compensation for property plundered or lives taken. Next came a blockade, which was only effective under certain conditions. Finally—*ultima ratio*—came the military expedition. Needless to say, this was only resorted to when every other means of coercion had failed, and further forbearance could only be ascribed to weakness. It must, therefore, always be borne in mind that each punitive expedition was a judicial act, analogous to the enforcement of legal penalties. The blowing up of villages and towers in such expeditions was justified by the fact that each was in itself a fortress; the burning of crops as being the only way of calling the tribesmen to account for their depredations in our territory.

In recounting these expeditions it has been thought convenient to follow a geographical rather than a chronilogical sequence. We shall begin, therefore, with the coast, and thence follow the main tribal divisions northward.

For the purposes of this history, Baluchistan may be taken as covering all that tract of country lying between the sea and the Gomal Pass. The Indian authorities had first come into contact with the Khan of Kalat, the over-lord of the Baluch tribes, during the First Afghan War, but with the return of the army from Afghanistan all troops had been withdrawn from Baluchistan. That great Warden of the Marches—and good gunner—John Jacob had pressed upon the Governor-General the importance of holding Quetta, but he died in the year before this narrative commences, and it was not until 1877 that his plan was acted upon. Its value will be shown in the next chapter, where we shall see our army in Southern Afghanistan dependent upon the camel caravans passing unmolested through the dreaded defiles of the Bolan and all the tribal territory beyond. With the establishment of the Sandeman system of tribal responsibility there are no punitive expeditions to chronicle here.

It is impossible to imagine a greater contrast than that presented by the state of affairs in Waziristan, next neighbour on the north. Inhabited by the most turbulent of

CHAPTER.
I.

Punitive
Expeditions.

Baluchistan.

Waziristan.

tribes, only roughly divisible into Mahsuds and Wazirs, the territory between the Gomal Pass and the Kurram Valley was the source of a long series of forays into the plains. Curiously enough the first expedition to be recorded has a special regimental interest in that it was undertaken in punishment for the murder of an artillery officer, Captain R. Mecham, commanding No. 3 Punjab Light Field Battery. The force detailed for the purpose under Brig.-General Chamberlain was a strong one, over five thousand men, with the following artillery :—

No. 1 Punjab Light Field Battery—Captain J. R. Sladen.
No. 2 do.   do.   do.   do. —Captain G. Maister.
The Hazara Mountain Train   —Captain F. R. Butt.
The Peshawar do.   do.   —Captain E. R. de Bude.

21st Dec.   So little was known of the country that on the very day after leaving Thal, the 21st December, the enemy were found in position, holding the narrow gorge giving entrance to the valley they occupied. The slopes on both sides of its mouth were very steep and sangared in terraces, but the ground was just practicable for mules, and in the attack the mountain guns proved their value by accompanying the infantry in their climb along the ridges which flanked the position. With a battery on each the enemy's defences were enfiladed in turn, and the position captured with little loss. General submission followed.

Undeterred by the punishment inflicted upon the
1860.   Wazirs, the Mahsuds gave a characteristic display of their habitual truculence a few months later by coming out into the plain with the avowed intention of sacking the town of Tank. They were driven off, but such an insolent defiance could not be allowed to go unrequited, and General Chamberlain was again called upon to assemble a force for the purpose. It was about the same size as before, but No. 3 Punjab Light Field Battery (Captain T. E. Hughes) took the place of No. 1.

17th April.   On the 17th April, General Chamberlain crossed the border, and, meeting with no opposition, pressed on with most of the infantry and the mountain batteries into the enemy's fastnesses, leaving the field batteries, which found the difficulties of the ground beyond their powers, with the cavalry and remaining troops, fifteen hundred in all, in camp at Palosin Ziarat. A day or two later this camp was surprised by a night attack, in which the outlying picquets were overpowered by some thousands of Mahsuds, and 500 of the bravest dashed in sword in hand. For a time the confusion was great, but the inlying picquets rallied, and the camp was cleared at the point of the bayonet. The batteries, which had come into action as soon as the alarm was given, rendered valuable assistance.

None knew the enemy and the country better than the troops and their commander, and every precaution known at the time had been taken. But "perimeter camps" were not yet, and Palosin Ziarat was long famous on the frontier.

As soon as Chamberlain returned the advance into the enemy's country was continued with the whole force, and was unopposed until the 4th May, when the Mahsuds were found in a similar position to that held by the Wazirs in December. The gorge through which the route lay was closed by a massive barrier, while along the cliffs on each side stretched a double line of sangars. As before, the infantry were directed against the flanking ridges, a mountain battery with each column, and the two field batteries in support in the centre. The leading infantry of the right column had nearly reached the defences when a counter-attack drove them back in panic upon the supporting battalion, which also gave way. The reserve and the mountain battery stood fast, however, and under their fire the Mahsuds fell back, hotly pursued by the reserve. The other troops rallied, the right of the position was carried, and the left attack put in. Enfiladed by the guns of the right attack the defenders offered such a feeble resistance that ridge after ridge was cleared with trifling loss. This was the end of the Mahsud resistance, and General Chamberlain led his force unopposed to Razmak, and thence back to Bannu.

Twelve years were to elapse before it was found necessary to despatch another expedition into Waziristan. On this occasion the force was under the command of Brig.-General Keyes, and included two howitzers of No. 3 Light Field Battery. The first pass was, however, too steep for their teams, and guns and wagons had to be dragged up by the gunners and infantry—to show their value when the first village was approached, and the inhabitants opened fire in spite of the agreement which had been come to with their leaders. The infantry stormed the gate, and the defenders fled, only to find themselves charged by the cavalry. This was too much for the remainder, who rushed for the guns, throwing down their arms. A hitherto independent tribe had been compelled to recognize that even their walled villages did not protect them from chastisement.

The Afridis.

Neglecting, for the time, the Miranzai Valley, leading from Kohat to the Kurram, and occupied by the Orakzais who gave no trouble during this period, we can pass on to the great tribe of the Afridis, occupying the whole country as far north as the Khyber Pass. With the main body our relations remained satisfactory, and their engagements regarding the safety of the pass were observed. But with the Kohat Pass branch of the tribe it was necessary to resort on one occasion to punitive measures.

The Kohat Pass Afridis occupy a peculiar position, for the road from Peshawar to Kohat passes through their territory. With occasional disagreements, an amicable arrangement had existed for many years until, in 1877, the Jowaki section took alarm at a proposal to make the road fit for wheeled traffic. Eventually joint expeditions had to be despatched from Peshawar and Kohat under Brig.-Generals Ross and Keyes. Colonel W. J. Williams was C.R.A., and the batteries with the two columns were :—

The Jowaki
Expedition,
1877.

PESHAWAR COLUMN.

I/C Horse Battery—Major G. R. Manderson.

13/9 Heavy do. —Major C. W. Wilson.

No. 4 Indian M.B.—Captain E. J. de Lautour.

KOHAT COLUMN.

No. 1 Indian M.B.—Captain J. A. Kelso.

No. 2 do. do. —Captain G. Swinley.

The horse battery had its guns carried by elephants, those of the heavy battery were drawn by them. The only operation of interest from an artillery point of view was the capture of the Sangasha Pass by the Peshawar column on the 4th December. The enemy's position was on a high and rugged ridge, its crest bristling with sangars, against which the 40-prs. of 13/9 opened fire with shrapnel at 3,000 yards. Under this support the infantry cleared the foreground, and the horse artillery then pushed forward and opened fire on the main defences with common at from 1,500 to 2,000 yards. The position was carried, and the horse artillery, "packing up" their guns, crowned the ridge, and drove off such parties of the enemy as were still disputing the advance.

4th Dec.

The use of elephants, both for pack and draught, is discussed in the "comments" at the end of this chapter, but mention may be made here of an interesting example of the early use of signalling communication between the guns and the infantry. Under General Ross's orders signalling stations were established with each battery, and when, unfortunately, a shell from the heavy battery burst among the infantry, the fire was at once corrected by heliograph. Next morning, before continuing the advance, what came later on to be known as a "panorama" sketch was made, showing the various villages and other localities of tactical importance. These were all lettered and numbered so that when the infantry were checked the general could at once communicate with the batteries that were supporting the advance.

During the rest of the winter the two forces were broken up into smaller columns, which traversed the country until

1878.

the enemy had had enough. An agreement was then come to, which survived the temptations afforded by the Afghan War and the rising of 1897.

The Mohmands.

North of the Khyber Pass come the Mohmands, a tribe dwelling partly in British territory, partly in the border hills, and partly in Afghanistan. During the first ten years of British rule they gave more trouble than almost any other tribe, but at the end of that decade matters were arranged, and the border was not disturbed until 1863, when, during the Ambela fighting, the Ahkund of Swat succeeded in rousing the fanaticism of the Mohmands.

1864.

The action of the tribesmen took its usual form of a threat to Shabkadr, but in this case a very harmless one, for the garrison had recently been reinforced, and had no difficulty in driving them off. The affair is chiefly notable as having afforded British cavalry their only chance of a charge in frontier fighting. Taken in flank by the hussars, who rode three times through their ranks; enfiladed by three guns of D/5 R.H.A. (Captain J. R. Sladen); and assailed in front by the infantry, the Mohmands fled back to the protection of their hills.

The Hindustani Fanatics.

Beyond the country of the Mohmands the border hills circle round to the east, the principal tribal territories bordering on British India being those of Swat, Buner, and Chamla. With the inhabitants of these the Government had no direct quarrel, but there had taken refuge in the valley of the Indus a curious community known as the Hindustani Fanatics, and it was essential for the peace of the borderland that it should be rid of this nest of outlaws. To clear them out a "Yusafzai Field Force", was assembled under the command of Brig.-General Sir Neville Chamberlain. It consisted of two infantry brigades and the following artillery :—

C.R.A.—1st Captain J. S. Tulloh.
C/19 Field Battery—Captain F. C. Griffin.

No. 3 Punjab Light Field Battery—Captain T. H. Salt.
The Peshawar Mountain Train—Captain T. E. Hughes.
The Hazara   do.   do. —Captain E. R. de Bude.

The plan was to advance through the Chamla valley
so as to come upon the Hindustanis from the west. The
country was unknown, but no opposition was expected,
and it was only when the troops had reached the crest of
the Ambela Pass that their descent into the Chamla valley
was found to be blocked by the Bunerwals. The fact that
this powerful tribe was prepared to dispute the advance
altered the whole situation. Instead of having to deal with
the one or two tribes in whose territory the Hindustanis
had taken refuge, with a view to their expulsion, General
Chamberlain found himself confronted by a general
coalition of all the tribes between the Indus and Swat
rivers. To add to this embarrassment the ascent of the
pass had proved far more difficult than had been anticipated—so much so that the field guns had to be packed on
elephants.

Fortunately there was some tolerably open and level
ground about the crest of the pass which allowed room for
the whole force to bivouac, and there it settled down after
picquetting the hills commanding the camp. On the left
these rose to a height of several thousand feet, but
plateaux and knolls on their slopes afforded convenient
sites for the picquets. On the right the hills were lower,
and the picquets were on their summits. In front the pass
widened out during its descent to the Chamla Valley. The
hill-sides were wooded in parts, in others covered with low
brushwood, and, though steep, were perfectly practicable
for good infantry.

A reconnaissance next day into the Chamla Valley was
only withdrawn with difficulty, and was followed by a
general attack on the picquets. This showed plainly that
any idea of advancing must be abandoned—the force was
in truth besieged in the pass. There ensued a war of
picquets, waged with the greatest gallantry on both sides,

22nd Oct.

and fluctuating fortune. The story of the "Crag Picquet" —thrice lost and won—was for long famous in the annals of frontier fighting. In these affairs the two mountain trains intervened on many occasions with effect, while the field battery, holding a breastwork in front of the camp, had more than once to stand a charge right up to its guns.

It was a month before re-inforcements arrived. These included the two howitzers of No. 3 Light Field Battery, whose value in the Tochi has already been recorded, and with them came two staff officers deputed by the Commander-in-Chief (Sir Hugh Rose) to report on the situation —both gunners, and both destined to rise to the highest distinction—Lieut.-Colonel J. M. Adye and Major F. S.

Roberts. It was not, however, until the 15th December that the offensive could be assumed. After clearing the hills, during which a very strong position known as "The Conical Hill" was stormed under the fire of the mountain batteries, two columns, a mountain battery with each, advanced into the Chamla Valley and captured the village of Ambela. A dramatic incident of the day was the furious onset of the Hindustanis, sword in hand, but the fire of the Enfields was too much for even their fanatical valour, and their annihilation proved the last straw. There was no further resistance, and the Bunerwals agreed to destroy the fanatics' refuge at Sitana themselves. In order to ensure the thorough completion of this work British officers had to see it done, and among the party selected for this dangerous mission figure again the names of Colonel Adye and Major Roberts. How dangerous it was is plain from their vivid accounts.

The North-West Frontier has now been traced from the sea to the Indus, but to complete the task that river must be crossed. For on its left bank, opposite the Sitana settlement, rose a mountain ridge the stronghold of a frontier tribe. Under the name of "The Black Mountain" it was to be very familiar to more than one generation of frontier soldiers.

The "Mountain" consists of a narrow ridge, its length 25 to 30 miles, its height some 8,000 feet, its upper part thickly wooded. Numerous precipitous spurs jut out from the main ridge, enclosing deep glens in which lie some of the smaller villages : the larger ones are situated on the banks of the Indus which washes the northern and western faces. It is only 30 or 40 miles from the military station of Abbottabad, and a punitive expedition had forced its way into its fastnesses as early as 1852, but in 1867 a recurrence of the raids into British territory necessitated another visit—the first of three which occurred during the period covered by this volume.

In order to overawe the neighbouring tribes, and prevent interference, a force of ten thousand men was assembled, drawn from down country so as to leave the troops on the frontier intact. The command was entrusted to Major-General Wilde, and he was given the following artillery :—

C.R.A.—Colonel E. Atlay.

Orderly Officer—Major T. E. Hughes.

Adjutant—Lieut. R. McG. Stewart.

D/F, Horse Battery—Lt.-Colonel G. A. Renny.

E/19, Field Battery—Lieut. A. L. Pringle.

2/24, Garrison Battery—Captain C. S. Jackson.

The Peshawar Mountain Train—Captain M. Elliot.

The Hazara Mountain Train—Lieut. E. J. de Lautour.

The horse battery had its guns carried on elephants, the garrison battery was armed with six $5\frac{1}{2}''$ mortars carried on mules, but neither it nor the field battery appear ever to have come into action.[1] The horse artillery covered the advance on the occasion of the attack of the main ridge, but otherwise the two mountain trains were the only artillery actively engaged. On every occasion they were

---

[1] Chamberlain wrote—"Some of our guns and the $5\frac{1}{2}''$ mortars have to be sent back as useless after having taken the pick of men and animals to equip a half-battery of R.A. Our 1st Light Field Battery have to be stripped to make a half-battery R.A. efficient." The Frontier Force did not welcome trespassors on its preserves.

able to afford the infantry close support in spite of the extreme ruggedness of the ground, and their effective fire contributed materially to the rapid success of the operations and the small loss of life. All was over in a week.

COMMENTS.

The
Mountain
Trains.
    The expeditions chronicled in this chapter mark the development of the mountain batteries as a definite branch of the artillery. The first Indian mule battery had been formed in 1841 during the Afghan War, only to be disbanded as soon as hostilities ceased. Five years later the conquest of Sind brought the Bombay Army up against the Hills and Jacob formed his "Mountain Train". With the annexation of the Punjab in 1849 the Bengal Army in its turn came into touch with the border tribes, and three "Light Field Batteries" were raised for the Frontier Force. But the country was too rough for any wheeled artillery, and so we find two "Mountain Trains" taking the field in the 50's. In his *Forty-one Years in India* Lord Roberts tells how in 1853 he arrived just too late for a frontier expedition "very disappointed at missing my first chance of active service, and not accompaning the newly raised 'Mountain Train' (as it was then called) on the first occasion of its being employed in the field." We have seen in this chapter the way in which the light field batteries which originally formed the artillery of the Frontier Force failed on several occasions to overcome the difficulties of the ground, and how the army came more and more to rely on the two mountain trains. The natural consequences followed. No. 1 Light Field Battery dropped out, Nos. 2 and 3 became Nos. 1 and 2 Mountain Batteries, and the Mountain Trains became Nos. 3 and 4 Mountain Batteries, all of the Punjab Frontier Force. Similarly in the Bombay Army

---

   [1] Later on the term "Mountain Train" was used to denote the equipment, the native establisment, and the animals which were attached to a garrison company of the Royal Artillery when it became a mountain battery for a time.

two mountain batteries were formed from the Sind Mountain Train.

The mountain trains soon developed tactics suited to their special attribute of mobility on the hill-side. It will be remembered how, in Waziristan, two batteries accompanying the infantry advance on parallel spurs, were able to play into each other's hands by enfilading the defences in turn. It was, however, in the Ambela expedition of 1863 that they established their value beyond all cavil. In those hard-fought picquet combats they were used with the greatest boldness to cover the infantry attacks. We read of the guns of the Peshawar Mountain Train being man-handled into action within six hundred yards of the enemy; and in defence they were equally freely used to sweep the approaches. A notable example is the night attack on the Crag Picquet, when a battery did good shooting at 250 yards, the fire directed by the defenders from the parapet of the post. During the assault of the Eagle's Nest the Hazara Mountain Train similarly repelled a determined effort to capture the neighbouring picquet.

It was not without a struggle, however, that the horse and field batteries surrendered the hillside to the mountain trains. As long ago as the Nepalese War of 1814 elephants had been used to carry the guns, and in the Ambela Expedition the field guns were transferred to the backs of elephants as soon as the track up the pass became impracticable for wheeled vehicles. When, on the other hand, the force advanced, the guns were brought down the pass on their elephants, but their teams were hooked in as soon as they got into the Chamla Valley. During the period of picquet warfare in the pass the field guns were not only used in defence of the camp, but, as soon as time had allowed of some improvement in the communications, a 9-pr. and a 24-pr. howitzer were sent up to the picquets. Again in Jowaki the horse artillery were able to crown the position—although it was on a high and rugged ridge—owing to their having been provided with elephants for the carriage of their guns.

The value of the heavier shell of the horse and field guns, and especially of their howitzers, was very apparent, and this was especially demonstrated when the horse and field artillery received the 9-pr. R.M.L. with its longer range. So much was this the case that before the outbreak of the Afghan War in the following year another horse battery had been provided with alternative elephant equipment.

It must be remembered that the 3-pr. guns and 12-pr. howitzers with which the mountain trains were armed were very feeble weapons. It was intended to replace them with 6-pr. Armstrongs, but these were found too heavy for mules, and so had to be relegated to elephants—as we shall see in Bhutan—or employed as light field guns as in New Zealand. As a result of the Ambela Expedition some of the 3-prs. were bored out to 3″ and rifled, but the first regular rifled mountain guns were the 7-prs. of 150-lb., brought into the service with a great flourish of trumpets for the Abyssinian Expedition.

LIST OF UNITS.

(With designation in 1914.)

I/C, R.H.A.	—W, R.H.A.
D/5, do.	—⎫
D/F, do.	—⎬ S, do.
C/19, R.A.	—35, R.F.A.
E/19, do.	—59, do.
13/9, do.	—Reduced as 53, R.G.A.
2/24, do.	— do. 103, do.
6/7, do.	—1, M.B. R.G.A.
12/11, do.	—2, do. .do.
No. 1 Punjab Light Field Battery	—Disbanded.
No. 2 do. do. do. do.	—21, Kohat M.B. (F.F.)
No. 3 do. do. do. do.	—22, Derajat do. do.
The Peshawar Mountain Train	—23, Peshawar do. do.
The Hazara do. do.	—24, Hazara do. do.

MEDAL & CLASPS.   CHAPTER
                                   I.

In 1869 the Indian General Service Medal of 1854 was granted to the survivors of the troops who took part in the expeditions narrated in this chapter. Special clasps were given for "Jowaki" and "Umbeyla" : for the remainder the "North-West Frontier."

CHAPTER II.

THE SECOND AFGHAN WAR—THE CAPTURE OF THE PASSES.

Introduction—Mobilization—The Peshawar Valley F.F.—Ali Masjid —The Kurram Valley F.F.—The Peiwar Kotal—The Kandahar F.F.—The Treaty of Gandamak.

MAPS 1 & 2.

NOTE.—*Questions of organization, armament, and employment are discussed in Chapter V, where will be found, also, lists of units and of officers, with some particulars of the Afghan Artillery.*

CHAPTER II.

Introduction.

1869.

ON the death of the Amir Dost Mahomed, so famous in connection with the First Afghan War, his son Shere Ali succeeded. In 1869 he paid a visit to India and was entertained by Lord Mayo at a great Durbar at Ambala. But a few years later relations became strained, and, failing to obtain from the British Government the assurances he desired, Shere Ali determined to throw in

1878.

his lot with the Russians. In 1878 he received at Kabul a Russian Mission, and, while this was still at his capital, turned back a British Envoy at the entrance to the Khyber Pass. Such a studied insult could not be allowed to pass, and an ultimatum was despatched, giving the Amir two months in which to make amends or have his country invaded.

The war that ensued has a special interest for the Royal Artillery since it was the first campaign subsequent to the adoption of rifled guns in which the enemy possessed an artillery worthy of consideration. On the personal side it will always be memorable in the annals of the Regiment from the fact that, as in the First Afghan War, the commander who gained the greatest distinction was an artilleryman.

It will be considered in three distinct campaigns or phases. The first, the "Capture of the Passes," ended

Field-Marshal Lord Roberts of Kandahar, V.C., G.C.B., G.C.S.I., G.C.I.E.

with the flight of Shere Ali, and the conclusion of the CHAPTER
Treaty of Gandamak with his son Yakub Khan. The II.
second, the "Occupation of Kabul," occasioned by the 1878.
murder of the British Resident, ended with the deportation Introduc-
of Yakub and the recognition of Abdur Rahman as Amir. tion.
The third, the "Relief of Kandahar," originated, like the
second, in a tragedy, and ended with the evacuation of
the country.

During the two months grace granted to the Amir the Mobiliza-
military authorities in India prepared to put in force the tion.
measures threatened by the ultimatum. An "Army of
Invasion" was assembled in three columns, with orders
to move forward on the 21st November. Owing to the fact
that batteries in India were kept virtually at war establishment, mobilization did not throw any excessive strain
upon the artillery. The batteries of the Royal Artillery
detailed for the campaign numbered in the first instance
only 3 horse, 6 field, and 7 garrison, with 5 Indian
mountain batteries. To complete these units for active
service batteries remaining in India (and not quartered
near the frontier) were called upon to supply the required
officers and other ranks, as well as horses. The numbers
at the outset, however, were not more than could be spared
without inconvenience, and, owing to the two months grace,
ample time was available for the transfers. The creation
of ammunition columns—the real crux of artillery mobilization—was not considered necessary.

Later on, owing to the wastage at the front, and the
mobilization of further forces, the drain on the supplying
batteries became more severe, and they suffered also, and
perhaps more seriously, from the demand for officers for
service on the staff and in the departments.

Although officially styled an "Army" the three
columns were quite independent of each other, and their
action will be considered separately. They were based
respectively on Peshawar, Kohat, and Quetta, and their
tasks were the seizing of the passes which penetrated the

CHAPTER II.

1878.

The Peshawar Valley Field Force.

huge mountain barrier separating the rugged territory of the Amir from the plains of India.

The first of these columns, the "Peshawar Valley Field Force," consisted of one cavalry and three infantry brigades, some ten thousand in all, under the command of Lieut.-General Sir Sam Browne, with the following artillery :—

C.R.A.	—Colonel W. J. Williams.
Adjutant	—Captain G. W. C. Rothe.
Orderly Officer	—Lieut. W. G. Knox.

I/C—Horse Battery.
E/3—Field   do.
11/9—Mountain do.
13/9—Heavy   do.
No. 4—Indian Mountain do.

The first duty of this force must obviously be the capture of Ali Masjid, the frontier fort some six miles within the Khyber Pass, the scene of the affront to the British Mission. The Afghan position here stretched right across the valley, with the fort on a detached hill in the centre. It was formidable by nature, and had been skilfully strengthened by the erection of sangars and other defensive measures. In the fort were eight guns, two or three more were on the cliffs below, a dozen or so were distributed along the entrenchments, with mountain guns at commanding points. The garrison consisted of some three thousand regular troops, supported by a considerable force of irregulars.

Ali Masjid

At sunset on the 20th November Sir Sam Browne despatched two of his infantry brigades, with No. 4 Indian M.B., to make a long detour through the mountains which would bring them on to the flank and rear of the enemy's position. With the remainder of his force he advanced directly against it on the morning of the 21st.

21st Nov.

About 11 a.m. the advanced guard batteries (I/C & 11/9) came into action on the Shagai Heights from which the fort and its defences were clearly visible at a range of about 2,500 yards. The Afghans returned their fire with the guns in the fort, and also opened fire upon the infantry with those below the fort, which were invisible from the heights on which the advanced guard batteries had taken up their position.

As the main body came up the field and heavy batteries relieved the horse and mountain, which pushed forward in close support of the infantry. By this time a way into the valley from the heights had been made practicable for guns, and I/C dropped down into the bed of the stream and advanced up it until a position was found from which four guns could support the left attack at a range of 1,200 yards. The infantry were under a continuous fire of round shot and musketry, and the other section, crossing to the right bank of the stream, moved forward to within 1,000 yards of the fort. Unfortunately no ammunition wagons had been allowed to accompany the advanced guard batteries, so the guns were withdrawn as soon as the limber ammunition had been expended.

In support of the right attack 11/9 found a position on a ridge within 1,500 yards of Ali Masjid. But the enemy scrambling along the cliffs to within four or five hundred yards made things unpleasant in spite of the efforts of the escort to dislodge them. When it became apparent also that an assault on this side would be impracticable owing to the inaccessible nature of the ground occupied by the enemy, the battery was withdrawn to a safer position on the ridge behind.

Meanwhile the 40-prs. of 13/9 had been pounding the fort at a range of 2,850, but there were no signs of the turning movement which should have been making its influence felt. Sir Sam Browne decided not to risk an assault, and as dusk fell the attack was broken off, the troops bivouacking where they stood—I/C on the left bank of the Khyber stream, 11/9 on its ridge on the right, E/3 and 13/9 on the Shagai Heights.

The threat of being taken in flank and rear had, however, produced its effect, although the difficulties of the ground had delayed the movement of the troops. When the guns reopened fire at daybreak it was soon apparent that the enemy had abandoned his position during the night, leaving 25 guns. The removal of these, entrusted to the heavy and field batteries, proved an unexpectedly hazardous occupation owing to the activities of the Afghans. On several occasions the officers had to use their revolvers to keep them off.

The ammunition expended was :—

I/C— 35 common, 168 shrapnel, total 203.

E/3— 48 common, 98 shrapnel, total 146.

11/9— 59 common, 31 shrapnel, total 90.

13/9—128 common, — shrapnel, total 128.

The advance was not further interfered with, and after a pause at Dakka, Jalalabad was occupied before Christmas. This was the limit of the task allotted to the Peshawar Valley Field Force,[1] and no further advance was attempted. But the second division (under General Maude) was added to the force, so as to allow the first division to concentrate at Jalalabad, and during the winter various punitive expeditions were undertaken. E/3 and 13/9 which had been left on the Shagai heights with details of cavalry and infantry (all under Major Wilson of 13/9) were heavily attacked; and forces had to be sent into the Bazar Valley in December and January. For these D/A was sent up from Peshawar with its guns on elephants, and 11/9 also took part. There was also a good deal of fighting with the Mohmands. In these the batteries chiefly engaged were I/C, and the two mountain batteries. The most serious affair was at Kam Dakka in April 1879, when a small body of troops sent to protect that village was attacked by a very superior force. It was only after

[1] Major-General J. E. Michell, R.A., commanded the Cavalry Brigade.

desperate fighting that its withdrawal was effected by reinforcements under Major Dyce of 11/9.

CHAPTER
II.

1879.
The
Kurram
Valley Field
Force.
1878.

The next force to be considered, that based on Kohat, was entrusted to Major-General F. S. Roberts. Between five and six thousand strong, its artillery was :—

C.R.A. —Lieut.-Colonel A. H. Lindsay.
Adjutant—Lieut. E. G. Osborne.
F/A —Horse Battery.
[1]G/3 —Field do.
No. 1—Indian Mountain Battery.
No. 2— do. do. do.

The duty of this force was to seize the passes between the Kurram and Logar Valleys, the latter of which led direct to Kabul. The Afghans instead of opposing the advance from its commencement, as at Ali Masjid, selected to make their stand at the very head of the Kurram, where a range of rugged, almost precipitous, heights, clothed with dense pine forest, rose some 2,000 feet above the valley. This ridge was crossed at the Peiwar and Spingawai Kotals some three miles apart, and the Afghan position covering both was held by a force of over three thousand regulars with 18 guns, besides large numbers of irregulars. Leaving No. 2 Indian M.B. at the frontier station of Thal, General Roberts advanced without opposition until he arrived at the foot of the Peiwar Kotal where the direct road crossed. A reconnaissance in force on the afternoon of the 28th November convinced the general that a direct attack would at the best be very costly, and he decided to halt for a few days in order to rest the troops, while he conducted further reconnaissances of the Afghan position.

28th Nov

The enemy soon showed that they were alert, for while the troops were awaiting the arrival of their baggage, they

---

[1] Half G/3 was left in garrison at Kohat.

were somewhat disagreeably surprised by the arrival of shell—fortunately all blind—from a mountain gun which had been pushed forward to the end of a spur overlooking the camping ground. Two guns of F/A were hastily brought into action in reply, but a safer site had to be selected for the camp : it was fortunate that the Afghan gunners had not reserved their fire until it was pitched.

The Peiwar Kotal.

The Peiwar Kotal itself was a narrow depression in the ridge, approached up a steep, zigzag path. The ascent was not only naturally difficult, but was completely commanded by the spur referred to above, which became known as "one-gun spur" and played a prominent part in the subsequent fighting. The Spingawai Kotal was of a very different character, being accessible by a valley broad enough to admit the movement of troops, although the actual track generally followed the bed of a dry watercourse. General Roberts decided to take the bulk of his force by this route under cover of darkness, hoping to surprise the defenders by an attack at dawn. Meanwhile the troops left in camp would do everything possible to conceal the movement, and to give the impression that the main effort was being made against the Peiwar. With the main body went the C.R.A., 4 guns of F/A, and No. 1 M.B., leaving in camp a section of F/A and the half-battery of G/3.

1st Dec.

The main body started on their long and laborious detour at 10 o'clock on the night of the 1st December, having to march for some miles down the valley before turning up the bed of the Spingawai. It was then that the difficulties began, for this was found to consist of loose boulders from which the earth had been washed by the floods, and, to make matters worse, they were coated with ice. Long before the enemy's defences were approached the elephants of F/A had fallen far behind, but the mules of the mountain battery made light of such difficulties, and the leading section came up just as the first order to charge was given. Lord Roberts has related how, when he told Captain Kelso to bring his guns into

action as soon as he could find a position, "I was struck
by the smile of satisfied pride and pleasure with which he
received this order. He was delighted, no doubt, that the
opportunity had arrived to prove what the battery—to
perfect which he had spared neither time nor labour—
could do; but it was the last time that gallant soldier
smiled, for a few seconds later he was shot dead."

The pass was defended by barricades of tree-trunks
behind which two mountain guns were posted, but these
defences were stormed with impetuous gallantry, and by
6.30 the summit had been gained. A flank attack along
the ridge was commenced, but the move through the forest,
against ever hardening opposition, proved a tedious and
expensive task, even when helped by the heavier shells of
F/A crashing through the trees. The back of the ridge
was found, however, to fall away in terraces instead of
being precipitous like its face. When, therefore, it became
apparent to Roberts that the Peiwar was really inaccessible
along the ridge, he determined upon another turning
movement, this time right round the rear of the position.
But, before embarking upon it, the mountain guns had
found a place from which they could shell the Afghan
camp behind the Peiwar Kotal. Unknown to Roberts
the sight of their tents blazing, and their followers and
baggage animals bolting, had been too much for the
defenders, who were already abandoning their position.

To return to the force left in camp. In accordance
with their mission the five guns were brought into action
against the Peiwar, and, since the Afghan guns commanded the approaches, the movement was made before
dawn. The position selected was in a field to the left of
the road, about 1,600 yards from the "one-gun" spur, and
here, as soon as daylight discovered their presence, they
were attacked by this gun and also by two 12-pr. howitzers
and a 6-pr. gun from the main ridge. For the next three
or four hours an artillery duel was kept up, without much
damage upon either side. Owing to the plunging nature
of the fire the shell of the gun on 'one-gun spur' burst

CHAPTER
II.

1878.

The Peiwar
Kotal.

2nd Dec.

too high to be effective, while the British position was almost out of range of those on the main ridge. And, although the Afghan guns were well within the range of the 9-prs., the skill with which they had been placed rendered the fire of the batteries almost innocuous.

During the artillery duel the infantry had been working their way forward, and about noon the guns were ordered to advance in support. While in column of route, crossing a ravine, they suddenly came under rifle fire from a party which had crept down through the wood from 'one-gun spur' on seeing the movement. Fortunately the escort was close at hand, and soon cleared the Afghans out of the way.

It was their last effort. Although their gunners had served their guns gallantly under the artillery fire they could not stand up against the infantry when they had worked their way to within 800 yards. So when the latter reached the kotal after a stiff climb, they found it completely deserted.

The ammunition expended by the artillery was :—

9-pr.—34 common, 442 shrapnel, total 476.

7-pr.—7 common, 45 shrapnel, total 52.

Next day General Roberts moved on to Ali Khel, finding the way strewn with abandoned arms and baggage. From there, with a small party, he reconnoitred the Shutargardan Pass, which led into the Logar Valley and opened the road to Kabul. In spite of its 11,000 feet it was found to present no defensive position as strong as the Peiwar, and no insuperable obstacles. It had been decided, however, that it was to be the limit of the force's activities, so the party returned without pushing their exploration beyond the summit, and Ali Khel became the most advanced post.

Meanwhile the collection and removal of the captured guns had proved almost as tough a task as at Ali Masjid, but here the elephants, and not the enemy, were the danger. Some were only half-broken, and their cradles did not fit the Afghan equipments. The gunners of G/3

CHAPTER
II.

The
Kurram
Valley Field
Force.

1878-79.

on the ridge had the awkward job of loading them up; to
the horse artillerymen fell the task of unloading them at
the foot of the pass, and then of adapting the guns to
bullock draught for the march to Kohat.[1]

To hold the hardly-won ridge during the winter a
garrison was left on the kotal, and this included the three
guns of G/3. The ammunition boxes were carried up on
elephants, but in spite of the precipitous nature of the
path Major Parry insisted on the guns being taken up by
their own teams. The gunners settled down to make
themselves safe and comfortable for the winter, with
blockhouses for the guns, and log-huts to shelter themselves from the snow, while the drivers and horses were
sent down to be attached to F/A.

The troops were in winter quarters, but the question of
supplies became a very serious one, especially for the
horse and field artillery, and the battery horses were in
poor condition when they were called upon to share with
the mountain battery mules the work of carrying up to the
front from Kurram the reserve ammunition of the
ordnance field park.

This was all part of a forward concentration of the force
at Ali Khel so as to be ready for an advance on Kabul,
the place of G/3 at Kohat and on the Peiwar being taken
by C/4. So serious, however, was the shortage of forage
at Ali Khel that all horses except two gun-teams per
battery had to be sent back until the grass began to grow
at that high altitude. This brought an interesting sequel,
for, in the absence of the gunners, the defence of the horse-
lines at Kurram fell upon the drivers, who were not armed.
They were served out with whatever was available in the
ordnance stores—Snider carbines discarded on the issue
of Martinis, and Enfield rifles captured from the enemy.

---

[1] The artillery with the Afghan force consisted of one field and two
mountain batteries. The former was armed with four 6-pr. bronze
smooth-bore guns, and two 12-pr. howitzers, but only the howitzers
and one gun were brought into action. The mountain batteries were
armed with 7-pr. rifled muzzle-loaders, but only one battery was
brought into action—the other was found complete, abandoned by the
road-side.

CHAPTER
II.

The
Kandahar
Field Force.

1878.

We must now turn to the last of the invading forces, that destined for the capture of Kandahar. It was far the strongest, consisting of two divisions with a total strength of over twelve thousand. The divisions were at first far apart, the 1st assembling at Multan under Lieut.-General D. M. Stewart, the 2nd under Major-General M. A. S. Biddulph, R.A., at Quetta.

The following was the artillery[1] :—

C.R.A.—Colonel C. G. Arbuthnot.

Brigade-Major—Captain A. D. Anderson.

1st Division.

Horse & Field Artillery.	*Mountain & Heavy Artillery.*
Lt.-Colonel A. C. Johnson.	Lt.-Colonel A. H. Dawson.
Lieut. F. J. W. Eustace.	5/11—Heavy Battery.
A/B—Horse Battery.	6/11— do. do.
L/11—Field do.	11/11—Mountain do
D/2— do. do.	
G/4— do. do.	

2nd Division.

Lt.-Colonel G. B. le Mesurier.

Lieutenant F. H. G. Cruikshank.

E/4	—Field Battery.	
No. 3	—Indian Mountain do.	
No. 2 (Bombay)—	do.	do. do.

On the 21st November General Biddulph commenced his advance from Quetta, and entered the Pishin Valley (then part of Afghanistan) without opposition. The troops of the 1st division had a long and weary march after they had left the railway at Sukkur, over the "put" to Sibi and through the Bolan Pass to Quetta. Meanwhile the 2nd division were collecting supplies in the Pishin Valley and making arrangements for the crossing of the Kwaja-Amran

---

[1] A Siege Train was also assembled at Sukkur, and started on the march, but was broken up on receipt of the news of the occupation of Kandahar.

Range. This last was an undertaking which aroused much interest at the time, especially as regards the measures taken to make the descent from the Khojak to Chaman practicable for wheeled artillery. In order to economise time it was decided to convert the bed of a watercourse into a ramp down which the guns and wagons could be lowered by ropes, thus leaving the road free for the innumerable camels carrying the baggage and supplies. The most elaborate arrangements were made—parking places for the vehicles waiting their turn, tackles, bollards, holdfasts: the lowering of each vehicle took just under ten minutes.[1]

This is to anticipate. It was not until the end of December that the 1st division was collected in the Pishin Valley ready to advance. General Stewart then assumed command of the combined force, and decided that it should enter Afghanistan in two columns by the Khojak and Gwajha Passes. Beyond some cavalry skirmishes, in which A/B played its part, no opposition was encountered, and Kandahar was occupied on the 8th January.

After a few days rest the two divisions moved on again, the 1st to Kalat-i-Ghilzai on the road to Ghazni and Kabul, the 2nd to the Helmund at Girishk on the road to Herat. These reconnaissances in force lasted a month or so, and entailed considerable suffering on the followers and animals from cold and lack of forage, but the only fighting was a sharp attack on the rear guard of the 2nd division. Before the end of February all were back in Kandahar, and a considerable reduction of the force was being made by the return of the 2nd division to India.

With the gates of his country all in the hands of the invaders, the Amir realised that further resistance was futile, and took refuge with the Russians; leaving his son, Yakub Khan, to make the best terms he could. At

Chapter II.

1878.

The Kandahar Field Force.

December.

1879.

8th Jan.

The Treaty of Gandamak. 26th May.

---

[1] Contemporary records give glowing accounts of the ingenuity of the arrangements, with pictures of the road, its zigzags marked by an endless stream of camels.

CHAPTER
II.

1879.

The
Treaty of
Gandamak.
26th May.
the end of May he signed the Treaty of Gandamak by which he agreed to the establishment of a British Resident at Kabul, and the occupation of the Khyber, Kurram, and Kandahar by British troops.

This involved the withdrawal of all troops between the Khyber and Jalalabad, and they suffered terribly during the march from the burning heat, the dust, and the scarcity of water—and the cholera was always with them. It was not until the end of June that Dakka was evacuated, and 11/9 and No. 4 were left at Landi Kotal to hold our newly acquired territory in the Khyber. Amongst the losses they suffered was that of Major Dyce commanding 11/9, who had been prominent in all the fighting on the Khyber line.

So ended the first phase of the war.

CHAPTER III.

THE SECOND AFGHAN WAR—THE OCCUPATION OF KABUL.

The Murder of the Resident—The Kabul F.F.—Charasia—The Bala
Hissar—Killa Kazi—The Conical Hill—The Sherpur Cantonment—
The Lines of Communication—The Ghazni F.F.—Ahmed Khel—
The Installation of Abdur Rahman.

MAPS 1, 3, & 4.

In accordance with the Treaty of Gandamak Sir Louis
Cavagnari was appointed Resident in Kabul. He reached
Kabul in safety, and was received with all due ceremony
by the Amir, but in little more than a month there was a
tumult in the city, the Residency was attacked, and Sir
Louis and his staff were murdered.

CHAPTER
III.

1879.

The murder
of the
Resident.

No time was lost in taking steps to avenge his death.
Unfortunately the Khyber Force had been broken up, and
the units dispersed to their stations, so nothing could be
done on that line for the time being. In the south, too,
the force had been reduced, the only batteries left at
Kandahar being 6/11 and 11/11, though G/4 was still
within reach at Chaman and was promptly ordered back.
All that Sir Donald Stewart could do was to make a demonstration, and he despatched a small column to Kalat-i-
Ghilzai, which included two guns each of G/4 and 11/11
—the latter at the special bidding of the Commander-in-
Chief in India who impressed upon Sir Donald how much
the presence of the heavy guns would add to the effect of
the movement by inducing the belief that an advance to
Kabul was contemplated. The Kurram Field Force was
the only one "in being," and General—now Sir Frederick
—Roberts was ordered to march on Kabul.

The constitution of the force detailed for the advance
on Kabul was little changed from that of the Kurram

The Kabul
Field Force.

Field Force of the last chapter. Its artillery was as follows :—

C.R.A. —Lt.-Colonel B. L. Gordon.

Adjutant—Captain J. W. Inge.

F/A—Horse Battery.

G/3—Field do.

No. 1—Indian Mountain Battery.[1]

No. 2— do. do. do.

There was also a new unit "The Gatlings," consisting of two Gatling guns organized on mountain battery lines.

On the day that news of the murder of Cavagnari reached Simla, orders were despatched to seize the Shutargardan with a couple of battalions and a mountain battery,[2] so that Sir Frederick Roberts might set about the preparation of his force with the pass by which he must enter Afghanistan in his hands. By the end of September —within a month of the receipt of orders—he was ready. The first obstacle was the Shutargardan, 11,000 feet high, with the last thousand feet so steep that the guns had to be man-handled up, and similarly down the other side. This took three hours in spite of the willing assistance of the infantry, but it was safely accomplished without interference by the Afghans. There were signs, however, of ever-increasing hostility as Kabul was approached, and all doubts as to their intentions were set at rest when the

village of Charasia, eleven miles from the capital, was reached.

Just beyond the village a horse-shoe shaped ridge, rising some thousand feet above the plain, barred the

approach to Kabul, and on this, at dawn on the 6th October, large numbers of troops were seen to be taking up their position. Owing to the shortage of transport one of his brigades and the horse battery were still a day's

---

[1] No. 1 M.B. was left in garrison on the Shutargardan and Ali Khel.

[2] Curiously enough the telegram used the old expression "Mountain Train."

march in rear, thus reducing his available strength to a bare four thousand men, but Sir Frederick Roberts decided to attack at once before the enemy's strength could be increased by the accession of waverers.[1]

The horse-shoe ridge occupied by the Afghans was cleft by a deep gorge through which flowed the Logar River, with the road to Kabul on its bank. Here the enemy's main strength appeared to be concentrated, and General Roberts decided to demonstrate against this point, where he was evidently expected, and to throw the weight of his attack against the enemy's right, where the end of the horse-shoe approached the village of Charasia. The former he entrusted to Major White, with half of G/3 for artillery, the latter to Brig.-General Baker, with No. 2 M.B. and the Gatlings. The remaining guns were left to guard the camp.

Baker's infantry had to fight hard to gain the first Afghan position, but the second line offered a weaker resistance, and from this a rapid advance carried the attackers on to the main ridge. General Baker was now able to bring up his left, and attack the main position at the gorge along the ridge.

Meanwhile White's small force had been working their way along the series of detached hills by the river which covered the approach to the gorge, supported by the field guns. The first of these hills had prevented the guns getting within 4,000 yards of the main position, but directly it was captured Major Parry advanced to a second position from which he was able to engage the twelve Afghan guns posted above the gorge. The defenders were now seen to be hurrying towards their right, in haste to oppose Baker's attack, and White, detecting the weakening of the force opposed to him, changed his rôle from a demonstration to a determined attack. His infantry

---

[1] Since the writing of this account Sir Philip Chetwode, Commander-in-Chief in India, when warning the students of the Staff College against hesitation to face risks, quoted this decision as an example of justifiable boldness.

CHAPTER III.

1879.

Charasia.

6th Oct.

pressing on succeeded in getting on the flank of a battery of Armstrongs in action in the mouth of the gorge, enabling G/3 to push in still closer. By the time Baker's troops were approaching the gorge from the left White's infantry had crowned the heights on the right. A rear-guard held up the cavalry in the gorge, but Baker's mountain battery finding its way on to the ridge crushed this last vestige of resistance. The road to Kabul was open.

The ammunition expended by the artillery was :—

G/3 — 6 common, 71 shrapnel, total 77.

No. 2—10 common, 94 shrapnel, total 104.

The Gatlings jammed after firing 150 rounds.

The whole of the 20 Afghan guns, specially brought out from Kabul, were captured—12 6-pr. Armstrongs, 6 8-pr. muzzle-loaders, and a couple of smooth-bores. The rifled guns had been left where they stood in action, the two smooth-bores (one an 8" howitzer) were found abandoned on the road.

The Bala Hissar.

October.

After spending a few days clearing the surrounding country, during which 33 more guns were captured, Sir F. Roberts took formal possession of the Bala Hissar, the historical fortress which dominates the city of Kabul.[1] The approaches were lined with troops of all arms, and, as the Union Jack was run up over the great gateway, the first gun of the Royal Salute was fired by F/A. But a sad fatality was to cloud the triumph. The Bala Hissar was stored with ammunition : hundreds of tons of gun-powder were in earthern pots; thousands of fuzes, friction tubes, percussion caps, littered the floor; hundreds of thousands of Enfield and Snider rifle cartridges filled the godowns. Captain E. D. Shafto, R.A., was detailed to take such precautionary measures as were possible, and he had been at work some days when a violent explosion

---

[1] Major-General Hills, R.A., was appointed Military Governor.

startled the army. It was followed by the rattle of rifle cartridges, but the profoundest anxiety was felt for the main magazine, reputed to contain two to four hundred tons of gunpowder. No time was lost in withdrawing the regiments quartered in the fortress, and all were clear before the last terrific explosion. A great disaster had been narrowly escaped, but the death of Shafto was deplored by the whole force.

At first all seemed peaceful about Kabul, but early in December it became apparent that the Afghans were gathering in ever-increasing numbers in the neighbourhood, and with hostile intent. Sir Frederick determined to break up the combination before it got too large, and sent out his brigadiers—Baker, Macpherson, Massy—with columns, each including one or more batteries, horse, field, or mountain. There was stiff fighting on several occasions, but the space available does not permit of following these minor operations in detail. One unfortunate incident must be described, however, for it caused much comment at the time, and for many years afterwards.

On the 11th December, Brigadiers Baker and Macpherson with their columns were within a few miles of Kabul, and Brigadier Massy was despatched with three squadrons and four guns of F/A to co-operate. Unfortunately, instead of following the way indicated to him, he took a short cut which brought him face to face with a large gathering of the enemy before he could get into touch with the other columns. As the little force approached the village of Killa Kazi the Afghans swept down upon them. Three times were the guns brought into action,[1] but four muzzle-loading 9-prs. could have no appreciable effect upon such numbers, and Massy ordered the battery to change

CHAPTER
III.

1879.

The Bala
Hissar.

October.

December.

11th Dec.

Killa Kazi.

---

[1] At 2,900, 2,500, and 2,000 yards, the first two positions being considered by Colonel Gordon, the C.R.A. (who was with the column) as being at too great a range. The next time the guns came into action it was 1,700.

front right back to a position on the bank of the river.
At this critical moment Sir F. Roberts appeared on the
scene. He had ridden out to take command of Macpherson's and Massy's columns as soon as they should
unite, and he has left a vivid picture of the extraordinary
spectacle presented to his view as he galloped forward
towards the firing. "An unbroken line of Afghans,
extending for about two miles, and formed of not less than
between nine thousand and ten thousand men, was moving
rapidly towards us, and to meet this formidable array there
were four guns and three hundred cavalry." Sending
back word for infantry to be hurried out, he ordered
Massy to retire the guns, covering the movement by a
charge. But the ground was so cut up with watercourses
that the cavalry could make little impression in spite of the
gallantry of their efforts, and two of the guns came into
action again to cover their withdrawal, while the other
section continued its retirement, having to abandon one of
its guns *en route* in a deep ditch.

Roberts now ordered Smyth-Windham to make for
the village of Bhagwana with the three remaining guns as
the only chance of saving them. Here they came into
action again, firing from behind the wall of one of the
enclosures. But the country was so intricate that the
Afghan footmen were able to move faster than the guns,
and as they closed on the village there was nothing for it
but to limber up again. At the further side was a ditch
fully twelve feet deep, forming part of the village
defences, and the leaders of the first gun had just begun
to scramble up the far side when one of the wheelers
fell, breaking the shaft. The gun was hopelessly stuck,
blocking the only point where there was any chance of
getting the others over. With a faint hope of saving the
guns Sir Frederick called on the cavalry to charge again,
but to no avail; and in the meantime the order to unhook
and spike the guns had been given.

It only remains to say that later in the day, when the advance of Macpherson's brigade had cleared the ground of the enemy, Colonel Macgregor, the Chief Staff Officer, collected a small party of cavalry and artillerymen, and picking up some of Macpherson's baggage guard, brought the guns in. They had been stripped of removable fittings, but were otherwise undamaged, and were fit for use next day.

There was much fighting during the next three days, and another loss of guns has to be recorded. The clearing of the Asmai Heights overlooking Kabul had been decided upon, and the force placed at Brigadier Baker's disposal included G/3 and No. 2 M.B., while further support was given by F/A and No. 1 M.B. During the operations No. 2 M.B. was left with a small party to hold a detached hill—the "Conical Hill" as it was generally termed. The heights had been cleared, and all seemed going well, when the Afghans who had been driven off, joining hands with fresh bodies who suddenly appeared, surrounded the Conical Hill in overwhelming strength and rushed its defenders. Two of the guns were got away after using their case shot with effect, but the mules of the other section were shot down, and the guns fell into the hands of the enemy.[1] In his despatch, quoted by General Roberts, Brigadier-General Baker stated explicitly that no blame for the loss was in any way to be attached to the officers or men of the battery, but that, on the other hand, every credit was due to them for the gallant manner in which they stood to their guns to the last.

The fighting on this day, which was in full view of the cantonment, showed unmistakably how strong and determined were the forces which had gathered with such dramatic suddenness. With the history of the First

---

[1] One of the guns which was saved was carried off the hill by Dr. Duke who was attached to the battery.

Afghan War vivid in his mind General Roberts decided to lose no time in concentrating his whole force, and ordered up as reinforcement the nearest brigade on the Khyber Line. Before nightfall all the troops at Kabul were safely gathered within the walls of the Sherpur Cantonment.

December. The Sherpur Cantonment, a mile or so north of the city, was a large fortified enclosure which had been built by Shere Ali to house the "New Model" army he was creating. It was enclosed on three sides with a bastioned wall nearly 20 feet high, although the eastern side had never been completed to the full height, and there was a gap on the western caused by an accidental explosion. The fourth side was protected by the Bimaru Heights Its perimeter, nearly five miles, was too extended for the number of men available for its defence, but the shelter was invaluable with the bitter winter to be faced, and soon after arrival Roberts had decided upon its occupation, and promptly set about the collection of supplies within its walls.

Instead of attacking at once, the Afghans let some precious days slip by, a respite which enabled the garrison to strengthen the weak spots in the walls, and perfect their arrangements for defence. The guns of the regular batteries were distributed—two of F/A and No. 2 M.B. on the bastions to sweep the approaches : four of G/3 and No. 1 M.B. on the Bimaru side : four of F/A and two of G/3 in reserve, free to be moved wherever required. Particularly interesting to gunners is the use made of the captured guns, of which there were so many available. Major C. A. Gorham, the Deputy Judge-Advocate-General was an old gunner, and he put aside his law books when the call came,[1] to be entrusted with the task of rendering the Afghan guns serviceable, and commanding them in action. The most efficient were the four

---

[1] His text-book on military law was in general use previous to the bringing out of the official manual. The Fourth Edition is dated "Kabul, 1st July 1880."

18-pr. smooth-bore guns and two 8″ howitzers of the siege train presented by Lord Mayo to Shere Ali during his visit to India in 1868-9. For the former nearly 400 rounds were found, though the shell had to be used as shot owing to the uselessness of the Afghan fuzes. For the howitzers there were 140 shell, for which service fuzes were used wrapped round with wax-cloth to fit the fuze-hole. Their huge explosions, so different from those of the little 7 and 9-lb. shell, were the delight of the garrison, as well as being very effective in dispersing any gatherings of the enemy within range. A species of case-shot was also manufactured which became popularly known as "Gorham's Mixture"[1] after its inventor. And it was not only by their fire that the Amir's guns were found useful. Their ammunition wagons made admirable movable barriers at the entrances, their wheels were used in hundreds to strengthen the obstacles.

It was not until the 23rd that the long-expected attack in force was delivered, heralded by a beacon-fire on the Asmai Heights. Heavy firing announced the enemy's approach before there was light enough for the defenders to see anything to aim at. But the star shell of No. 1 M.B. soon lighted up their masses, and their fire was returned by the infantry lining the parapets and every gun that could bear. Under this fire repeated efforts at assault melted away. A pause about 10 o'clock was followed by renewed attempts, but these lacked the determination of the first, and about mid-day there were signs of wavering. Roberts, with true tactical instinct, saw that the moment had come to launch his cavalry and artillery in counter-attack. A cavalry regiment and G/3 passing out through the Bimaru gorge took the main gathering in flank, and all was over. The great attack had failed absolutely, and one such lesson was enough. Christmas saw the country clear.

---

[1] In allusion to "Gregory's Mixture" a domestic medicine of universal use in Victorian nurseries.

CHAPTER
III.

1879.

The Lines of
Communi-
cation.

October.

December.

It is time now to say something of events on the Lines of Communication.

Taking the Kurram first. Small forces had been left on the Shutargardan and at Ali Khel, but no sooner had the Kabul Field Force left than the neighbouring tribes gathered to their attack, and the little garrisons had to fight hard to maintain their positions. As soon as the occupation of Kabul was effected Sir F. Roberts sent back a brigade for their relief, and since the winter snows would soon render the Shutargardan impassable, it was decided to take the opportunity to close that pass, leaving the Peiwar as the most advanced post. We need not linger over events in the Kurram during the winter except to mention an expedition against the Zaimukhts in December. This is noteworthy from the fact that a newly converted mountain battery (1/8) had just come up armed with a new mountain gun, the 2·5″ R.M.L., the "screw-guns" of popular parlance. It was their first appearance and the reputation they there gained had important results.

By the abandonment of the Kurram Line the troops at Kabul became dependent upon the Khyber for their communications with India. All troops beyond the pass itself had, as we have seen, been withdrawn under the Treaty of Gandamak, but among the steps taken in connection with the despatch of Roberts to Kabul was the collection of a force to reopen the Khyber Line. The command of this was given to Major-General R. O. Bright, with Brigadier-General C. G. Arbuthnot, R.A.,[1] in command of one of his brigades, and the following artillery :—

---

[1] He had been C.R.A. of the Kandahar Field Force during the first phase. In 1880 he was called home to take up the appointment of D.A.G., R.A., at the War Office. He held in succession all the highest appointments in the Regiment.—Inspector General R.A. in India and at Home, Deputy Adjutant-General R.A. at the War Office, President of the Ordnance Committee. Then, returning to India, he commanded first the Bombay Army, and then that of Madras when Burma was added to it.

C.R.A.   —Colonel C. R. O. Evans.

Adjutant   —Captain R. A. Lanning.

Orderly Officer—Captain R. H. S. Baker.

I/A—Horse Battery.

C/3—Field   do.

11/9—Mountain do.

No. 4—Indian Mountain do.

CHAPTER
III.

1879.

The Lines of
Communi-
cation.

There was no opposition, and before the end of November convoys were passing freely. The opportunity was taken to deport to India the Amir Yakub Khan, who had come into Sir F. Roberts' camp at the commencement of his advance upon Kabul. But the sight of their Amir passing down the line under guard tried the temper of the tribes, and the news of the shutting up of the British troops at Kabul spread like wildfire. Although things quieted down round Kabul after the failure of the great attack on the Sherpur Cantonment, there were sporadic outbreaks along the Kyber Line all through the winter and spring of 1879-80, and the garrisons of the posts had to fight hard to beat off attacks and keep the road open. The troops were kept busy with punitive expeditions, movable columns, etc., and in all these activities the batteries took their full share, usually by sections.

November.

In spite of the disturbing events in the north the winter passed without serious trouble about Kandahar. As a preliminary to the eventual evacuation of the country it was decided that all troops originally drawn from the Bengal Presidency should march to Kabul, and that their places in Southern Afghanistan should be taken by a Bombay Division. Accordingly Sir Donald Stewart left Kandahar at the end of March 1880 with a force to which the name of the "Ghazni Field Force" was given. Its strength was slightly over seven thousand, formed in one cavalry and two infantry brigades. The artillery consisted of :—

The Ghazni Field Force. 1880.

March.

C.R.A. —Colonel A. C. Johnson.

Adjutant—Lieutenant R. Bannatine.

A/B—Horse Battery.

G/4—Field   do.

11/11—Mountain do.

6/11—Heavy   do.

Ahmed
Khel.

19th April.

At first there was no resistance, but soon after leaving camp on the 19th April the enemy was seen to be occupying a range of hills which ran parallel to the road. In the order of march on that day the cavalry brigade formed the advanced guard, and the baggage and supplies were placed between the two infantry brigades. The horse battery was with the cavalry, the field battery with the leading infantry brigade, the mountain and heavy batteries with the rear brigade. On the appearance of the enemy the cavalry and leading infantry brigades were ordered to move off the road, to the right and left respectively, so as to leave the way clear for the guns, and the march was continued in this formation until the head of the column reached a spot where the track turned directly towards the hills on which the enemy's standards were planted. Here A/B and G/4 were ordered into action, one on each side of the road, under escort of a squadron and a company respectively. The line was thus formed with the horse battery and the cavalry on the right, the field battery next slightly in advance, then the infantry practically parallel with the road. On their left again there were a squadron or two.

A/B fired the first round at a range of 1,500 yards, and scarcely had the guns opened fire when the hills were crowned by a great mass of Afghan troops. From the centre of this mass successive waves of swordsmen on foot swept down on the infantry and guns, its full force falling upon the two batteries thrust forward so dangerously. The onslaughts of the footmen were made with fanatical desperation, and so rapid was their advance that within twenty minutes both horse and field batteries were firing case, and, after that was exhausted, reversed shrapnel. The guns were served with the utmost steadiness although the Ghazis charged to within five and twenty yards of the muzzles, but when the infantry were forced back by weight of numbers both batteries were ordered to retire to a hillock some hundred yards in rear.[1] From here A/B was able to do great execution upon the Afghan troops on our extreme right, while Colonel Johnson sent G/4 by sections to take post in the infantry line as soon as it was reorganized. He sent orders also for the heavy battery to come into action against the enemy's cavalry threatening an attack upon the baggage column, and a few well-directed shrapnel from their 40-prs. quickly checked any such intention. By 10 o'clock the battle was over. The fighting had lasted hardly an hour, but the Afghans had lost in that hour more than in any previous battle. The ammunition expended was :—

A/B— 1 common, 142 shrapnel, 23 case, total 166.

G/4—18 do. 138 do. 24 do. do. 180.

6/11— — do. 8 do. — do. do. 8.

Ghazni was occupied without further opposition, and there were found two of the guns lost by 1/1 Bengal Horse Artillery in the First Afghan War, when that troop gained so much glory.[2] But resistance was not quite over, and on the 23rd April Sir Donald Stewart sent out Brigadier Palliser with a force which included A/B and 11/11 to break up the hostile gathering. The two batteries commenced by shelling the villages of Argu and Shalez with common, changing to time shrapnel when the Afghans showed themselves, and advancing to successive positions

margin notes:
CHAPTER III.
1880.
Ahmed Khel.
19th April.
23rd April.

---

[1] In his despatch General Stewart wrote—"The gallantry with which the batteries maintained their ground until the last moment, and the orderly manner in which the retreat was effected, reflected the greatest credit on officers and men."

[2] The guns were taken on to Kabul and eventually given to A/B the lineal descendant of their original owners.

at 1,800, 1,700, and 1,600. But the cover in the villages was too good for the 9-pr. and 7-pr. shell to have much effect, and the brigadier considered that he was not strong enough to push an attack home. So the guns were withdrawn to a position on a ridge some 2,500 yards from the villages.

The news brought out Sir Donald with the remainder of the force and an immediate resumption of the attack was ordered. The horse and field batteries, on the right and left respectively, advanced as before, coming into action at intervals, while the mountain battery gave close support to the attack of the infantry, moving forward with them to successive positions from 1,400 to 900 yards. The villages was then rushed, and the enemy fled, closely followed by the infantry, accompanied by the field and mountain batteries, while the horse artillery joined the cavalry.

5th May.

There was no further trouble, and Sir Donald Stewart proceeded to Kabul, where on the 5th May he assumed supreme command in Northern Afghanistan. The Ghazni Field Force became the 3rd Division under Sir James Hills, R.A.

The
Installation
of Abdur
Rahman.

The military situation had been cleared up, but, anxious as the Government were to withdraw the troops, this could not be done until a ruler strong enough to maintain order after their departure had been found. It seemed that Abdur Rahman Khan, another grandson of Dost Mahomed, who had been exiled by Shere Ali, might be capable of filling the part. Negotiations with him,

22nd July.

which had been in progress for some time, were now brought to an issue, and on the 22nd July he was publicly recognized by Her Majesty's Government at an imposing Durbar.

The second phase had ended, and it only remained for the troops to march back to India.

CHAPTER IV.

THE SECOND AFGHAN WAR—THE RELIEF OF KANDAHAR.

The Position in Southern Afghanistan—Maiwand—The Defence of
Kandahar—The Measures for Relief—The Kabul-Kandahar March—
The Battle of Kandahar—The Evacuation of Afghanistan.

MAPS 1 & 5.

After the departure of the ·Bengal troops under Sir
Donald Stewart affairs had settled down so satisfactorily
in the south that it was decided to detach the whole
Province of Kandahar from Afghanistan, and to make it a
British Protectorate. With this change in view, an Afghan
Governor, or "Wali", was appointed, with power to raise
levies. But as the summer approached there were disquieting rumours from Herat, where Ayub Khan, a
younger brother of the deposed Amir Yakub Khan, was
reported to be stirring up trouble. The Wali took his
levies to the Helmund to guard the frontier, and in June
there came word from him that Ayub had succeeded in
forming a confederacy of the chiefs in the west for an
attack on the British. Under the leadership of Ayub they
were now advancing on the Helmund, and the Wali asked
for British troops to stiffen his levies.

There had been considerable changes among the troops
in Southern Afghanistan, and the five to six thousand
men now under the command of Major-General Primrose
included the following artillery :—

C.R.A. —Colonel W. French.
Adjutant—Lieut. W. A. Plant.
E/B—Horse Battery.
C/2—Field   do.
5/11—Heavy   do.
No. 2 Bombay—Mountain do.

In response to the Wali's appeal General Primrose
despatched Brigadier Burrows to the Helmund with two

CHAPTER
IV.

1880.

The
Position in
Southern
Afghanistan.

CHAPTER
IV.

1880.

The
Position in
Southern
Afghanistan.
regiments of Indian cavalry, three infantry battalions (one British, two Indian), and E/B. When, however, they reached Girishk on the Helmund, where the levies were in camp, the latter deserted *en masse*. They were followed across the river, and although they showed fight the well directed fire of the horse artillery obliged them to abandon their guns. These consisted of a field battery of four 6-pr. guns and two 12-pr. howitzers (all smooth-bore) which had been presented to the Wali by the British. General Burrows' position was now entirely changed. Instead of having a loyal force to co-operate with him, these troops had either dispersed or gone over to the enemy, and the Wali himself was a refugee in his camp. The river Helmund was no protection, for it was now fordable everywhere, and the supplies expected at Girishk were not forthcoming. He decided, therefore, to fall back to a position where supplies were obtainable, and whence he could intercept Ayub whether he marched on Kandahar or Ghazni. Here he remained for over a week, and during that time, at the suggestion of Captain Slade of E/B, who was acting as galloper to the Brigadier, the smooth-bore guns were handed over to him to be formed into a battery. As section commanders he had two gunner subalterns who were serving with the Ordnance and Transport, and one subaltern from the 66th.[1] Non-commissioned officers from E/B provided Nos. 1, and volunteers from the 66th the gunners.

Maiwand.
At last, on the afternoon of the 26th July, information was received that the Afghans were making for the Maiwand Pass some few miles distant. Burrows decided to march against them at once, hoping to break up the advanced troops before help could reach them from the main body.

27th July.
Early on the 27th the force moved off, but it was not until after 10 o'clock that there was any sign of the Afghan

---

[1] The 2nd Battalion Royal Berkshire Regiment.

army. Horsemen were then seen moving across the front, and Lieut. Maclaine dashed out with his section and opened fire. He was followed by the cavalry and another section, which came into action about half-a-mile further back. On this last a line was gradually formed, with the guns[1] in the centre, two battalions on the right, one on the left.

For half-an-hour or so nothing occurred to disclose the fact that the little force was facing the Afghan main body, for the heat haze, mirage, and dust prevented any accurate estimate of the enemy's strength being made. Then their artillery began to come into action, until eventually the fire of thirty guns[2] was concentrated on the British line. For two hours this artillery duel continued while the Afghan cavalry and infantry endeavoured to work round the flanks, and then an incident occurred which was to have fatal effect. The smooth-bore battery had no ammunition wagons, and when it ran short of ammunition Captain Slade ordered it back to the reserve to get what was left there. The retirement of the guns was followed by a general development of the enemy's attack, under which the battalion on the left, next to which the smooth-bores had been, was borne back, mixed up with swarms of Ghazis, upon the guns of E/B. The gunners, who had borne the brunt of the Afghan fire throughout, made a gallant stand, until Captain Slade gave the order to limber up. Fortunately the limbers and detachment horses were formed up, according to the drill of the period, only ten yards from the trails, so there was no delay, and two of the sections got safely away.

After this brief outline we can turn to Captain Slade's own account. It was addressed to Captain Saward, who had only left E/B a year before on appointment to

CHAPTER
IV.

1880.

Maiwand.

27th July.

---

[1] The smooth-bore battery was distributed by sections during the course of the action, but had been reformed before ordered to retire for ammunition.

[2] 24 horsed field guns, and 6 mountain guns.

the staff, and was now holding the office of D.A.A.G., R.A.
in India. He was thus very well placed to ensure the
true story of the battery's action becoming known, and
we may assume that this was in Captain Slade's mind when
he wrote. The letter was certainly looked upon as of
importance by Saward, for he kept it carefully until his
death, and left directions for its preservation as a historical
document. It has been handed over to the Institution.[1]

COPY.

Camp, Kokeran,
9th Sept.

"My dear Saward,

The 27th of July was certainly an unfortunate one for
British arms—but I think when the truth is known Gunners
will be found to have done their duty. Nothing c^d. have
been steadier in my opinion than the behaviour of both
N.C. Officers & men of E/B both in the action and in the
retreat, & I have already brought to the notice of the Lt.
Genl. Comdg. the distinguished and conspicuous conduct
of 5 or 6 of the men, & I trust if you can further their
interests that you will do so, as considering the panic
striken state of 9 out of every 10 individuals present it was
all the more praiseworthy on their part.

Sergt. Major Paton's conduct was everything to be
desired, his conduct under a very heavy fire was as cool
and collected as if on parade—& in the retreat he stuck
by me & assisted me, most efficiently during the whole
night. Sergt. Mullane (I was in hopes w^d. have got the
V.C., but unfortunately I could not collect sufficient
evidence, but his) behaviour was most gallant, as when I
gave the order to limber up, he ran back under heavy fire
& to within some 15 yards of the enemy's infantry & picked
up one of our wounded men (who unfortunately was then

---

[1] Major-General M. H. Saward, Colonel-Commandant, R.H.A.,
d. 1928. Major Blackwood had been badly wounded in the thigh, and
handed over the command to Slade.

dead) & placed him on the limber. Corporal Thorogood, Trumpeter Jones, & Gunner Collis, are also mentioned for individual acts, so I trust they may be rewarded with the distinguished conduct medal.[1]

(You will have seen the various accounts in the papers of the action, but I w^d. not pin my faith to their accuracy— but w^d. be inclined to form my opinion fr. the official Account.

Maclaine opened the ball by galloping out to the left front & coming into action on the enemy's right at a range of about 1,700 yds. & in place of withdrawing him, we shd. have moved up to his position, as it was he was withdrawn & formed up with the battery in the centre of the line—in echelon of divisions.

For 3 hrs. we were exposed to a very heavy artillery fire, and our horses and carriages suffered greatly, almost all our men were killed by artillery fire, in fact I don't know of any individual being killed by infantry fire, 2 or 3 were wounded by sabre cuts when we were retiring, and one man had his left arm smashed by a Snider bullet.

The enemy had to advance over a distance of about 600 yds. & during this time were exposed to a very heavy fire of both musketry & artillery but tho' they fell in hundreds they were not to be deterred—and poor Maclaine waited a moment too long & lost his guns—they were within 15 yards of us when I limbered up—besides being in our rear. I then formed close interval & retired to a position about 400 yds. back where I came into action again to cover the retreat.

Owing to the artillery fire being so heavy I had to leave 67 horses dead or severely wounded on the field besides 3 wagons completely disabled. Poor Osborne was shot dead just as we were limbering up to retire. Blackwood was wounded in the thigh early in the day and never

---

[1] The following awards were made :—

V.C.   —Serjt. Mullane and Gunner Collis.
D.C.M.—Both Staff Serjeants and one of every rank from Serjeant to Trumpeter.

came back to the battery, & is supposed to have been killed in the garden with the 66th. Sergt. Wood of No. 4 was shot dead whilst laying the gun. Gunners Swinnerton, Roberts, Br. Lowe, Collar-maker Cumings, Gr. George Smith, Wheeler Dix—Drivers Gray, Istead, Webster, Richard Jones, Loughlin, Macalister, Mathewman, Macdonald, Dewley, killed. Of the men who were wounded I think you know—Mangan—Edwards lost his left arm, Sergt. Burridge, Sergt. Guffin, Br. Clarke.

Yours very truly,
John R. Slade."

To this may well be added a few graphic touches from the diary of Mr. W. M. Williams, then a gunner in No. 4 sub-division.

"By this time our loss had been great, the enemy's cavalry advanced opened out & came within 400 yards of us, & to our surprise we found each horse had 2 riders, for they halted and dropped down each an infantryman, who at once opened fire on us . . . These Ghazis advanced placing small coloured flags in the ground as they came up . . . they were repulsed by us several times, but came on again with renewed vigour . . . Lt. Osborne superintended the fire of No. 4 gun, and under his direction much damage was done to the enemy's advancing columns. After several repulses the enemy's infantry again advanced so near that we were compelled to resort to case shot. The case shot proved very effective, rows & rows of their infantry falling before us . . . The report of musketry was deafening, the enemy's artillery firing slowly and with good aim. The detachment horses of Nos. 3 & 4 subdivisions left in charge of Grs. Davis & Carver & myself, remained with the gun limbers some 12 or 14 yards from the guns in action. One after another the detachment horses fell until 3 only of them remained . . . many draught horses were kicking & plunging in the last agonies of death, while the drivers were engaged in removing their harness. . . . The enemy's entrenched infantry finding our firing slacker owing to the amount of shell used during the engagement, became more daring, & led on by their chiefs who carried large silken banners of various colours, they charged down on our guns, yelling & shouting as they came on. After firing a couple of rounds of case each gun, some one gave the order to mount. I cannot say whom. The two guns of the left division under Lt. Maclaine had to be abandoned at the moment when they were surrounded by the fanatics, the gunners escaped only by a miracle. . . . The right & centre divisions with the smooth-bores, limbered up, & commenced to retire. After proceeding about a hundred yards at a good gallop, I was suddenly brought to the ground, my horse, shot in the off hind leg, rolled on top of me, and tossing & writhing he set me free & I arose to my feet. The battery were some distance in front and were again coming into action."

Of the horrors of the retreat all that need be said is that the horse artillery played a part of which the Regiment will always be proud, bringing in as many of the wounded as could be carried, and suffering themselves a sad loss in the capture of Lieut. Maclaine while on an errand of mercy to get water for them.

The casualties in the battery were :—

Killed—2 officers, 19 n.c.o's. & men, 63 horses.

Wounded—2 do.   14 do.   do.   ?   do.

The news of the disaster reached Kandahar early on the 28th, and a small force, which included a section of C/2, was at once despatched to bring in the survivors. General Burrows, with the remnant of his brigade, was met at Kokeran and escorted back to Kandahar, but not without some sharp fighting in which the fire of C/2 proved very effective in clearing the road.

The defence of Kandahar is not an episode to be lingered over, but the Royal Artillery may always remember with satisfaction that the Regiment was a notable exception to the despondency which afflicted so large a part of the garrison. Lord Roberts tells how one and all bore testimony to the unfailing good behaviour and creditable bearing of the artillery.

The cantonment occupied by the troops had been built during the First Afghan War, and was quite untenable if attacked. The troops were therefore all withdrawn within the walls of Kandahar before nightfall on the day after Maiwand.

The city was in form quadrilateral, with a total perimeter of slightly over six thousand yards. It was surrounded by a solid wall twenty to thirty feet high, and included a separately enclosed citadel. The effective strength of the garrison was three thousand Indian and a thousand British troops, with fifteen guns. The artillery was :—

C.R.A. —Colonel W. French.

Adjutant—Lieut. F. J. Fox.

E/B—Four 9-prs.

C/2—Four 9-prs.[1]

5/11—{ Four 40-prs. R.B.L.
       { Two 8″ S.B. Mortars.
       { One 6-pr. S.B. Gun.[2]

Platforms for the guns were constructed on the walls, the 9-prs. at the gates, the 40-prs. on the bastions from which a wider field of fire was obtainable. The mortars were kept in the citadel ready for service wherever required.

On the south and east sides of the city groups of

---

[1] The other two guns of C/2 were at Kalat-i-Ghilzai.
[2] The sole survivor of the smooth-bore battery.

CHAPTER
IV.

1880.

The
Defence of
Kandahar.

August.

villages came close up to the walls, and these, with their
vineyards, orchards and walled enclosures, afforded ample
cover not only for assembling troops but also for concealing
guns. In the making and masking emplacements for the
latter the Afghan gunners showed much ingenuity, and,
when two 12-pr. Armstrongs were unmasked, the danger
of allowing the establishment within 1,000 yards of guns
of such power became evident. The possibility of an
attempt to rush a gate by troops collected under cover of
the villages had also to be remembered, and it was determined to clear the village of Deh-i-Kwaja opposite the Kabul
gate by a sortie. This was unwisely preceded by a bombardment by all the guns that could bear from their
positions on the wall, reinforced for the occasion by the two
mortars.[1] The fire could have little effect, however, on
such a mass of houses, and served only to warn the
enemy of the attack to follow. Entangled in the narrow
streets, fired on from all sides, the infantry suffered
severely. The brigadier[2] was killed, and the regiments
were only withdrawn with difficulty after losing heavily.

While this fighting was going on to the east of the city
the enemy threatened an attack from the west, bringing
guns into action on Picquet Hill above the cantonment,
and deploying infantry in the latter. But "such was the
steadiness and accuracy with which the 40-prs. were
served that the infantry were unable to form up, and
within an hour their artillery was silenced and one gun
dismounted."

Investment continued, however, and there was daily
artillery fire from Picquet Hill and from six guns in the
villages near the walls, until the 24th August, when great
commotion was observed, and guns moving off. The news

---

[1] The orders were for the artillery to open fire at 4.45 a.m. and
for the infantry to debouch from the gate at 5.0. They did not reach
the village of Deh-i-Kwaja until 5.30.

[2] His orderly was one of the trumpeters of C/2. He received the
D.C.M. for clearing the way, sword in hand, when the general was
being brought in and some Afghans tried to block the streets.

that Abdur Rahman had been proclaimed Amir, and that the relieving force was at Kalat-i-Ghilzai, had reached the Afghans.

On the receipt at Headquarters in India of the news of the disastrous defeat at Maiwand, and the withdrawal of the Kandahar Force within the walls of that city, measures for its relief were at once put in hand. The garrisons of the various small posts on the line of communication were concentrated at Chaman and the Khojak, at each of which there was a section of No. 2 Bombay Mountain Battery, and to further strengthen these important points detachments from 14/9 at Quetta were sent up, taking two 25-prs. to the Khojak, and to Chaman the two 9-prs. which had arrived to replace those lost by E/B.

At Quetta meanwhile a division was being organized and equipped under the command of Major-General Phayre. His artillery consisted of :—

C.R.A.   —Brig.-General Denis-de-Vitré.

Brigade-Major—Captain E. Blaksley.

at Quetta.

D/B—Horse Battery.

F/2—Field   do.

14/9—Heavy   do.[1]

en route to Quetta.

H/1—Field Battery.

5/8—Mountain do.

at Sukkur.

A/4—Field Battery.

15/9—Mountain do.

Their fate was hard, for after toiling through Sind and the Bolan Pass at the hottest time of the year, retarded not only by endless difficulties with the transport, but also by the attitude of the tribes, they received the despatches

---

[1] 14/9 was really a garrison battery in the fort at Quetta. It took on the 25-prs. and was re-equipped as a heavy battery with 40-prs. on arrival at Kandahar.

containing the news of the relief of Kandahar when within thirty miles of that city.

When the news of Maiwand reached Kabul it was at once apparent to Sir Donald Stewart that relief by way of Sind and Quetta at that time of year must be a matter of difficulty and delay, and he recommended the immediate despatch of a picked force from Kabul under Sir Frederick Roberts.

The sanction of the Government of India to Sir Donald's proposal was received on the 3rd August, and the selection of troops, followers, and baggage animals, and the survey and completion of equipment was at once proceeded with. The success of such an enterprise demanded much organization and forethought, for the force would be cast loose in a hostile country, with no base to fall back upon, and with a hard pressed fortress as its goal, the siege of which must be raised in the face of an enemy far superior in strength and flushed with victory. The reasons which led to the decision to take no wheeled vehicles must always be of interest to gunners. Their full discussion requires, however, some space, and has, therefore, been deferred to the "Comments" in the next chapter, so as to avoid interrupting the narrative. Here it need only be said that the force consisted of one cavalry and three infantry brigades with a total strength of ten thousand fighting men. The artillery was :—

C.R.A.　　　　—Colonel A. C. Johnson.

Adjutant　　　—Captain H. Pipon.

Orderly Officer—Lieut. R. Bannatine

　　6/8—British Mountain Battery.

　　11/9— do.　　do.　　　do.

　　No. 2—Indian Mountain　do.

The reasons for bringing up the two British batteries from the Khyber are interesting—6/8, a newly converted battery armed with 2·5″ jointed guns, owing to the reports of

CHAPTER
IV.

1880.

Measures
for the
Relief.

August.

CHAPTER
IV.

1880.

The Kabul—
Kandahar
March.

the effect of those of 1/8 in the Zaimukht expedition; 11/9 owing to the reputation it had won at the capture of Ali Masjid and in all the fighting on the Khyber Line since.

This force left Kabul on the 9th August, reaching Ghazni on the 15th, and Kalat-i-Ghilzai on the 23rd. Here news from Kandahar permitted a day's halt, and the garrison (which included a section of C/2 under Lieut. Mercer) were picked up. Another day's halt when within a march of Kandahar brought the troops to their goal on the morning of the 31st August, fresh and fit for anything. The total distance covered was just over 300 miles, and the average day's march therefore just under fourteen and a half. But for the 230 miles to Kalat-i-Ghilzai, during which no news of the position in Kandahar could be received, the average was close on seventeen miles a day. And this, be it remembered, through an enemy's country where all precautions had to be observed, where no sick or stragglers could be left behind, and with a temperature ranging from freezing point at dawn to over 100° at noon. The strain on the infantry was intense, and the transport animals suffered severely, but witness to the care taken of their mules by the mountain gunners is borne by the horse artilleryman, Gunner Williams of E/B, who records in his diary going to see them on their arrival in Kandahar.

A reconnaissance in force in the afternoon of the day on which Sir Frederick Roberts reached Kandahar found Ayub Khan's army busy entrenching a position on a broken and precipitous ridge some two and a half miles north-west of the city. This ridge was crossed by the road from Kandahar to the Arghandab valley at the Baba-Wali Kotal, while the road to the Helmund skirted its southern end, where it drops aburptly to the plain at the village of Pir Pamial.

So aggressive was the reception accorded to the reconnaissance that the troops in camp had to be called out to cover its withdrawal, and Roberts decided not to

CHAPTER
IV.

1880.

The
Battle of
Kandahar.

2nd Sept.

risk delaying until the arrival of the Quetta column, but to
attack next morning. His plan was to threaten the Afghan
centre and left about the Baba-Wali Kotal with the troops
of the Kandahar garrison, and to turn the right of their
position on the ridge with those of the Kabul column
directed on Pir Pamial, the cavalry making a wide sweep
across the plain in the hope of cutting off the retreat to
the Helmund. It is of a particular interest to note that
he attached the horse and field batteries of the Kandahar
garrison to the cavalry and infantry of the Kabul force
which had brought only mountain artillery with it.

Breakfasts were at eight o'clock, and shortly after nine
the 40-prs., whose turn it now was to occupy Picquet Hill,
opened the ball. Their segment shell did great execution
among the mass of Afghans in the Baba-Wali Kotal, but
the latter replied with much spirit, the shells proclaiming
the presence in the pass of the 9-prs. lost at Maiwand. At
the same time the Kabul cavalry brigade moved out into
the plain on the left, and E/B had the satisfaction of
clearing out of the way some of the Afghan batteries under
whose fire they had suffered so severely only a month
before. But the intense heat and want of water proved
a greater obstacle to the cavalry than the enemy, and the
hope of cutting off the latter's retreat was doomed to
disappointment.

Between the Kandahar troops on the right and the
cavalry on the left, the three brigades of Kabul infantry
with their mountain batteries, supported by C/2[1], assembled
behind the Karez Hill. The 3rd was to remain for the
present in reserve, the 1st and 2nd to carry out the
turning movement. Before, however, they could advance
against Pir Pamial they must capture the two villages of
Sahibdad and Gandigan perched on their detached knolls,
surrounded by orchards and vineyards. From a position
on Karez Hill, firing over the infantry, C/2 and 6/8

---

[1] More than half-a-century later the three subalterns of C/2—then
generals Sir F. Mercer, H. L. Gardiner, and G. B. Smith—presented
the battery, then 75/R.F.A., with a silver model of the guns they had
fought at Kandahar.

CHAPTER
IV.

1880.

The
Battle of
Kandahar.
1st Sept.

supported the attack of the 1st brigade by sweeping with shrapnel the roofs of Sahibdad. No. 2 Indian M.B. did the same for the 2nd brigade in their attack on Gandigan. The villages captured, E/B took up a position on the hill by Gandigan to cover the left flank while the two brigades advanced, side by side. It was slow enough work for the infantry and mountain guns through the tangle of enclosures, but the field battery found still greater difficulty in threading its way between the walls, with no sort of road to help until it got past the villages and found a track by the canal—though here it came under a pretty hot fire from the guns on the Kotal. It was not until noon that the southern extremity of the ridge was rounded, and the Afghan flank turned. But even after the capture of Pir Pamial, the infantry found themselves faced with a second position. From the fire of the guns in this position, and of those on the Kotal which were turned round to bear upon them, men began to drop fast. The only course was to storm the entrenchment without delay. Covered by the fire of 6/8 the 1st brigade charged with the bayonet and rushed a couple of guns, and the 2nd brigade, coming up on their left with No. 2 M.B. in support, captured the remainder. The enemy broke and fled in confusion, but the danger from artillery fire was not done with, for the 40-prs. of 5/11 were still bombarding the Kotal, and their "overs" were falling among the infantry. It was all important to inform the battery of the progress of the attack in the valley behind, though to do so involved great risk. Captain Straton, superintendent of signalling, undertook the task, and met his death at the hands of a lurking Ghazi.

There remained only to count the spoil. All the Afghan guns which had proved so fatal at Maiwand were captured, and also the two of E/B[1] which had been lost there. Ayub Khan's camp was standing just as he had

---

[1] These were presented to Sir Frederick Roberts by the Government of India.

CHAPTER
IV.

1880.

The
Battle of
Kandahar.
1st Sept.

left it. But the victor's triumph was marred by the discovery of Maclaine's body, with the head nearly severed, lying by the side of the track. The body was still warm, and he had evidently come out of the tent in which he had been kept prisoner to meet the British troops, and been cut down by a straggler.[1]

The expenditure of ammunition was :—

    40-pr.—161 rounds.
    9-pr.—152 do.
    7-pr.—161 do.

The captured guns were :—

6-pr. smooth-bore bronze	16
12-pr. do. do. howitzers ...	2
9-pr. rifled breach-loaders, iron	6
3-pr. do. do. do.	4
4-pr. do. do. do.	2
	30

With the battle of Kandahar the third and final phase of the war came to an end, for Ayub Khan fled and his army dispersed. A change of Government at home brought also a reversal of policy, and it was decided to abandon the advanced positions in Afghanistan which had been acquired under the Treaty of Gandamak. During the winter of 1880-81 the Khyber and Kurram were evacuated, and in the following summer the Province of Kandahar was handed over to the Amir.

It remains only to record that the services of the artillery during the final phase were not overlooked. The fine work of 5/11 during the defence and the battle of Kandahar[2] was acknowledged by the selection of this

---

[1] Lieut. Bannatine, who had been sent on to report as to the captured guns etc. in the camp, found the body, and identified it.

[2] This was the last occasion on which one of the old "Bail Batteries" was in action.

CHAPTER
IV.

1880.

The
Evacuation
of
Afghanistan.

battery to convey to India the guns captured in the battle. And the return of E/B was a veritable triumph. At Jacobabad their gallantry in the battle of Maiwand and their self-sacrifice during the retreat were extolled by the Viceroy. At Bombay all ranks were entertained by the citizens to a banquet, with the Governor presiding. At Kirkee the Honours they had gained were presented by the Commander-in-Chief at a full-dress parade of the whole division. We may note how often disasters to the other arms have brought opportunities of service to the artillery.

CHAPTER V.

THE SECOND AFGHAN WAR—COMMENTS.

ORGANIZATION.

Allotment of Guns to Formations—Command and Staff—Mountain and
Heavy Artillery—Siege Train—Ammunition Supply—Elephants.

ARMAMENT.

Guns—Ammunition—Range-finders—Gatlings.

EMPLOYMENT.

Tactics—Gunnery—Casualties—Medal & Clasps.

APPENDIX.

Regimental Commands and Staff Appointments—Army Commands and
Staff Appointments—The Afghan Artillery.

ORGANIZATION.

In considering the organization of the artillery in the
Second Afghan War, the first point that strikes one is the
small proportion of guns to bayonets. In the three
columns prepared at leisure for the invasion of Afghanistan
the proportion works out at approximately 2·8 per thousand
in the Peshawar Valley Field Force, 2·7 in the Kurram,
and only in the combined force under Sir Donald Stewart
does it rise to four per thousand. In the later campaigns there was little change. At Ahmed Khel the proportion was under three; in the force so carefully organized
for the Kabul-Kandahar march it was under two; and in
the Battle of Kandahar it only just reached that figure.

But, however small the total number of guns might be,
every nature was generally represented. For anything
approaching the strength of a division the usual allowance
was one horse, one field, one heavy, and one mountain
battery. If a second mountain battery was allotted it

*Chapter
V.*

*Allotment of
Guns to
Formations.*

would be an Indian one if the first were British, or vice
versâ, and so complete the diversity.

The most interesting experiment in this respect was
undoubtedly the decision to take only mountain artillery
with the force to relieve Kandahar. Much discussed at the
time and pregnant with results, it is well to recall the chief
arguments adduced at the time.

For the composition of his force, Sir Frederick Roberts
was given full freedom of selection from the whole of the
troops under Sir Donald Stewart in or about Kabul. These
included horse, field, and heavy batteries which had just
marched from Kandahar to Kabul, and knew the road, as
well as having proved their worth in the most closely con-
tested battle of the war. And yet he decided to take only
mountain batteries. The Commander-in-Chief in India
was naturally surprised, particularly in view of the fact
that Ayub Khan was known to have thirty guns, including
a battery of Armstrong breech-loaders "admirably served."
In reply to his expostulation, Sir Frederick pointed out
that his object was to reach Kandahar in the shortest
possible time, and that it was not improbable that he would
have to leave the main road and find his way across
country. With his recollection of the day of Killa Kazi,
he went on to explain that the ground throughout
Afghanistan was such that artillery could never be safely
employed with cavalry alone, and that, in any case,
rapidity of movement was not so much required of artillery
as the power of being able to operate over the most difficult
ground without causing delay to the rest of the troops. No
doubt he counted also on the fact that while the absence
of wheeled artillery would give him freedom of movement
on the way to Kandahar, he would find there horse, field,
and heavy batteries upon whose services he could call if he
had to fight a battle in order to raise the siege. It will be
remembered that on arrival he took this course—attaching
E/B to his cavalry, and C/2 to his infantry in the battle
of Kandahar.

Summing up the experience of the war in after years,

Sir Frederick, horse artilleryman as he was, came to the conclusion that "for service beyond the N.W. Frontier, the most useful armament would be the best gun that can be carried upon mules, and the heaviest gun that can be dragged upon wheels."

As a general rule a lieut.-colonel was appointed to command the artillery of a force, with an adjutant and often an orderly officer as well. His command was sometimes referred to as the artillery "brigade," sometimes as the artillery "division," but these terms were used indiscriminately without any suggestion of a permanent formation. The only case of a higher staff is that of the Kandahar Field Force of two divisions, which had a brigadier as C.R.A., with a brigade-major. Under him lieut.-colonels commanded "Horse & Field" and "Heavy & Mountain" respectively.

Colonel W. J. Williams pointed out in his usual incisive style the drawbacks of this system of command, showing how, owing to their junior rank and their position with the staff of the general, the divisional C.R.As. never commanded their artillery, which was in consequence frittered away, the batteries attached to cavalry and infantry brigades. He stressed the fact that owing to the want of unity of command the full power of the artillery was not developed in battle, and he urged the appointment of brigadiers, with the same status as those of cavalry and infantry, to command the artillery. To be with their guns in action and not "hanging about with the staff." It was to take thirty years to get them!

The system of converting garrison companies of the Royal Artillery into mountain batteries by the attachment of a "mountain train" has been explained in the Introduction and need not be repeated here. On the first outbreak of the war there were two of these mountain trains, and the batteries serving with them (11/9 and 11/11) were the only two British mountain batteries to

CHAPTER
V.

Mountain
and Heavy
Artillery.

take part in the first phase of the war. When, later on, more such batteries were required their creation, or rather conversion, proved a difficult business. Thus 6/8 was moved to Peshawar in December 1879 and provided with guns, but their completion with mules and drivers was such a tedious affair that the battery was not able to start for the front until March 1880.

Indian mountain batteries, on the other hand, had been permanent units for some years, but only took four guns into the field. The other two were kept at their depots in battery charge but no men or mules were provided for them. It was not until the conclusion of the first phase of the war that orders were issued for the batteries to be brought up to a six-gun strength. Volunteers from Indian infantry regiments were called for to supplement the number of recruits at the depôts, and such suitable mules as could be found in the Transport were taken to eke out the inadequate supplies procurable in the market.

Heavy batteries were in much the same position as British mountain batteries in that they were temporary formations, a garrison battery being "converted" by the issue of the necessary equipment, and the native establishment with the animals.

Siege Train.

Mention must in conclusion be made of the Siege Train although it never reached Afghanistan in that form. With the possibility of having to lay siege to Kandahar, three garrison companies were assembled at Sukkur in the autumn of 1878, and were formed into a Siege Train, under a Colonel, with a Staff Officer and an Adjutant. It was divided into three "divisions," armed respectively with 40-prs., 25-prs. and 6·3″ howitzers. But it was still on the march when news came that Kandahar had been occupied, so the colonel, the adjutant, and the companies were sent home, and the staff officer took the guns on to Quetta.

Ammunition
Supply.

The total number of rounds to be taken into the field was supposed to be 500 per gun. Owing, however, to want of transport this had frequently to be cut down. Thus, in his advance on Kabul, General Roberts took only 300, and in his march to Kandahar only 120,—90[1] of which were carried by batteries, 30 by the Park.

It will have been noticed that there has been no mention of ammunition columns in the narrative. The organization of the Army in India at that time contained no such link between batteries and the Ordnance Field Park. The latter was a mobile unit, usually close up to the front—necessarily so for protection—and the batteries had two lines of wagons. Of these the first line only was horsed, and even they found great difficulty in getting along the mountain tracks. Everyone was agreed as to the drag on the mobility of batteries presented by the ammunition wagons, but they could not be left behind, for, small as the expenditure of ammunition was, it was more than could be carried in the gun limbers. So there arose an agitation for two-wheeled carts, or limbers, instead of wagons. There were even some who advocated the adoption of pack transport, as in a mountain battery, so as to confine the wheeled vehicles to the guns and limbers. It was urged that it was always possible to get the guns along, but that the wagons were quite another matter, and if they were not permitted to accompany the guns it was almost impossible to get them up through the block of traffic in a mountain pass.

To return to the question of the bullock-wagons. Up till the Mutiny period bullock-draught had been the normal system for field batteries in India. But after the great development in the mobility of field artillery which followed the Franco-German War, its retention for any portion of a battery was certainly an anachronism. Sir Donald Stewart commented bitterly on its results before crossing the Khojak in his invasion of Afghanistan.

"*1st Jan., 1879.* The artillery have simply collapsed, owing to the complete failure of the bullocks.

---

[1] 44 common, 10 double, 24 shrapnel, 8 case, 4 star.

They have died in large numbers, and from sore feet and other causes are hardly able to drag themselves, much less loaded wagons, along, even an easy road. . . . . In these days an artillery that is practically tied down to bullocks is simply an encumbrance, and we should have been in Candahar a fortnight ago if the guns could get along like the other branches—and field artillery ought to do this".

Gradually the ammunition from the second-line wagons was transferred to camels. Special boxes were designed to carry nine rounds, two making a load, but for the most part gunny bags were utilized, carrying twelve rounds on each side.

Elephants.

The central figure of the scene depicted on the reverse of the Afghan medal is an elephant carrying a gun, and since this war was the last in which elephants were used by the artillery, this would seem an appropriate place for some reflections on the value of their services.

These come under two distinct catagories :—

(i) In pulling the guns of heavy batteries.

(ii) In carrying the guns of horse and field batteries.

Taking the heavy batteries first. On the march, and on all occasions until coming under fire, the guns were drawn by two elephants, tandem fashion. But owing to the elephants' objection to being shot,[1] alternative teams of bullocks had to be provided for taking the guns into action. With the native attendants required for elephants and bullocks the establishment of a heavy battery thus became enormous.

Yet on occasions elephants were found invaluable for getting a heavy gun out of difficulties. On the sharp hair-pin bends of the mountain roads they could exert their strength where there was no space for a team of

---

[1] In accounts of the Peiwar Kotal it is specially noted that when the section of F/A was brought into action against the gun shelling the camp, the elephants did *not* bolt.

horses or bullocks. In descending steep places an elephant in his breeching had complete control and could bring a gun down where any other "wheeler" would have been helpless.

To turn now to their use in pack. The early mountain guns were feeble weapons, and the value of the heavier metal of horse and field guns was clearly recognized. But there were many occasions on which the difficulties of the ground proved too much for wheeled vehicles. So field guns were put on the backs of elephants, and these demonstrated their value as good climbers, even when so loaded, in many an Indian campaign as well as in Abyssinia. Only a year before the Afghan War the guns of I/C had been so carried in the Jowaki expedition with admirable results, and a second battery of horse artillery was, in consequence, provided with elephant equipment.

In the first phase of the Afghan War I/C managed to get their guns into action at Ali Masjid without calling upon their elephants, although these were on the spot. But if the Afghans had held out for another day they would certainly have been required. In the Bazar Valley expeditions that followed D/A were similarly equipped, and in the Kurram besides carrying the guns of F/A up the rough track to the Spingawai Kotal they performed many incidental services.

But, as the war went on, it was found that in a country where the cultivation was sparse, the ground rocky, and the climate rigorous, the use of elephants had its drawbacks. Their appetite was too voracious, their feet too soft, their constitution too delicate. And so, before the winter of 1879, all the elephants of the Kabul Field Force were sent back to India.

ARMAMENT.

The whole of the horse and field batteries were armed with the 9-pr. R.M.L. gun, which had been introduced on

CHAPTER
V.

Guns.

the reversion to muzzle-loaders in 1870, to supersede the Armstrong breech-loaders. In accordance with the same decision the armament of the heavy batteries was being changed from 40-pr. R.B.L. guns and 8″ mortars to 40-pr. R.M.L. guns and 6·3″ howitzers. The batteries first employed had the old equipment, those sent up later the new, and exchanges were sometimes effected.[1]

The mountain batteries, like the heavy, were in process of re-armament. At the beginning of the war all had the old 7-pr. of 200 lb., but shortly afterwards the new 2·5″ jointed gun arrived in India, and was issued to the British batteries converted to mountain during the war—1/8 and 6/8. The first occasion on which it was used was in the Zaimukht expedition in the winter of 1879-80 by 1/8 when its superiority over the 7-pr. in range, accuracy, and shrapnel effect attracted general attention. Thus arose that enthusiasm for the "screw-guns" which found expression in prose and verse.[2]

The gun was noteworthy, however, for more than its jointed construction. It was the first of the longer type, having a length of 26½ calibres, in place of the 17 calibres of the 9-pr., and its projectiles were studless, the rotation given by a gas-check.

Ammuni-
tion.

As regards ammunition, there is little to note except that the R.B.L. guns had "segment" instead of shrapnel shell. This, like the "ring" shell much used on the continent, was a common shell arranged to break up into a large number of fragments, and was used with a percussion fuze. Possibly for this reason it was very popular : there was a considerable school in the Regiment who did not believe in the possibility of time fuzes being

---

[1] For instance, 5/11 was re-armed when it returned to Quetta temporarily, but after getting back to Kandahar it was ordered to hand over its new equipment to 6/11 which was to accompany Sir Donald Stewart on his march with the Bengal troops to Kabul in April, 1879.

[2] The innovation did not meet with general approval at first. The following is an extract from the diary of a distinguished staff officer : "17 April 1880. Went out to see the new screw-guns : one of them stuck from dust, which is just what I said."

set accurately under fire. On the other hand the war saw the greatest step ever taken in the evolution of such fuzes in the trial of the "time and concussion" fuze afterwards adopted for the service—the first in the long series of metal time and percussion fuzes which were so soon to supersede the old "Boxer" wooden time fuzes. The time fuzes would not act at less than a thousand yards, and so when the case shot had all been fired away, all that could be done was to load the shrapnel reversed, with their plugs removed.

At the time of the 2nd Afghan War there were no such things as range-finders in the service. The first practice camp at Okehampton in 1875 had drawn attention to the want, and the necessity had been recognized, but no satisfactory instrument had been forthcoming. Just about the time of the Afghan War, several attempts to solve the problem had been made. Of these the Weldon, the invention of an officer in Madras, had been adopted in India, and gained many adherents in the war where its simplicity, in comparison with the Watkins and Nolan also under trial, greatly recommended it. The long base required in order to get accurate results with the Watkin was also found a great disadvantage in a mountainous country, but it was eventually adopted as the service instrument.

There remains only to notice an interesting experiment in armament, in the shape of the issue to artillery of "Gatling" machine guns. During the summer of 1879 there arrived in the Kurram, direct from England, cases containing a couple of these newly invented weapons, with orders that an artillery unit was to be formed on mountain artillery lines. Gunners were sent up to man them, ponies were purchased to carry them, and the British infantry were called upon for drivers. But when the cases were opened it was found that there were many deficiencies. Spare parts were manufactured by the armourers, saddlery was improvised from that of the Afghan battery captured on the Peiwar, but when it came to the first trial on the ranges the shooting was found to be hopelessly inaccurate. This was a great disappointment not only to the gunners but to the whole of the troops who had turned out to watch the practice. By degrees, however, the cause was found, and this and other defects were eliminated, so that the unit was in a fairly serviceable condition when it started with the Kabul Field Force. But the arrangement of the ammunition drums was always liable to cause jams unless the drill was absolutely accurate, and at Charasia, the first occasion on which the guns came into action, they failed from this cause, and were little used afterwards. The last heard of them is in the defence of Sherpur Cantonment.

EMPLOYMENT.

It will have been plain from what has been said under the head of organization that no brigading of batteries for the control of their fire was contemplated. Before the advent of range-finding or signalling such control was, of course, out of the question unless the batteries were in action side by side, and such instances of massing of guns as can be found appear to have been due to accident rather than design. The idea of the brigade as a tactical unit had not yet occurred to any except the more advanced thinkers like Colonel Williams. The tactical unit was the battery.

And yet it was rare to find even a battery in action as a whole. It was common to take out only four guns, sometimes only three. Camps had of course to be protected, but even when whole batteries were in the field there was a tendency to split them up by sections.[1] It has, of course, to be remembered that "fire discipline" of a battery had not yet been invented. The fire-unit was the gun.

---

[1] This does not refer to the splitting up of batteries in posts along the lines of communication, nor to the allotment of guns to the small columns employed in many minor operations, which was unavoidable since a couple of guns was often all that was required or available.

The evil results of separating the ammunition wagons from the guns was well exemplified. It will have been noted that on these occasions it was the custom to send back the guns. Prince Kraft's admirable remarks on the ill-effects of this practice had not yet inspired our manuals, but Maiwand pointed the same moral.

The action of the artillery was usually simple, the batteries supporting the brigades to which they were attached, advancing to successive positions side-by-side with them, never hesitating to go in to close range. Their fire was thus in many cases of great value in keeping down musketry fire from the Afghans crowded upon house tops or mountain ridges. Where the enemy brought their guns into action this advance was generally preceded by an artillery duel, in which the 40-prs. were often employed with great effect.

Escorts appear to have been told off to each battery for the whole day as a matter of routine—a squadron to a horse battery, a company to a field. Several cases will have been noticed where this precaution was amply justified. As pointed out by Sir F. Roberts, the nature of the ground in Afghanistan was such that wheeled artillery were liable at any moment to be held up by some impassable obstacle which would afford an opportunity for an active enemy.

The No. 1 commanded and layed his gun, although he had been promoted on account of aptitude for command and not for his eyesight. So much indeed was the laying of the gun looked upon as a function of command that it seems to have been common for not only section but battery commanders to undertake it. It will be remembered that Captain Kelso was killed while laying a gun at the attack on the Spingawai Kotal, Lieut. Montanaro was another case in this war, and many will be found in other campaigns.

The range was always found with common shell and percussion fuze before proceeding to time shrapnel. The small number of rounds expended bears witness to the slow rate of fire possible with the muzzle-loading guns then in use but it has also to be remembered that the smoke of the gunpowder was a fruitful cause of delay.

If the casualties in action were comparatively slight the rigours of the climate and the privations suffered took a heavy toll, and there was ever-present the risk of a fanatic's knife.[1] The deaths from cold among the followers were very numerous, and the sufferings of the battery animals—elephants, camels, oxen, horses, mules—from semi-starvation and overwork were terrible.[2] The horrors of the march to Quetta, across the desert and up the Bolan, lingered long in the recollection of all that took part in them. Many accounts have come down to us of the gunners and the infantry of the escort assisting with drag-ropes the exhausted animals—elephants and bullocks so enfeebled from want of food and water, and so footsore from the shingle of the river-bed, that they lay down in many cases never to rise again. The agonized demands of battery commanders for supplies for their starving animals elicited nothing but orders to "push on with all speed," but it took some batteries three months to reach Kandahar from their Indian stations although they railed as far as Sukkur.[3]

And with this may be compared the return of the troops on the Khyber line to India at the end of the first phase of the war—in the hottest time of the year, and with the cholera always with them.

MEDAL & CLASPS.

The Afghan War Medal was granted to all troops who served over the Afghan Frontier between the 22nd November 1878 and the 26th May 1879; and between the

---

[1] E/4 lost an officer and two men from such murderous attacks.

[2] The horses that stood it best were the "Gulf Arabs" or "Persians," with which C/2 was half horsed.

[3] It was supposed to be dangerous to take elephants by train so those of the heavy batteries had to march the whole way—a thousand miles from Gwalior to Quetta.

3rd September 1879 and the 15th August 1880 on the
Khyber and Kurram Lines, and the 20th September 1880
in Southern Afghanistan.   Clasps were given for the
battles or actions of PEIWAR KOTAL, ALI MASJID, CHARASIA,
KABUL, AHMED KHEL, and KANDAHAR; and a Bronze Star
to the troops which accompanied Sir F. Roberts from Kabul
to Kandahar, including the garrison of Kalat-i-Ghilzai
which he picked up on the way.

APPENDIX.

REGIMENTAL COMMANDS AND STAFF APPOINTMENTS.

Major-General W. French	G.O.C., R.A.
do. do. A. H. Lindsay	do.
Brigadier-General C. G. Arbuthnot	do.
do. do. C. R. O. Evans	do.
do. do. W. Denis-de-Vitré	do.
Colonel B. L. Gordon	C.R.A.
do. A. C. Johnson	do.
do. W. J. Williams	do.
do. A. H. Dawson	do.
do. C. B. le Mesurier	do.
do. H. M. G. Purvis	do.
do. T. P. Smith	do.
Lieut.-Colonel W. Carey	do.
Captain A. D. Anderson	Brigade Major.
do. E. Blaksley	do.
Major J. W. Inge	Adjutant.
do. J. C. Robson	do.
do. G. W. C. Rothe	do.
do. H. Pipon	do.
Captain F. H. G. Cruickshank	do.
do. F. J. W. Eustace	do.
do. C. H. Hamilton	do.
do. R. A. Lanning	do.
do. J. R. Slade	do.
do. J. Keith	do.
do. H. T. Lugard	do.
do. W. A. Plant	do
Lieut. Hon. F. E. Allsop	do.
do. R. Bannatine.[1]	do.
do. J. W. Dunlop	do.
do. F. J. Fox	do.
do. W. G. Knox	do.

---

[1] Changed later to Bannatine-Allason.

ROYAL HORSE ARTILLERY.

BATTERY.	BRIGADE.	BATTERY COMMANDER.	1914.
D	A	{ Major P. E. Hill { Lieut.-Colonel S. Parry	} — O R.H.A.
E	—	Major W. W. Murdoch	— E do.
F	—	{ Lieut.-Colonel W. Stirling { Captain H. Pipon { Major J. C. Smyth-Windham	} — D do.
I	—	do. M. W. Ommanney	— I do.
A	B	{ Colonel D. MacFarlan { Captain R. G. S. Marshall { Major H. de G. Warter	} — F do.
D	—	Major F. W. Ward	— 57 R.F.A.
E	—	{ do. G. F. Blackwood { Captain J. R. Slade { Major J. A. Tillard	} — 58 do.
H	C	do. C. E. Nairne	— No. 1 Depôt R.F.A.
I	—	{ do. G. R. Manderson { do. Hon. A. Stewart	} — W R.H.A.

FIELD BATTERIES.

H	1	{ Major F. H. Pritchard { do. C. Crosthwaite	} — 29 R.F.A.
I	—	do. H. C. Lewes	— 16 do.
C	2	do. P. H. Greig	— 75 do.
D	—	do. E. Stavely	— 14 do.
F	—	do. J. R. J. Dewar	— 64 do.
C	3	do. H. C. Magenis	— 33 do.
E	—	do. T. M. Hazlerigg	— 42 do.
G	—	{ do. S. Parry { Captain R. Purdy { Major W. R. Craster	} — 6 do.
C	4	do. J. C. Auchinleck	— 25 do.
D	—	do. J. F. Free	— 18 do.
E	—	do. T. C. Martelli	— 66 do.
G	—	{ do. Sir J. W. Campbell { do. P. K. L. Beaver	} — 69 do
L	5	do. W. R. C. Brough	— 74 do.

GARRISON COMPANIES.

1	8	Major J. Haughton	— Reduced as 92 R.G.A.
6	—	{ do. T. Graham { do. J. C. Robinson	} — No. 3 Mtn. Batty.
11	9	{ do. J. R. Dyce { do. J. M. Douglas	} — 65 R.G.A.
12	—	do. H. L. Gwynn	— 64 do.
13	—	do. C. W. Wilson	— Reduced as 53 R.G.A.
14	—	do. G. A. Crawford	— 42 R.G.A.
5	11	{ do. C. Collingwood { Captain G. M. B. Hornsby	} — 86 do.
6	—	Major J. A. Tillard	— Reduced as 87 R.G.A.
10	—	do. C. D. A. Straker	— 89 R.G.A.
11	—	do. N. H. Harris	— 84 do.
Gatling Battery		do. A. Broadfoot	— —

SIEGE TRAIN.

Commander—Colonel E. J. Bruce.

Staff Officer—Major W. H. Noble.

Adjutant —Captain R. A. Lanning.

13	8	Major E. S. Burnett	— 40 R.G.A.
16	—	do. J. H. Blackley	— No. 8 Mtn. Batty.
8	11	do. H. H. Murray	— 91 R.G.A.

INDIAN MOUNTAIN BATTERIES.

No. 1 M.B., P.F.F.	{ Captain J. A. Kelso { do. H. N. Jervois { Major H. R. L. Morgan	} 21 Kohat M.B., (F.F.)
No. 2 do. do.	{ Major G. Swinley { do. A. Broadfoot { Captain H. F. Smyth	} 22 Derajat do. do.
No. 3 do. do.	Major J. Charles	23 Peshawar do. do.
No. 4 do. do.	do. E. J. de Lautour	24 Hazara do. do.
No. 5 Garr. Bty., do.	{ Captain H. F. Smyth { Lieut. R. A. C. King	} The Frontier Garr. Arty. do.
No. 1 Bombay M.B.	Captain J. D. Snodgrass[1]	25 Bombay M.B.
No. 2 do. do.	Major R. Wace	26 Jacob's do.

[1] Changed later to Douglas.

Army Commands and Staff Appointments.

Lieut.-General Sir F. S. Roberts	Cmdg. Division.
do. do. Sir M. A. S. Biddulph	do. do.
Major-General Sir J. Hills[1]	do. do.
do. do. Sir C. W. Arbuthnot	do. Brigade.
do. do. J. E. Michell	do. do.
Colonel E. F. Chapman	D.A. & Q.M.G.
Lt.-Colonel C. A. Gorham	D.J.A.G.
Captain R. McG. Stewart	A.Q.M.G.
do. A. B. Stopford	D.A.Q.M.G.
do. W. Law	Brigade Major.
do. G. T. Pretyman	A.D.C.
do. D. C. Dean-Pitt	do.
do. F. J. W. Eustace	do.
do. Hon. W. C. Rowley	do.

The Afghan Artillery.

After his visit to India in 1869 the Amir Shere Ali determined to create an army on European lines. The most difficult problem was of course the artillery, but he carried back with him as a nucleus the complete equipment of two batteries presented by the Indian Government. These were a siege battery consisting of four 18-pr. S.B. guns and two 8″ S.B. mortars, and a mountain battery of six R.B.L. (Armstrong) 6-prs. These could be copied, and the Amir despatched his best gun-smiths to Peshawar to be instructed in the arsenal there. They learnt not only how to cast bronze guns but how to build up iron ones, and brought back complete models, so that they were soon able to construct at Kabul excellent imitations of Armstrong guns, utilizing water-mills for driving the boring machines, and rifling them by hand.

The English patterns of shot and shell were skilfully copied but it was beyond their skill to imitate successfully our time fuzes. None were turned out until the year of the war, and these were not a success.

A great many of the old brass guns were broken up, and improved pattern brass mountain guns made from their material. A return made in 1879 shows the

---

[1] Changed later to Hills-Johnes.

following number of guns in possession of the Afghans at
the commencement of the war :—

Siege Train.

     Presented by the Government of India   6

     Afghan imitations with elephant draught   ... 10

     do.   do.   do.   bullock   do.   ... 18

Field Guns.

     Iron R.B.L.   89⎫

     Bronze S.B.   56⎭ 145

Mountain Guns.

     Iron R.B.L.   6⎫

     Iron R.M.L   48⎬ 150

     Bronze S.B.   96⎭

Small Guns of Position 50

                                                 379

The artillery units of the Afghan Regular Army consisted of :—

     2 Elephant Batteries.

     22 Horsed Batteries.

     18 Mule Batteries.

     7 Bullock Batteries.

Their guns were often skilfully placed, generally in
concealed or protected positions at wide intervals, much
on the same lines as adopted by the Boers twenty years
later. Great care was devoted to taking advantage of the
conditions of the site. Thus on the Peiwar Kotal the
guns were at least 30 yards apart, and their platforms were
cut in the reverse slope and ramped so that the guns could
be quickly run down into shelter. On the Baba-Wali
Kotal similar positions were occupied, and these were
screened by the natural lines of rocks against which even
the 40lb. shell could make no impression.

They were inclined to open fire at ranges where no effect was attainable, but the great obstacle to their obtaining serious effect was the want of reliable time-fuzes. The gunners stood to their guns well, and on several occasions their fire proved an important factor in the action—at Maiwand it was undoubtedly the deciding factor.

CHAPTER VI.

THE NORTH-WEST FRONTIER—AFGHANISTAN TO TIRAH.

The Zhob Valley—The Samana—The Black Mountain—The "Roof of the World"—Hunza-Nagar—Chitral—The Gilgit Relief Column—The Indian Relief Force—The Durand Line—Wana.

COMMENTS.

Organization—Armament—Employment—Medals and Clasps—List of Units.

MAPS 1 & 6.

THE Zhob River, rising in the neighbourhood of Quetta, and flowing in a north-easterly direction to join the Gomal, forms a link between Baluchistan and Waziristan. Attention seems to have been first directed to it by the hostile attitude of the inhabitants during the Afghan War, and when the war was over it was decided to send an expedition into the country. The force detailed for the purpose under the command of Brig.-General Sir Oriel Tanner was provided with the following artillery :—

CHAPTER VI.

The Zhob Valley.

1884.

C.R.A.—Lt.-Col. Sir J. W. Campbell.

9/1 Northern, British M.B.—Lt.-Col. T. Graham.

No. 1 (Bombay) Indian M.B.—Captain A. Keene.

The operations lasted less than a month, but in that time the leader's forts were captured, his supporters routed, and general submission enforced. The "great execution" effected by the "screw-guns" of the British battery appears to have excited general admiration.

Again in 1890 General White marched through the country, bringing with him from Quetta :—

1890.

No. 7 British M.B.—Major F. Beaufort.

and being joined by troops from the Punjab including :—

No. 1 Indian M.B.—Captain G. F. W. St. John,.

No. 7 Indian M.B.—Captain C. P. Triscott.

The valley was traversed in every direction, Fort Sandeman was established and the venerated "Takht-i-Suliman" was scaled by Sir George White and a hundred men—half British, half Indian—fully accoutred. Then, after a Durbar at which the terms imposed were announced, the force broke up.

In the general survey of the frontier with which this Part commenced, the Miranzai valley was passed over, for little trouble had been given by the Orakzais, the tribe that dominated it, previous to the Afghan War. That war was followed, however, by a series of outrages, and, as soon as more pressing affairs were settled, a Miranzai Field Force was assembled under Sir William Lockhart for the infliction of the long-delayed punishment. His artillery consisted of :—

No. 3 Indian M.B.—Captain F. H. J. Birch.

No. 4 do. do. —Captain W. J. Honner.

Starting from Kohat in January, 1891, this force occupied, with negligible opposition, the Samana Range, a steep rocky ridge running parallel to the important road from Kohat to the Kurram. By February all sections of the Orakzais had made submission and agreed to the establishment of British posts on the Samana. Working parties were left to construct these under the protection of a battalion, and the rest of the force was withdrawn.

They were not left in peace for long. On the 4th April the tribesmen swarmed on to the ridge with no sign of hostile intent, and then treacherously turned on the guards over the working parties. Temporary abandonment of the ridge was unavoidable, but no time was lost in restoring the situation. Sir William Lockhart was once again called upon to take command, and troops were collected as rapidly as possible.[1] Within a fortnight he was advancing with 8,000 men, and the following artillery :—

---

[1] No. 3 British M.B., after detraining at Khushalgarh, and crossing the Indus, marched the 34 miles to Kohat between 9 p.m. that night and noon next day.

C.R.A.   —Lieut.-Colonel E. J. de Lautour.
Adjutant   —Lieut. F. H. F. R. McMeekan.
No. 3 British M.B.   —Major J. D. Cunningham.
No. 2 Indian do.   —Captain J. L. Parker.
No. 3   do.   do.   —Captain F. H. J. Birch.
Punjab Garrison Battery—Captain T. W. G. Bryan.

The British, and No. 3 Indian, mountain batteries had
2·5″ guns, No. 2 Indian battery 7-prs., and the garrison
battery 12-prs. B.L. taken off the ramparts of Kohat Fort,
and drawn by bullocks. This unusual step was taken on
the strength of reports of large sangars having been
erected to bar the ascent of the Samana, but though the
guns were brought into action to cover the advance on the
17th they found no target and were sent home. On
that and the following day the mountain batteries
accompanying the infantry did useful work in driving the
enemy out of their strongly walled villages. On the 20th
the tribesmen were caught between the columns and completely routed, survivors complaining that they could not
stand against the shrapnel fire of the three mountain
batteries. The Orakzai valley was then traversed by
columns as far as the head waters, hitherto looked upon
as inaccessible, and our occupation of the Samana insisted
upon. But this time the garrison left upon it was three
battalions with No. 3 Indian M.B.[1]

Twenty years after the first Black Mountain expedition
the lesson had to be repeated in consequence of an unprovoked attack within British territory upon a detachment
of Gurkhas from Abbottabad, in which two British officers
were killed. A "Hazara Field Force," nearly 10,000
strong, was assembled in September, 1888, under Brig.-
General McQueen, with four mountain batteries :—

> CHAPTER VI.
>
> The Samana.
>
> 1891.
>
> 17th April.
>
> The Black Mountain.
>
> 1888.

---

[1] Next year during a tribal feud one side applied for the services of
two guns of the battery offering to feed men and animals and pay for
all ammunition expended, and even to try and raise money for the
hire if the government demanded it.

G

C.R.A.   —Lieut.-Colonel C. J. Deshon.

Adjutant   —Lieut. F. R. Drake.

2/1 Scottish, British M.B.—Major H. R. L. Morgan.

3/1 South Irish do.   do. — do. R. A. C. King.

No. 2 Indian M.B.   —Lieut. W. Moore-Lane.

No. 4   do.   do.   —Captain W. J. Honner.

Gatling Section   — do.   N. D. Findlay.

   do.   do.   —Lieut. M. S. Eyre.

The British batteries had 2·5″ guns, the Indian batteries 7-prs. of 200 lb.

This force traversed the territory of the Black Mountain tribes in a number of columns, meeting with but slight resistance except on one occasion. While three columns were toiling up the eastern slopes, the fourth—with which were 2/1 Scottish and Eyre's Gatling Section—moved up the Indus valley, thus taking the mountain in reverse. The tribesmen were quick to recognize the danger of such an encirclement, and took up a position to block the way between the hills and the river, with sangars on the slopes and sharpshooters on the opposite bank. The guns came into action at six hundred yards against a wood in front of their centre, and after the whole had been well searched by artillery and machine-gun fire, the infantry started to rush the position. And then occurred a repetition of the charge of the Fanatics in the Ambela Expedition of 1863. A body of Ghazis who had been concealed in a nullah made a desperate attempt to break through the line —only to be shot down by the infantry and gatlings.

With the repulse of this forlorn hope the enemy's resistance collapsed. But the idea of celebrating it by a military promenade along the crest of the ridge roused such fierce opposition that it had to be abandoned, and another expedition substituted.

The ammunition expended by the batteries was :—

2/1 Scottish—	55 Shrapnel,	80 Common,—Double, 3 Case.
3/1 S. Irish—129	do.	132 do. — do. — do.
No. 2 Indian— 9	do.	24 do. — do. 2 do.
No. 4 do. — 23	do.	20 do. 3 do. — do.

CHAPTER
VI.

The Black
Mountain.

1888.

1891.

For the expedition of 1891 the troops under the command of Major-General W. K. Ellis amounted to a couple of thousand less than before, with only three batteries :—

    C.R.A.   —Lieut.-Colonel J. Keith.
    Adjutant   —Captain G. C. Dowell.
    No. 1 British M.B.—Lieut.-Colonel H. R. L.
                      Morgan.
    No. 9   do.   do. —Major F. A. Bowles.
    No. 2 Indian do. —Captain H. B. Brownlow.

Owing to a wide-spread coalition of the tribes the operations were more protracted than on the previous occasion, although again there was no serious fighting. The advance began in March, but although the main body was withdrawn in June, a brigade—which included No. 9 M.B.—remained in occupation until the end of November.

There was still another expedition in 1892 by the "Isazai Field Force" under Sir William Lockhart, but the tribes offered no resistance to the march of the troops through their country.

As pointed out in the general survey at the beginning of Chapter I, the North-West Frontier turns eastward after the Mohmand territory is passed, and circles round the Peshawar plain until it strikes the Indus opposite the Black Mountain. Beyond this semi-circle of hills there extends a block of tribal territory into which, up to the last decade of the XIXth century, British troops had never set foot. No mention of the tribes, or of their country, has, therefore, yet been made. As it was now to become the scene of hostilities, some description is necessary.

North of the Mohmands lie the Pathan countries of Dir and Bajaur occupying the basin of the Panjkora River.

To the east of them the rich valley of the Swat River runs behind the border hills. Further north still lie the little congerie of states, no longer Pathan, with which we are now concerned—Chitral, Hunza, and Nagar. Their country, stretching right away to the Hindu Kush, where "three Empires meet," is on a far grander scale than the border hills along which we have hitherto wandered. Here we are in the heart of mighty mountains, whose fastnesses are only accessible by mountain paths—little more than goat tracks. These cross passes of vast altitude, free from snow for only a few months in the year, or thread a tortuous course through gorges of immense depth. Sometimes they follow the bed of the torrent between stupendous walls of rock, made doubly perilous by prepared "shoots" for avalanches; sometimes they are carried giddily across the face of cliffs along narrow shelves in the rock, or even on galleries shored up on wooden brackets, crossing and recrossing the ravines on bridges of twisted twigs.

1885.
Strained relations with Russia in 1885 made it advisable to investigate the country adjacent to the Pamirs, for although it was hard to imagine such a country possessing any military value, the northern passes afforded a route not impracticable for a force large enough to cause dangerous excitement along the frontier. Further missions followed, and in 1889 a British Agency was established at Gilgit.

1889.
It is unnecessary to enter upon the troubles that disturbed the Agency. We are only concerned with the military operations to which they gave rise.

Hunza—
Nagar.
1891.
The first of these was with two tribes, living on opposite banks of the Hunza river, whose importance was due to their possession of the passes leading to the Pamirs, where the Russians had been indulging in a military promenade. It was decided that a mule road must be made from Gilgit to Hunza, and that we must have free access to their territory in order to ensure the safety of the frontier.

The tribes would have none of it, and even threatened the
Kashmir frontier fort at Chalt. There could be no
hesitation, and Colonel Durand, the British Agent,
collected a force just over a thousand strong, the greater
part Kashmir Imperial Service Troops, but including a
couple of hundred Gurkhas and a section of No. 4 Indian
M.B. With this force he attacked the frontier fort of
Nilt, a famous stronghold, reputed impregnable, which
formed part of a very strong defensive position. The
fort itself could not be seen until within a few hundred
yards, but the section was brought into action directly
a clear view could be obtained. It was soon obvious,
however, that the walls of solid stone were quite im-
pervious to the little 7-lb. shells although the range was
only 250 yards, and that if the fort was to be captured it
must be taken by assault. The only possible entrance
was by a gateway protected by walls and ditches, yet a
gallant little party blew in the gate, and took the place by
a *coup-de-main*—in the literal sense. Unfortunately
Colonel Durand was wounded in the moment of victory,
and the opportunity of capturing the rest of the defences
while abandoned was missed. By next morning they had
been re-occupied, and the guns again proved powerless to
destroy the sangars. An attempted advance failed somewhat disastrously, the one British officer with the guns
(Lieut. R. St.G. Gorton) being wounded. For between two
and three weeks the antagonists faced each other across
the ravine which formed the enemy's main defence, but
night after night search was being made for a way by
which this might be turned. At last a path was found,
up which an adventurous party made a perilous ascent,
and got above the sangars, while the main body and guns
kept their defenders under fire. As soon as these realized
that their flank was turned, they abandoned their position
and all resistance was at an end. An advanced force,
which included the guns, made a forced march of 27 miles
to Nagar, and peace was soon concluded with both that
state and Hunza.

There were risings in the Indus valley in the following years, but these do not concern us for the guns had no part to play. It was not until 1895 that they were called upon again.

Chitral.

1895.

The death of the old Mehtar of Chitral had been followed by the usual fratricidal feuds, but on this occasion the situation was complicated by the intervention of a redoubtable adventurer, Umra Khan, who had made himself master of the neighbouring state of Dir. The British Agent happened at the time to be at Chitral with a small escort, and he decided to occupy the fort. Umra Khan immediately advanced against it, drove back the portion of the escort which attempted to check him, and shut up the Agent and his little force in the fort.

The
Gilgit Relief
Column.

1895.

22nd Mar.

The troops in the Gilgit Agency at this time consisted of a battalion of Indian infantry (Pioneers), three battalions of Kashmir infantry, some Sappers and Miners, and No. 1 Kashmir Mountain Battery. Lieut.-Colonel Kelly, the commanding officer of the Pioneers, was the senior military officer in the Agency, and on the 22nd March a telegram from India directed him to assume military command, and to make such dispositions and movements of the troops as he thought fit.

Although the total number of men in the Agency amounted to nearly 3,000 the safety of Gilgit and of the long lines of communication had to be provided for as well as the relief of Chitral. Colonel Kelly decided, therefore, to limit his relieving column to a wing of his own regiment, and of the Kashmir infantry, a section of the mountain battery,[1] forty sappers, and a hundred Hunza-Nagar levies, very expert cragsmen.

The story of their march has been told many times :[2]

---

[1] The right or Dogra section was taken.

[2] *With Kelly to Chitral* by Lieut. (now Major-General Sir William) Beynon, K.C.I.E., C.B., D.S.O., whose reconnaissance sketches have been utilised here.

here it is only possible to summarize the struggles of the gunners under Lieut. C. G. Stewart.

A start was made at the end of March, and the march up the Gilgit valley was uneventful except for the constant struggle against the difficulties of the mountain road. But it was early in the year for such altitudes, and towards the head of the valley they got into the snow. A few days later it was in deep drifts where avalanches had swept across the track, and it deepened everywhere as the Shandur pass was approached. Where it was frozen hard on the steep slopes it caused dangerous falls, where it was soft it bothered the mules by balling, and as fresh snow fell they were floundering up to their girths. An attempt to trample down a track by driving some yaks in front proved of little value, and leading the mules in the bed of the stream had to be abandoned, for men and animals were soon numbed by the ice-cold water. Further progress with laden animals was impossible. Rough sledges were made from broken-up ration boxes, but these constantly upset, and their loads had to be fished out of the snow with much labour. It was evident that no animals could possibly pass over the snowfield in its then condition, so there was nothing for it but to shoulder the guns,[1] and this could never have been accomplished if the men of the Pioneers and Kashmir infantry, heavy laden as they were, had not volunteered to help the gunners in their desperate struggle. Worst of all, the intense glare of the sun on the snow soon began to affect all eyes. Slowly the little band struggled up the slopes towards the Shandur Pass (13,000 feet), everyone taking their turn at carrying. The sun went down on a waste of snow, the pace got slower and slower, and by 9 o'clock Lieut. Stewart realized that if they were not to be frost-bitten they must have rest and food. The loads were stacked by the way-side, but it was not until midnight that the men reached the camping ground at the foot of the pass. There was

The
Gilgit Relief
Column.

1895.

March.

April.

---

[1] The gun and carriage loads were each 200-lb., the ammunition boxes 125-lb.

Chapter
VI.

The
Gilgit Relief
Column.

1895.

9th April.

only one little hut, but there was wood, so fires could be made.

Next day the guns were brought up, and the day after the pass was tackled. It was a great effort, for there was an ascent of two thousand feet, and the snow was fresh-fallen and three to five feet deep. But by 7 o'clock in the evening the section struggled into Laspur, carrying their guns, amidst the cheers of the infantry. The presence of guns meant much to that little advanced guard of a couple of hundred men, for the pass behind them might be closed at any moment by a snow-storm, and the defile in front was certain to be held by the enemy. The next day a reconnaissance of eight miles down the Mastuj valley found them in position sure enough. But since the men were all, more or less, snow-blind, and no one could see to lay a gun, it was decided to have a couple of days rest and treatment before attacking. And so they climbed back to Laspur, where rigorous confinement in darkened huts with drastic treatment to the eyes worked wonders.

The position at Chakalwat, blocking the road to Mastuj, was a strong one, stretching right across the valley, but as soon as the guns had knocked a hole or two in the sangars their defenders began dribbling away.

Next day, the 10th April, Mastuj was reached, and the little garrison found in good order. Here a halt of a few days was made in order to collect transport and supplies, and to await the arrival of the remainder of the force, and of the ammunition and stores which had been jettisoned before tackling the pass. Stewart utilized the time to get some ponies for the guns in place of the mules left behind, impressing prisoners as drivers. But the ponies were not up to the work when the road was specially bad, and then the gunners had once again to carry everything—sometimes for a mile at a time. Twenty of the common shell were emptied of powder and filled up with sand.

Between Mastuj and Chitral stretched a long series of most difficult and dangerous defiles, and the enemy was

known to be holding the "Nisa Gol" the most famous
defensive position in the country. From Lieut. Beynon's
reconnaissance sketch, reproduced here, its general features
can be realized more clearly than from any verbal description, but it may be mentioned that the valley is about a
mile wide, the river is unfordable, and the steep rocky hills
rise several thousand feet on each side. The Nisa Gol
itself is a remarkable feature, a veritable cleft, two to three
hundred feet deep, cut through the plain between the foot
of the mountains and the river.

Colonel Kelly's force now consisted of 400 pioneers,
100 Kashmir infantry, 100 levies, 40 sappers and the
guns. The latter's orders were first to take on the
sangars on the enemy's left. Owing to the slope of the
ground this meant going in to closer range than was
agreeable to the impressed drivers, and it was only with
pistols at their heads that the two lead drivers[1] could be
persuaded to bring up the guns. Opening fire at 500
yards a hit was obtained with the second round, and the
plugged shell cut down the wall, so that the defenders
could be got at with shrapnel. This soon cleared them
out, and the process was then repeated with the next
sangar, but this time at 300 yards which allowed of a
round or two of case. Meanwhile the levies had climbed
the precipitous slopes and turned the enemy's left. Seeing
their retreat thus threatened, the Chitralis abandoned
their whole position and streamed down the valley, followed
up as far as range allowed by shrapnel. The section had
had a hot time, for the enemy concentrated their fire on
the guns. Their casualties were severe—eight killed and
wounded out of the little band.

It was the last of the fighting. A week later the
column marched into Chitral. The tale of the gallant
defence of its Fort cannot be given here, for it is one of
those rare feats of arms in which gunners can claim no
share—either collectively or even individually. And this

---

[1] They were the owners of the ponies, and both were hit while the
guns were in action.

CHAPTER
VI.

The
Gilgit Relief
Column.

1895.

was unfortunate, for, in the absence of anyone with knowledge of artillery, no use was made of the old guns in the Fort.

In 28 days the relief column, drawn for the most part from the plains of India, had crossed 200 miles of mountains and had captured two positions of great natural strength. It was a great achievement, and the Kashmir gunners had borne a part worthy of the best traditions of the Regiment with which they were so closely connected. Hindus and Mussulmans, Gunners and Infantry, Officers and Men, working shoulder to shoulder, had carried the guns, with their carriages and ammunition, across some twenty miles of deep soft snow, over a pass more than twelve thousand feet high, while suffering acutely from a blinding glare and a bitter wind. Handled in action with boldness and skill the effect of their fire on the enemy's defences had fully justified the efforts which their presence had required. At the parade held by Sir Robert Low on his arrival at Chitral, the section was generously accorded the place of honour on the right of the combined line—the guns still on their little ponies, and with their prisoner drivers.[1]

To return now to the steps taken in India on receipt of tidings of Umra Khan's incursion into Chitral. Although little hope was entertained of the possibility of succour over the mountains from Gilgit, instructions were, as we have seen, sent to Colonel Kelly giving him a free hand. At the same time steps were taken for the despatch of a relief force from India, and the mobilization of the 1st (Peshawar) Division of the Field Army was ordered. It was commanded by Lieut.-General Sir Robert Low, and the artillery was as follows :—

---

[1] The two Nos. 1 of the section—Havildars Bulwan Singh and Dhrm Singh both received commissions in the Imperial Service Troops, and in the Great War commanded Nos. 1 and 2 Kashmir batteries, the first in East Africa, the second with Malleson's force on the Persian Frontier.

Colonel-on-the-Staff—Colonel W. W. Murdoch.

Staff Captain   —Captain M. F. Fegen.

C.R.A.   —Lieut.-Colonel W. Aitken.

Adjutant   —Captain G. C. Dowell.

15th Field Battery —Major E. Cassan.

No. 3 British M.B. —Major J. D. Cunningham.

No. 8 do. do. — do. J. C. Shirres.

No. 2 Indian do. —Captain J. L. Parker.

The first step in the advance was the crossing of the Swat Valley, but when the troops commenced their advance from Nowshera it was found that the passes were held. Sir Robert Low decided to force the Malakand, and this was attacked on the 3rd April by the 2nd brigade supported by the three mountain batteries.

Brigaded under Major Cunningham the mountain batteries took up a position in the valley leading to the pass, but finding the range (4,000 yards) too long, advanced to a second position from which they could shell the enemy's defences at from 2,300 to 2,800 yards. Meanwhile the infantry were pushing on, two battalions scaling the heights on the left, the other two advancing directly up the valley. As these latter began to climb the spurs at its head which led to the pass, the guns were brought forward in close support, taking up a third position at 1,000 to 1,400 yards. From here they kept up a shower of time shrapnel on the sangars over the heads of the infantry. Fire was concentrated on every gathering of swordsmen, and under the support of this well-directed fire, which was a revelation to many present, sangar after sangar was carried at the point of the bayonet.

The ammunition expended was:—

No. 3 British M.B.—148 shrapnel, 48 ring.

No. 8 do. do. — 69 do. 35 do.

No. 2 Indian do. —114 do. 50 do.

On the next day the advance was continued into the Swat Valley, and the 1st brigade, with No. 3 M.B., had a

CHAPTER
VI.

The Indian
Relief Force
1895.

7th April.

sharp fight with the tribesmen who disputed the way, the battery coming into action again and again as fresh bodies of the enemy appeared, and, on one occasion, forming up to await an expected rush—fortunately frustrated. There was opposition also to the crossing of the Swat river at Chakdara on the 7th, and both Nos. 8 & 2 M.B's. were engaged. But that was the end of resistance in the Swat Valley.

No news had come in of the situation in Chitral, so Sir Robert took on a brigade, and finally, hearing that the garrison were hard pressed, 500 men with a couple of guns of No. 2 battery. But, as we have seen, he was anticipated by the Gilgit column.

While the above events were taking place in this far-away outpost, negotiations of great importance had been proceeding with Afghanistan which were to effect profoundly the situation on the frontier.

South of the Gomal Pass Baluchistan marched with Afghanistan and Persia, and there were no difficulties with semi-independent tribes. But north of the Gomal the respective responsibilities of the Indian and Afghan Governments for the behaviour of the tribes whose territory lay between was ill-defined. The Amir laid claim to certain rights of suzerainty : the Government of India pressed for the laying down of a definite line which should determine clearly where his claims began : the tribes were jealous of their independence, and suspicious of our motives, but quite capable of playing off one authority against the other. There were obvious inconveniences in such a situation, and in 1893 Sir Mortimer Durand succeeded in persuading the Amir Abdur Rahman to agree to the extension of the zone of British influence up to Afghan territory, and the demarcation of this line on the ground. Thus the "Durand Line" became the *Frontier*, the old frontier at the foot of the hills became the *Administrative Border*.

It was decided to start the demarcation from the Gomal, and the Delimitation Commission accordingly assembled at Wana in southern Waziristan, their escort a brigade which included

No. 3 Indian M.B.—Captain F. H. J. Birch.

Sufficient consideration had not, however, been given to the effect on the inhabitants of seeing the boundary pillars going up. The time had come to make their protest, and unfortunately the site of the camp had been chosen for political rather than military reasons. Even so the measures taken for its security were inadequate, and when the tribesmen swept down on the night of the 3rd November, 1894, the picquets were rushed, and a considerable number of fanatics broke into the camp. Fortunately the main body of the attack was shown up by the star shell and prevented from charging home by the heavy fire which could thus be brought to bear upon it. By dawn those who had succeeded in effecting an entrance had been driven out with the bayonet, and the whole were in full flight, with the cavalry riding hard on their heels.

Three brigades (including that forming the escort to the Delimination Commission) were placed under the command of Sir William Lockhart, and during December, 1894, these scoured the Mahsud country from end to end.

The batteries included in the three brigades were :—

No. 8 British M.B.—Major J. C. Shirres.

No. 1 Indian do. —Captain G. F. W. St. John.

No. 3 do. do. —Captain F. H. J. Birch.

COMMENTS.

Organiza-
tion.   During the period covered by this chapter a very
important step was taken towards improving the status of
the British mountain batteries. Up to, and during, the
Afghan War, these batteries had been ordinary companies
of garrison artillery temporarily transformed into mountain
batteries by the attachment of a "mountain train" consisting of the equipment, the mules, and their Indian
drivers. In 1885 it was ordered that garrison companies
once equipped as mountain batteries should retain their
status for eight years, and in 1889 they became permanent formations in a separate branch of the Regiment.

In view of the part played by the Kashmir troops in the
relief of Chitral, a word must also be said here as to their
organization. In consequence of the Russian scare following the "Penjdeh Incident" of 1885, the Indian Princes
formed from the masses of unorganized soldiery which
they had previously maintained certain corps known as
"Imperial Service Troops". These were set apart for the
service of the Empire if required. In return the Government furnished equipment, and deputed British officers to
train and inspect the various corps, although the command
was exercised by their Indian officers.

In pursuance of this scheme Nos. 1 and 2 Kashmir
M.B's. had been entirely reformed, and at the time when
we have been following its fortunes, No. 1 was organized,
clothed, and equipped on exactly the same lines as an
Indian mountain battery.

Armament.   Two points of considerable importance in connection
with the armament of mountain batteries were brought to
notice during the operations recorded in this chapter. The
first was the much bigger scale of the natural features in
some of the country traversed in the Black Mountain
Expeditions than any which had been previously experienced. In consequence the ranges were longer, and
the 7-pr. guns were found not to have sufficient accuracy

for fire over the heads of the infantry. In the Black
Mountain they were relegated in consequence to the duty
of keeping open the lines of communication, and the
batteries armed with 2·5″ guns used in the fighting line.
The latter nature, previously reserved for British batteries
was, therefore, gradually extended to the Indian batteries.

On the other hand, in the still greater mountain
features met with further north, the 7-prs. were probably
the only guns that could have surmounted the difficulties
of the way. But at Nilt they had failed in power against
the sangars, though all that was required was to displace
the stones, since these would then roll down the steep
hill sides on which they were built. It seemed possible
that the explosion of the small bursting charge in the shells
really reduced their effect, and that they would do better
if used as solid shot. In pursuance of this theory we
have seen Lieut. Stewart replacing the powder with sand
before attacking the position at Nisa Gol, and taking his
guns in to such close range that they could be depended
upon to strike the same spot when fired in salvos—with
the complete success described.

The gatlings used in the Black Mountain did not jam,
but their range was found insufficient, and the employment of artillery *personnel* with machine guns found many
objectors.

From a tactical point of view the outstanding feature
of this period is the first appearance of the brigade as a
tactical unit. In the attack on the Malakand in 1895 the
three mountain batteries were handled as a brigade by the
senior battery commander, who in action took up his
position with his own battery on his right and the other
two on his left, and controlled the fire by voice. The effect
of such concentrated fire was, as we have seen, a revelation
to those whose previous experience had been confined to
batteries acting independently.

MEDALS & CLASPS.

The special clasp given to the Indian Medal, 1854, for the operations recorded in this chapter were "Hazara, 1888," "Hazara, 1891," "Samana, 1891," "Hunza, 1891," and "Waziristan, 1893-95".

In 1895 a new India General Service Medal was approved, with clasps for "Relief" and "Defence" of Chitral.

LIST OF UNITS.

(With designation in 1914.)

15th Field Battery, R.A. —15, R.F.A.
9/1 Northern,   do.   — 3, M.B., R.G.A.
2/1 Scottish,   do.   — 1,   do.   do.
3/1 South Irish,   do.   — 6,   do.   do.
1 Mtn. Battery,   do.   — 1,   do.   do.
7   do.   do.   do.   — 7,   do.   do.
8   do.   do.   do.   — 8,   do.   do.
9   do.   do.   do.   — 9,   do.   do.
1 Indian Mtn. Battery —21, Kohat M.B. (F.F).
2   do.   do.   do.   —22, Derajat do.   do.
3   do.   do.   do.   —23, Peshawar do.   do.
4   do.   do.   do.   —24, Hazara do.   do.
7   do.   do.   do.   —27, Mountain Battery.
1 (Bombay) do.   do.   —25,   do.   do.
The Punjab Garrison Bty.—The Frontr. Gar. Arty. (F.F.)
No. 1 Kashmir Mtn. Bty. — 1, Kashmir Mtn. Bty.

CHAPTER VII.

THE NORTH-WEST FRONTIER—TIRAH AND AFTER.

Introduction—Maizar—The Defence of the Malakand—The Malakand
Field Force—The Mamunds—Nawagai—The Mohmand Field Force
—The Samana—The Tirah Field Force—Dargai—Tirah Maidan—
The Bara Valley—The Buner Field Force—Waziristan—"Will-
cocks' Week-end Wars"—The Makran Coast—Buner.

COMMENTS.

Organization—Armament—Employment—Medals & Clasps—List of
Units.

MAP 1.

CHAPTER
VII.

1897.

Introduc-
tion.

THE year 1897 will long be famous in the annals of the
North-West Frontier, for in it occurred the series of tribal
risings which came to be generally included under
the designation of "Tirah", although, strictly speaking,
that term belongs only to the final operations against the
Afridis.

Many causes have been suggested for such wide-spread
outbreaks, but any discussion of these would be beyond
our province. Suffice it to say that the tribes had some
excuse for suspicion regarding British designs. To a
people so jealous of their independence, the military
occupation of the Malakand pass and the Chakdara bridge
in the Swat valley; the location of political officers in the
Kurram and Tochi valleys; the building of forts on the
Samana; and the cantoning of a brigade at Wana overlooking the Gomal—taken in conjunction with the line
of boundary pillars advancing slowly but surely along the
Durand Line—must have seemed ominous.

Commencing in June with the treacherous attack on the
political officer's escort at Maizar in Waziristan, there
followed in July the totally unexpected gathering of the
tribes of Swat and Buner against the Malakand. A fort-

night later the Mohmands were pouring down into the plains not a day's march from Peshawar, their neighbours the Mamunds from the Panjkora valley making common cause. In September the Orakzais were swarming round the Samana forts, backed up by the Afridis, who had already overrun the posts in the Khyber.

After General Keyes' expedition of 1872 the tribe occupying the Tochi valley had given no trouble until the Afghan War. During that war, however, they had been implicated in some attacks on posts and minor raids, and as this lawlessness continued, the Government decided on the permanent location of a political officer in the valley. In June he had occasion to visit Maizar, a village only nine miles from his headquarters, to settle some local affairs. He took with him a small escort, including a section of No. 6 Indian M.B., but no danger was anticipated, and on arrival the party was received most amicably, the villagers even offering to provide a meal for the Mahomedans. So all settled down under the walls of the village.

Breakfast over, the inhabitants, who had been listening to the pipers of the Sikhs, drew off, and, without the slightest warning, a fusilade was opened from the walls and every vantage spot. The guns at once opened fire with case at a hundred yards on a crowd collecting for a charge, and drove them into the village, although both the British officers with the section had been wounded almost at once. This gave time to limber up, and, under cover of a stand by the infantry, the guns were brought into action again on a low *kotal* some three hundred yards further back. Here Captain Browne succumbed to his previous wound, and Lieut. Cruickshank received a third, and this time a mortal wound. Under their Indian officers, however, the gunners continued to serve their guns until all the shell were expended, and they had to fall back on blank. Successive withdrawals from ridge to ridge were carried out by units in turn with the greatest steadiness, the Indian officers nobly filling the places of the British officers[1], who had now all been hit.   Meanwhile the news had reached Datta Khel, and the arrival of Lieut. de Brett of the battery, with reinforcements and the much needed ammunition, turned the tables. They had done the nine miles from Datta Khel in ninety minutes.

There could be no delay in inflicting punishment for such a flagrant outrage, and during July a force under Major-General Corrie-Bird destroyed the guilty village, and then, splitting up into smaller columns, traversed the country, doing much useful work. Beyond a little sniping there was no opposition, and in due course full submission was made and the fines paid. The batteries employed were :—

No. 3 Indian M.B.—Captain F. H. S. Giles.
No. 6   do.   do. — do.   O. C. Williamson.

It will be remembered that after the relief of Chitral and the withdrawal of the Relief Force, garrisons were left to guard the bridge over the Swat river at Chakdara and to hold the Malakand pass. On the 26th of July officers from these posts, playing polo together in the valley, in perfect friendliness with the inhabitants, were warned of possible trouble owing to the arrival of a fanatical leader who came to be generally known as "The Mad Fakir."

The garrison of the Malakand, under command of Colonel W. H. Meiklejohn, consisted of the bulk of three battalions, a squadron, and

No. 8 Indian M.B.—Captain F. H. J. Birch.[2]
There were also two 9-pr. S.B. guns in the fort.

At ten o'clock that night the alarm was sounded, and the troops had hardly time to reach their posts before the

---

[1] On one occasion when the gunners were delayed in limbering up owing to mules being shot, an Indian officer of the infantry shouted to them not to hurry as he and his men would stand between the guns and the enemy to shield them from the bullets.

[2] Captain Birch was temporarily absent, Lieut. Wynter was in command.

CHAPTER
VII.

1897.

The
Defence
of the
Malakand.

26th—31st.
July.

attack began.   All through the night it was pressed with great vigour, and it was plain that, far from being a mere local disturbance, it was a combined attempt of the tribes to drive the British out of the valley.   Next day the North Camp, which was at some little distance from the remainder, was evacuated in order to contract the perimeter to be defended; and the dispositions for the defence of the remainder were revised, the battery being distributed by sections.

It is not necessary to follow in any detail the five nights of fierce assaults to which the garrison was subjected, or the steps taken during the day to counteract the enemy's attempts.   On the 29th a welcome supply of ammunition was received and reinforcements began to arrive.   By the end of the month all danger was over.

In order to crush the rising and punish the Swatis and the tribes who had joined them in their unprovoked attack, a "Malakand Field Force" was formed under Sir Bindon Blood.   It consisted of three brigades with the following artillery :—

C.R.A.   —Lieut.-Colonel W. Aitken.

Adjutant   —Captain H. D. Grier.

do.   — do.   H. Rouse.

10th Field Battery —Major C. A. Anderson.

No. 1 British M.B.— do.   G. F. A. Norton.

No. 7   do.   do. — do.   M. F. Fegen.

No. 8 Indian do. —Captain F. H. J. Birch.

General Blood turned his attention in the first place to Upper Swat where the enemy were holding a very strong position at Landakai on a spur running down from the
mountains to the river.   This was attacked on the 16th August by the 1st brigade with the field battery and Nos. 7 and 8 M.B's.   Three battalions climbing the hills,

accompanied by No. 8 battery, turned the enemy's left and threatened their line of retreat, while the other two batteries kept their position under a hot fire from the front at a range of 1,400 to 1,600 yards. It was the first time that tribesmen had been subjected to the shrapnel fire of the 12-pr. B.L., and under it, and the threat of the infantry, they began to waver and finally dispersed. The position was thus carried with little loss, and no further opposition was offered by the Swatis. The Indian battery, ordered to the top of the ridge to shell them as they made off across the plain, was climbing up the pretty steep slope, the guns loaded up on the mules, when a small band of about a dozen *ghazis* suddenly appeared from behind some boulders close to the track, and made a rush, sword in hand. They got within a few yards of the guns but were shot down by the escort before any damage was done. On the top of the ridge the battery came into action against the retreating enemy, and had a bird's eye view of the gallant but abortive charge which resulted in the death of two British officers and the bestowal of three Victoria Crosses.

The total ammunition expended by the three batteries at Landakai was 352 shrapnel and 112 ring shell. Of this the 10th Field Battery fired 132 shrapnel.

Meanwhile the Mohmands had attacked Shabkadr, and Sir Bindon Blood was directed to turn his attention to Dir and Bajaur in co-operation with a Mohmand Field Force which was being formed. Leaving his 1st brigade to hold the Swat valley, he took the remainder of his force to the Panjkora valley, where the attitude of the Mamunds—not to be confused with the Mohmands—was causing anxiety. During September both brigades had severe fighting.

Owing to scarcity of water in the Mamund country the 2nd brigade was established in camp at Inayat Kila as a centre for punitive operations, and one of these columns had an unpleasant experience on the 16th September.

Chapter
VII.
1897.
The
Malakand
Field Force.

The
Mamunds.

16th Sept.

The troops got separated as the evening was closing in, the darkness being accentuated by a heavy thunderstorm, and the brigadier, with No. 8 Indian M.B. and a few sappers and infantrymen, finding themselves cut off, sought refuge in the village of Bilot. The tribesmen, swift to seize such a chance, rushed into the village, which was little more than a hamlet, and from walls within close range poured in a destructive fire. The gunners dropped fast, many of them still continuing to serve the guns in spite of their wounds, among them being Lieut. Wynter, who remained at his post though shot through both legs. Every effort to dislodge the enemy failed, but the arrival at midnight of four companies, which had also been benighted, saved the situation. The battery had fared badly, losing one officer and nine men killed, two officers and 32 men wounded, besides 31 mules. Fortunately the 2nd line mules had all been sent back to camp.

An advance against the villages of Agrah and Gat on the 30th September found the Mamunds in considerable strength occupying the villages and precipitous ridge between them. No. 7 British M.B. opened fire on the ridge from a position about a mile and a half south of Gat, but the enemy were well protected by numerous sangars, and by the huge boulders with which the slopes were strewn, and offered a desperate resistance until driven out at the point of the bayonet. The battery then advanced to a second position from which they could shell the ground east of Gat, and after very hard fighting the villages were taken and destroyed. But the enemy, strongly reinforced, were seen advancing in very large numbers, and a general retirement had to be ordered.

Fresh troops were sent up—among them the 10th Field Battery, and its march is noteworthy. Early in September it had advanced as far as the Panjkora although the mountain track only just gave room for the wheels, and the least mistake would have precipitated gun and team into the river far below. Four guns and two wagons now moved on to Inayat Kila where no wheeled traffic had previously been seen.[1]

The Mamunds had gained a well-deserved reputation by their stubborn resistance, but they had the sense to recognize now that the Government were equally determined, and re-opened negotiations which were soon brought to a satisfactory conclusion.

Meanwhile the 3rd brigade and its battery were awaiting at Nawagai the arrival of the Mohmand Field Force with which they were to co-operate, and here they had to stand fierce night attacks, on the 19th and 20th September.   These may be described in Sir Bindon's own words :—

"They began with heavy firing, followed by rushes of swordsmen from several directions . . . . but no one could live under the fire with which our men swept the glacis surrounding our camp, especially as our gunners gave us great help with their illuminating star shell, together with their ordinary fire both shrapnel and case, which was most effective . . . . our losses included that gallant horse gunner Brig.-General Wodehouse of our 3rd Brigade, who was severely wounded—truly a loss to us."

At the time of the rising in the Swat valley there had been no sign of unrest among the Mohmand clans, but another pestilent priest—the "Hadda Mullah"—had no doubt been in communication with the "Mad Fakir," and on the 7th August he led the Mohmands in an attack upon Shabkadr—the usual objective of their incursions into British territory.

On news reaching Peshawar a small column, which

CHAPTER
VII.

1897.

The
Mamunds.

October.

Nawagai.

19th—20th.
Sept.

The
Mohmand
Field Force.

7th Aug.

---

[1] The Official Account says—"The advance of the 10th Field Battery from the Panjkora Bridge is worthy of notice.   After having successfully hauled his six guns over the ford of the Panjkora River without mishap, the O.C. took four of the guns on to Inayat Kila over ground destitute of any road of even the roughest description.   Overcoming all obstacles—man-handling the guns frequently—the battery without a single casualty arrived at the Watelai valley, and thus reached a point in the hills fifty miles further than any other wheeled traffic had ever proceeded."

CHAPTER
VII.

1897.

The
Mohmand
Field Force.

9th Aug.

included four guns of the 51st Field Battery, under Captain S. W. W. Blacker, was despatched post haste. The enemy had come out on to the plain between the fort and the hills, and here they were attacked. But it was difficult ground, the guns were late in getting into action, and a determined attempt to turn the left of the British line threatened to cut off the infantry from the fort. A withdrawal had commenced when General Elles arrived from Peshawar and assumed command. Seeing that the tribesmen had been tempted from the hills into the open, the general extended the infantry further to their left, ordered the guns into action again to support them, and sent the cavalry by a wide detour to come down upon the enemy's flank. With the artillery fire sweeping the ground in front of them the lancers rode along the whole line in extended order, repeating the exploit of the hussars in 1864. The Mohmands fled back to the hills where it was not deemed advisable to follow them with the small force present.

Captain Blacker was severely wounded and Battery Serjeant-Major Wallman took command, and brought the battery out of action. It had fired 53 shrapnel and 9 common.

Towards the end of August the Mohmands were gathering again, and a field force of two brigades was formed under General Elles to co-operate with the Malakand Field Force, as mentioned above. The artillery consisted of :—

C.R.A.   —Colonel A. E. Duthy.

Adjutant   —Captain W. K. McLeod.

No. 3 British M.B.—Lieut.-Colonel J. D. Cunningham.

No. 5 Indian do. —Captain F. R. McC. de Butts.

For various reasons action was delayed, and it was not until nearly the end of September that the two forces were in touch. The enemy were then holding a position on the

Bedmanai pass, and General Elles attacked this on the 23rd September with his own troops reinforced by the 3rd brigade and No. 1 M.B. from the Malakand Field Force.

The position was carried without difficulty, and little opposition was offered to the subsequent movement of columns throughout the country, in which the most inaccessible districts, never before visited by British troops, were traversed, and the towers of the leaders destroyed. Before the end of the month the tribe had made submission, and the majority of the units were on their way back—to be absorbed in the force being gathered for the invasion of Tirah.

While the operations described above were being carried out against the northern tribes along the border of the Peshawar plain, still more serious events were taking place south of the Khyber. The first sign of the unrest which was to spread so far had been the treacherous attack at Maizar, and not far away was the Samana Ridge, on which, as will be remembered, posts were established after the Miranzai expedition of 1891. There had not been wanting signs of smouldering discontent with our occupation of this commanding position, and the isolated situation of Kohat, 32 miles from railhead at Kushalgarh, was a cause for anxiety. The garrison was accordingly reinforced,[1] and Major-General Yeatman-Biggs, R.A., was sent to take command of all the troops on the Kohat-Kurram border.

During the latter part of August the position became daily more serious. The water supply on the Samana was not sufficient to allow of its occupation in any strength, but on the 31st August, General Yeatman-Biggs moved forward to Hangu, just below the ridge. Here he was kept busy with raids all along the border, until, on the 12th September, affairs took on a somewhat alarming aspect with the arrival of the main fighting force of the Afridis to join the Orakzais. When the news reached General Yeatman-Biggs he was engaged in dealing with raiders, and owing to the scarcity of water in the country

---

[1] The reinforcements included the 3rd and 9th field batteries. The latter was in action on the 27th August at the Ublan Pass, and on the 31st August and 3rd September against raiders, working by sections.

he was obliged to descend to his base at Hangu, thus leaving the tribes free to throw their whole weight against the Samana forts. The gallantry with which these were defended forms a proud page in the record of the regiments engaged—saddened as it is by the fate of Saragarhi. Steps were at once taken for their relief. No. 2 Indian M.B. marched from Thal to Hangu, 36 miles in twelve hours by night. The 9th Field Battery was sent out with the cavalry to the nearest point to which the guns could be taken to signal that relief would arrive on the morrow. Perhaps more cheering still to the beleagured garrison at Gulistan, the guns were brought into action and a few shell dropped among the enemy clustering round the walls.

The main force starting at midnight gained the ridge on the morning of the 14th, and drove the tribesmen head-long from it. The climb was beyond the power of a field battery, but the mountain guns, in line with the infantry, swept the front with shrapnel, and this brusque treatment, coming on the top of the heavy losses they had suffered in their attacks on the forts, was too much for the Orakzais. Thoroughly dispirited, they dispersed to their homes, and gave no further trouble.

During the Second Afghan War an arrangement had been made with the Afridis for the keeping open of the Khyber Pass, and this had worked satisfactorily for nearly twenty years. But the general unrest of 1897 was too much for the tribesmen. Their first overt act of hostility was the seizing of the posts in the pass, although these were held by their own levies, the "Khyber Rifles."

For the moment nothing could be done beyond sending "K" battery into the mouth of the pass to send a few shell among those attacking Fort Maude. But such unprovoked aggression could not be allowed to go unpunished. It had created a situation so subversive of our position on the Frontier that nothing short of humbling the pride of the tribe by dictating terms in the heart of their country would meet the occasion. Orders were accordingly issued

for the formation of a force under Lieut.-General Sir
William Lockhart for the invasion of Tirah.[1] The re-
opening of the Khyber could wait.

The "Tirah Field Force" consisted of a main body of
two divisions, with two subsidiary columns to operate from
Peshawar and The Kurram respectively.

G.O.C., R.A.	—Brig.-General C. H. Spragge.
Brigade-Major	—Captain C. de C. Hamilton.
Orderly-Officer	— do. F. E. Johnson.

1st Division.

C.R.A.	—Lieut.Colonel A. E. Duthy.
Adjutant	—Captain W. K. McLeod.
No. 1 British M.B.	—Major G. F. A. Norton.
No. 1 Indian do.	—Captain G. F. W. St. John.
No. 2 do. do.	— do. J. L. Parker.

2nd Division.

C.R.A.	—Lieut.-Colonel R. Purdy.
Adjutant	—Captain H. D. Grier.
No. 8 British M.B.	—Major J. C. Shirres.
No. 9 do. do.	— do. A. ff. Powell.
No. 5 Indian do.	—Captain F. R. McC. de Butts.
	— do. A. W. Money.
Rocket Battery	—Major A. H. Browne.

Peshawar Column.

C.R.A.	—Lieut.-Colonel W. W. M. Smith.
Adjutant	—Captain F. R. Drake.
57th Field Battery	—Major A. B. Helyar.
No. 3 British M.B.	—Lieut.-Colonel J. D. Cunningham.

Kurram Column.

3rd Field Battery	—Major H. Guise.

Lines of Communication.

9th Field Battery	—Major A. S. Wedderburn.
No. 1 Kashmir M.B.	—Commandant Khajur Singh.

---

[1] The troops under General Yeatman-Biggs were absorbed into the
2nd division of this force, and he himself took charge of the concentration until the arrival of Sir William Lockhart, who was on leave
at home.

The 2nd division was already on the Samana, and, as a preliminary to the general move forward, they cleared the enemy off a position at Dargai which commanded the descent from the ridge. But, unfortunately, arrangements did not permit of the troops remaining in occupation, so that when the advance commenced in earnest on the 20th October the operation of dislodging the enemy had to be repeated. And this time the position was held in force.

The line of steep cliffs, along the summit of which the Afridis were extended, was accessible at one point only, and to reach their foot at this point a narrow neck had to be crossed. This was approached by a spur of the Samana, and by keeping along the side of this spur a certain amount of shelter could be obtained. But the neck which intervened between the spur and the foot of the cliffs was absolutely bare, without a scrap of cover, and commanded at a distance of only five hundred yards by the enemy's position, which was three to four hundred feet above it.

20th Oct.

The divisional artillery, reinforced by No. 1 Indian M.B. from the 1st division, was disposed with No. 8 British and Nos. 1 and 5 Indian M.B.'s., massed under Colonel Purdy on the *kotal* facing the position at 1,800 yards range, No. 9 British M.B. by itself, further to the right.

At 10 o'clock an accurate and sustained fire was opened, under cover of which the leading troops reached the neck, and the first rush got safely across to the dead ground at the foot of the cliffs. But from then on such a devastating fire was concentrated upon this exposed space that although it was not bigger than a tennis court very few of those who gallantly attempted the passage reached shelter. It was clear that a fresh effort must be organized. Two fresh battalions were ordered up, the guns poured in a heavy fire, culminating in a rapid burst for three minutes : the two new battalions, joined by the survivors of the previous attempts, dashed across the neck and swarmed up the cliffs. The tribesmen did not await their assault.

The advance was resumed next morning, but a week's
halt in the Khanki valley followed, so that it was not until
the 29th October that the Sampagha—the first of the passes
on the way to Tirah—was reached. Here the whole of
the artillery of the two divisions worked by brigades under
the control of the G.O.C., R.A. The 1st divisional artillery
first drove the enemy from a knoll half way up the pass
which they had occupied as an advanced position. This was
at once occupied by the 2nd divisional artillery, and from
there the enemy's main position at the summit of the pass
was kept under fire until the fire was masked. Then a
couple of batteries were pushed forward to crown the
position, and clear the Afridis from some ridges to which
they were still obstinately clinging. By half-past eleven
all opposition had ceased, but the artillery had to mourn
the loss of one of the best-known mountain gunners of the
day, Captain de Butts, killed at the head of his battery
as he led it through the kotal.

Two days later it was the turn of the Arhanga Pass,
and here again the massed action of the artillery proved
decisive. A suitable position at the foot of the ascent
having been secured by the infantry, the artillery of the
leading division opened fire from it, to be shortly joined
by the remainder of the guns. Under cover of this concentrated fire the infantry seized the summit without
difficulty, and the way into Tirah Maidan, the summer
home of the Afridis, into which a European had never set
his foot, was now open.

By the end of November every settlement had been
visited, and winter was drawing in, so Sir William decided
to move his troops to the winter settlements of the Afridis
in the neigbourhood of Peshawar. On the 7th and 8th
December the move was commenced, and none too soon
for the snow was commencing. The 1st division
returned over the Sampagha pass and down the Mastura
valley, the 2nd by Dwatoi and the Bara valley. The
march of the former, though arduous, was uneventful as

far as fighting was concerned : the latter experienced for five consecutive days some of the hardest rear-guard fighting ever known on the Frontier.

The width of the valley down which the 2nd division had to find its way varied considerably, but the river banks were never out of rifle range from the hills on either side, and here and there were narrow gorges. The weather was bitterly cold, with incessant rain and sleet, making the paths slippery and turning the fields into morasses, so that there were constant delays with the baggage animals falling or floundering.

As a rule the leading brigade met with no opposition, but the other was harried all day, being fired on at times even before it moved out of camp. The enemy's efforts were particularly directed towards cutting off the rear-guards, and these were often unable to get in before dark, and so had to remain out all night without food or water.

As a general rule No. 9 British M.B. worked with the 3rd brigade, Nos. 8 British and 5 Indian with the 4th, but on occasions the whole of the divisional artillery was required to cover the withdrawal of the rear brigade.

The following figures show the total expenditure of ammunition during the campaign by the batteries of the 1st and 2nd divisions of the Tirah Field Force :—

Division.	Battery.	Shrapnel.	Ring.	Case.	Star.	Total.
1st	No. 1 British M.B.	124	101	22	28	275
	,, 1 Indian ,,	792	80	—	—	872
	,, 2 ,, ,,	502	129	—	—	631
2nd	No. 8 British M.B.	1350	57	—	25	1432
	,, 9 ,, ,,	775	96	—	—	871
	,, 5 Indian ,,	1427	527	—	2	1956

Before concluding this account of the series of operations generally known under the name of "Tirah", we must return to the Malakand Field Force, which we left in the Swat valley. Sir Bindon Blood had intended to complete his operations in Upper Swat in August by an advance into Buner, but the rising of the Afridis had induced the Government to postpone the punishment of the Bunerwals to a more favourable time. With the conclusion of the Tirah campaign this had now arrived, and in January 1898 the Malakand Field Force was reconstituted[1] as the "Buner Field Force." The artillery remained the same, except for the absence of No. 1 British M.B., viz :—

> C.R.A. —Lieut.-Colonel W. Aitken.
> Adjutant—Captain H. Rouse.
> 10th Field Battery —Major S. W. Lane.
> No. 7 British M.B.— do.   M. F. Fegen.
> No. 8 Indian do. —Captain F. H. J. Birch.

On the 7th January the Tangi Pass was attacked, the field battery opening fire at 2,200 from a knoll near the mouth of the pass.   Under cover of this fire the two mountain batteries, with a battalion as escort, climbed a spur from which they could bring fire to bear on the enemy's position at a range of under 2,000 yards, and the main body of the infantry then commenced their thousand foot scramble up the face of the ridge.   The three batteries firing over the heads of the attackers kept the crest under a steady fire of shrapnel under which the mass of standards so proudly displayed gradually wilted away.   Demoralized by the artillery fire, and realizing that they could not meet the frontal attack without further exposing themselves, the Bunerwals did not wait for the infantry to close.   The complete submission of the tribe soon followed, and the Field Force returned by the Ambela Pass, the scene of such heavy fighting in 1863.

With the capture of the Tangi Pass at the beginning of January 1898 the curtain was at last rung down on the long drawn-out drama which had opened at Maizar in June 1897.

---

[1] Except the 3rd Brigade and No. 1 British M.B. which as we have seen returned to India with the troops of the Mohmand Field Force, there to be absorbed into the forces being prepared for the Tirah Campaign.

For a few years after the lessons of 1897 all was quiet on the North-West Frontier, but the turn of the century brought trouble again in Waziristan. A dreary blockade of the Mahsuds followed, intensified by raids into the tribal territory by small columns, in which No. 7 Indian M.B. was chiefly employed, usually by sections. There was little serious fighting though there were some sharp encounters.

In a very different category come two brilliant little campaigns conducted by Major-General Sir James Willcocks some five years later, and christened by *Punch* "Willcocks' Week-End Wars"—to the delight of the General.

The first of these was against the Zakha Khels, the most persistently hostile of the Afridi clans, who dwelt in the Bazar valley immediately adjoining the Khyber. They had kept quiet for nearly ten years, but in 1907 their raids became so daring that the Government were forced to take action, and in 1908 Sir James Willcocks, who commanded the 1st division at Peshawar, mobilized two of his infantry brigades, with a mountain battery attached to each—the batteries being :—

No. 3 British M.B.—Major F. H. S. Giles.

No. 22 Indian do. — do. C. de Sausmarez.

Starting on the 13th February, and acting with great rapidity, Sir James secured all the principal entrances into the Bazar valley before the tribesmen could combine to oppose the advance. Henceforth they were given no rest, but were harried day after day by columns striking in every direction except the expected. Before the end of the month the clan had tendered its submission, and the force was back at Jamrud on the 1st of March.

This attack on the Afridis caused considerable stir among their neighbours on the other side of the Khyber, the ever pugnacious Mohmands. Not content with looting villages, they attacked border posts and were evidently spoiling for a fight. So the troops facing their territory were gradually increased, the artillery being :—

18th Battery, R.F.A.—Major W. G. H. Manley.

80th   do.   do. —Major W. Strong.

No. 8 British M.B.—Major F. W. S. Stanton.

No. 23 Indian   do. —Captain G. G. W. Corrie.

No. 28   do.   do. —Lt.Colonel R. W. Fuller.

CHAPTER
VII.

1908.

"Willcocks'
Week-end
Wars,"

The field batteries as divisional artillery, the mountain batteries with the brigades.

20th April.

On the 20th April the Mohmands were given a taste of long-range fire which is of some Regimental interest, and Sir James may tell the tale :—

"The enemy had driven in one of our cavalry patrols, and as I happened to be present on the spot, I ordered our 18-pr. field guns to disperse them. Three faultless shell very soon did this, but why I mention it that this was the first occasion on which this new pattern field gun was fired in action."

An attack in force by the 1st and 2nd brigades followed, and drove the Mohmands back into their hills helter-skelter. But before a further advance could be made the centre of activities was suddenly shifted to the Khyber, where hostile forces were reported to be moving on our post at Landi Kotal. Not a moment was to be lost, and the 3rd brigade with its mountain battery (No. 22 Indian) and the 80th field battery were despatched up the Khyber by forced marches. In spite of the fiery heat, they arrived in time to ward off the danger, and on the 4th May drove the Afghans—as they turned out to be—across the frontier.

Before concluding this account of operations on the North-West Frontier something must be said of the coastal region of Makran which lies between Baluchistan and the sea. This has not been mentioned previously, but in 1898 one of our survey camps was looted, and its *personnel* murdered. An expedition under Colonel

The Makran
Coast.

1898.

Mayne, was at once despatched by sea from Karachi, and had quite a hot fight after landing at Pasni. The tribesmen were holding the mouth of a defile, and it was not until the guns—a section of No. 4 Indian M.B., under Lieut. J. H. Paine—had been brought up into the infantry line to fire case that they broke and fled. One lesson was enough. The stronghold of the instigator at Twifat was blown up, and the troops returned, the cavalry and guns overland to Quetta, the remainder by sea to Karachi.

1901.

Again in 1901 the unruly tribes required punishment, and again they put up a stout resistance—this time at the strong fort of Nody. The guns—a section of No. 9 Indian M.B.—were told off to demolish the battlements as it was thought that the breaching of the walls was beyond their power. To do this they had to come into action at under six hundred yards, but fortunately the scrub offered a good deal of cover from view, and the mules found shelter in a nullah. A practicable breach was made, the stormers dashed in, the guns were brought up to blow in the interior defences, and the garrison threw down their arms.

1910.

In 1910 the "Gun-Running Operations" 'in the Persian Gulf took a section of Indian mountain artillery again to the Makran Coast—this time of the 32nd Battery. It formed part of a military landing force detailed to co-operate with the Navy patrolling the coast by raiding the depôts established some twenty miles inland where the guns taken off dhows were collected to await the arrival of caravans from Afghanistan. There was no serious fighting, but the section had the experience of living on board ship, with frequent disembarkations, in which all ranks, mules included, became very expert.

The long list of operations on the North-West Frontier between the Indian Mutiny and the Great War concludes, very appropriately, with a punitive expedition carried out by a well-known officer of the Regiment.

Buner.

1914.

By 1914 the Bunerwals had forgotten the punishment

inflicted upon them by Sir Bindon Blood in 1898, and taken to raiding and harbouring outlaws. Major-General Bannatine-Allason, R.A., commanding Nowshera Brigade, was entrusted with the conduct of a counter-raid by a couple of thousand men and the 30th Indian M.B. (Major H. J. Cotter). Leaving the manœuvre area on the 21st March, the white bands still round their head-gear and blank ammunition in their pouches, Rustam (37 miles) was reached on the next afternoon in spite of heavy rain and deep mud. Starting again at midnight the Malandri Pass—another 12 miles and a 2,000 foot climb—was secured by 8 a.m. on the 23rd. The offending villages, a few miles over the pass, were then destroyed, the flocks and herds driven off. By 2 p.m. the withdrawal commenced under cover of the guns, the pass was clear by 4 o'clock, and the troops were back in camp by night-fall. Owing to the precautions taken to secure secrecy, and the rapidity of the movements, the Bunerwals were taken unawares, and complete success was secured with trifling casualties.

CHAPTER
VII.

Buner.

1914.

21st March.

COMMENTS.

It will have been plain from the narrative how unexpected, and how widespread, were the risings in 1897. Under the mobilization arrangements then in force, the sudden and severe call upon the batteries detailed for service involved heavy demands upon the "supplying batteries" in India. These could be met by the field batteries, but were more than the mountain batteries could cope with owing to the large proportion of this branch in the field. A mountain artillery depôt was, therefore, formed at Rawalpindi for the purpose of supplying their

Organiza-
tion.

needs. It was maintained by drafts from garrison companies, from Indian infantry regiments, from the
Hyderabad Contingent,[1] as well as by calling up the Indian
reservists of mountain artillery.

When it came to taking stock of the experiences of the
year it was obvious that the time for such impromptu
arrangements had gone by, and the whole subject of the
mobilization arrangements for the artillery in India was
considered by a committee at Simla in 1898, from which
important changes resulted.

The Tirah Field Force was the first example in India
of a complete artillery organization. This organization
was, it is true, scarcely more than a skeleton, owing to
the small number of batteries, and the fact that they all
belonged to the mountain artillery. But the framework
was there, from the G.O.C., R.A., at Headquarters, and
the C.R.A.'s. of Divisions, to the ammunition column
officers.[2] And we shall see how valuable this proved when
discussing the tactical lessons of the campaign.

The decision to take nothing but mountain artillery
into Tirah was no doubt a wise one. But the value of the
heavier metal of field guns when the ground made their
employment possible was exemplified by the 51st Field
Battery at Shabkadr, the 10th with the Malakand and
Buner Field Forces, and the way in which, ten years later,
Sir James Willcocks made use of his field batteries. An
interesting attempt to revive the use of elephants for the
carriage of field guns, of which so much has been said in
previous chapters, was also seen in connection with the
preparation for the Tirah campaign.

A horse battery was moved to Peshawar, the old

---

[1] The Hyderabad Contingent was a force maintained under treaty
with the Nizam. It included four field batteries officered from the
Royal Artillery in much the same way as Indian mountain batteries.
They had done fine service in the Central Indian Campaign during the
Mutiny.

[2] These officers were appointed to each brigade to take charge of
the reserve ammunition mules of the batteries, and arrange for the re-
plenishment of the supply from the Ordnance Field Parks.

elephant equipment was drawn out of store, and a couple
of elephants were borrowed from the heavy battery at
Campbellpore.   No difficulty was found although the
elephants had heretofore been in the habit of drawing guns
not carrying them, but eventually it was decided that the
mountain guns could do all that was required.   This last
attempt to use elephants in the field is, it is thought,
worthy of record for by the turn of the century they had
disappeared from even the heavy batteries.

Another revival of ancient custom in armament was
scarcely more fortunate.   The rocket battery was formed
at the express wish of Sir William Lockhart, with gunners
from 5/Western and drivers and mules from No. 3 British
M.B.   The rockets proved ineffective, however, as well as
dangerous, and the battery was sent back from Tirah.

During the operations recorded in this chapter breech-
loading field and mountain guns, and finally quick-firing
field guns, made their first appearance.

The value of the 12-pr. B.L. (of 7 cwt.) was shown in
the operations of the Malakand and Buner Field Forces in
1897-98, and the only fault found with the equipment was
at the attack on the Tangi pass when the volute springs
gave some trouble.   But the battery had fired 476 rounds
from a position 900 feet below the target, so the trial was a
high one.   The excellence of the *matériel* was also proved
by the way it survived the march to the Panjkora, and
from there to Inayat Kila, over country into which no
wheeled traffic had ever previously penetrated.[1]

The 2·5″ R.M.L. gun—the famous "screw-gun" of the
Afghan War, and many a frontier expedition since—was
at the height of its reputation at the time of Tirah.   But
there had been whisperings for some time as to the poor
effect of its little 7-lb. shells against the frontier towers,

---

[1] It is to be noted that on this occasion, gunners were always carried
on the off lead and centre horses.

except at ranges which were becoming almost impossible with the modern rifles of the tribesmen. The plight of the two mountain batteries in South Africa sealed its fate. A muzzle-loader firing black powder was an anachronism in the XXth century, and in 1901 a 10-pr. B.L. took its place. Its smokeless powder, heavier shell, longer range, and greater rapidity of fire came as a rude shock to the Afridis in 1908. A month or two later the 18-pr. Q.F. made its début against the Mohmands, and we have had Sir James Willcocks' own account of its effect.

The withdrawal of star shell was under consideration in 1897, but its value was clearly shown on numerous occasions during the fighting of that year, especially in the defence of the Malakand and in the night attacks in Bajaur, where the tribesmen still used cold steel. At Nawagai the night was so dark that without the star shell it would have been impossible for the infantry to use their rifles with effect. Unfortunately batteries had not had sufficient practice in its use, and found in consequence some difficulty in obtaining the best effect.

Employ-
ment.

In the tactical handling of the guns the outstanding feature is the massing of the batteries in the attacks on the passes leading into Tirah. In the last chapter attention was called to the first brigading of batteries, which occurred in the attack on the Malakand. The value of such concentrated fire was immediately recognized, and the massing of the artillery of two divisions under the Brigadier-General R.A. in Tirah was a natural development. Here again the reports give due acknowledgment to the decisive contribution to the success of the attacks made by this imposing array of guns.

But it must be admitted that the share of the artillery in the opening action of the campaign met with considerable criticism. At Dargai, as has been related, the first attacks of the infantry were repulsed with heavy loss, and, as was only natural, there were some who attributed this rebuff to the failure of the artillery to render adequate support. Fortunately an artillery officer who was present during the attack, and was able to go over the ground afterwards, contributed a detailed account to the *Proceedings*[1] From this the following notes have been extracted.

The Afridi position was on the absolute edge of precipitous cliffs. Many of the defenders had found ideal rifle pits in clefts of the rock, and where sangars had been built they were not more than two feet high, and therefore presented practically no target, even if a direct hit with a 7-lb. shell at 2,000 yards would have done them any harm. As regards time shrapnel and ring shell, the splashes of the bullets on every square foot of sangar and of rock along the enemy's position bore elequent witness to the accuracy of the fire. It must have been plain enough to the defenders that any man leaning over to shoot down at men climbing up the face of the cliffs would himself be hit, and this consideration no doubt influenced their timely flight as soon as the attackers had passed the neck. That they did not escape without loss was plain from the patches of blood on the stones. Their casualties would probably have been much greater if it had not been that the main artillery position was some seven hundred feet below the top of the cliffs, greatly reducing the searching effect of the shrapnel bullets. The fire of the detached battery on the right was probably the most effective of all owing to its higher position and longer range.

The batteries engaged fired altogether some 1,300 rounds.

The Bara valley will no doubt long be quoted as an example of the work of artillery in covering a retreat. As always in frontier warfare the signal to withdraw was

---

[1] "The Artillery at Dargai"—Lieut. G. F. MacMunn, D.S.O. *Proceedings*, Vol. XXV.

also that for the appearance of hordes of creeping foes. The duty of the artillery was plain—to cover the rear-guard in its successive positions without risking the loss of guns or delaying the movement of the infantry. But to perform this duty successfully is a matter of nice judgment.

In the defence against night attacks the action of the Malakand Field Force at Nawagai illustrates the change of method since the camps had been rushed at Palosin Ziarat and Wana. No longer was reliance placed on outlying picquets. The front was cleared and the troops saw to their own defence, the guns taking their share on the perimeter.

The way the section at Maizar was fought is worthy of unstinted admiration, but its deficiency of ammunition cannot be passed over without comment. It was ordered out without its full normal ammunition equipment, because the mission was supposed to be entirely peaceful. But if it is worth while to take guns at all on such occasions they should be adequately equipped. There are, of course, exceptions to this as to every other rule, and one was provided in the last chapter, for, in order to get the guns of the Gilgit column over the Shandur pass, the ammunition was cut down to only fourteen rounds a gun. But the risk was deliberately faced, and minimized by taking the guns into action at such short ranges that no rounds were wasted in "ranging".

Attention may be drawn in conclusion to the number of officers of the Regiment who held army commands or staff appointments in the operations chronicled during this period. It was the beginning of the change in this respect which the Great War brought.

Major-General A. G. Yeatman-Biggs—2nd Division,

Tirah F.F.

do.   do.   E. R. Elles—Mohmand F.F.

do.   do.   R. Bannatine-Allason—Nowshera Brigade.

Brig.-General J. H. Wodehouse[1]—3rd Brigade,
Malakand F.F.

do. do. C. A. Anderson[1]—1st Brigade,
Mohmand F.F.

Colonel A. W. Money　　　—A.A.G.,
Peshawar Division.

Major C. P. Triscott　　　—A.Q.M.G., Tirah F.F.

,, E. A. P. Hobday　　　—D.A.A.G.,
Malakand F.F.

Captain H. E. Stanton　　— do.　　do.　do.

Lieut. F. L. Galloway　　　—A.D.C.

,, H. D. Hammond　　　— do.

CHAPTER
VII.

Employ-
ment.

MEDALS & CLASPS.

Medals.

For the operations recorded in this chapter the following Clasps were given to the India Medal 1895— "Malakand 1897", "Samana 1897", "Tirah 1897-8", "Punjab Frontier 1897-98", "Waziristan 1901-2."

In 1908 a new India Medal was introduced with clasp for "North-West Frontier 1908."

LIST OF UNITS.

(With Designation in 1914.)

K, R.H.A.　　　　　　—K, R.H.A.

3, Field Battery, R.A.— 3, R.F.A.

9,　do.　　do.　　do. — 9,　do.

10,　do.　　do.　　do. —10,　do.

18,　do.　　do.　　do. —18,　do.

51,　do.　　do.　　do. —51,　do.

57,　do.　　do.　　do. —57,　do.

80,　do.　　do.　　do. —80,　do.

---

[1] These officers had commanded "I" battery, R.H.A., in succession.

LIST OF UNITS—*Continued.*

1, Mtn. Battery, R.A.— 1, M.B., R.G.A.
3, do. do. do. — 3, do. do.
7, do. do. do. — 7, do. do.
8, do. do. do. — 8, do. do.
9, do. do. do. — 9, do. do.
1, Indian Mtn. Battery—21, Kohat   M.B. (F.F.)
2, do. do. do. —22, Derajat do. do.
3, do. do. do. —23, Peshawar do. do.
4, do. do. do. —24, Hazara do. do.
5, do. do. do. —25, Bombay M.B.
6, do. do. do. —26, Jacob's do.
7, do. do. do. —27, Mountain Battery.
8, do. do. do. —28, do. do.
9, do. do. do. —29, do. do.
22, do. do. do. —22, Derajat M.B. (F.F.)
23, do. do. do. —23, Peshawar do. do.
28, do. do. do. —28, Mountain Battery.
30, do. do. do. —30, do. do.
32, do. do. do. —32, do. do.

CHAPTER VIII.

The Northern and Eastern Frontiers.

Introduction—Sikkim—Bhutan—Akas and Daphlas—Tibet—The Naga
Hills—Manipur—Lushai.

COMMENTS.

Organization—Armament—Employment—Medals & Clasps—List of
Units.

MAP 7.

FROM the Black Mountain on the banks of the Indus, to CHAPTER
VIII.
Darjeeling, the hill station of Calcutta, the northern frontier of India marches with Kashmir and Nepal for more Introduc-
tion
than a thousand peaceful miles.   But eastward of Nepal we
return to a fringe of tribal territories—Sikkim, Bhutan,
and, beyond these, a series of smaller tribes, Akas,
Daphlas, Abors, Mishmis—occupying the hill country
between Tibet and the Brahmaputra.   Behind and above
these border hills and tribes, in a position somewhat
analogous to that of Afghanistan on the north-west, towers
Tibet.

The eastern frontier follows the range of hills that
runs from the valley of the Brahmaputra almost to the
mouths of the Irrawaddy, and along these hills we find
another succession of tribes—Nagas, Manipuris, Lushais,
Chins.

On these northern and eastern frontiers, as on those
of the north-west, the rich plains of India are overlooked
by mountain races who, like all mountaineers, have been
accustomed to raiding the lowlanders, and must be kept
under control.   So, once again, the story is of punitive
expeditions when the depredations of the tribesmen become

too outrageous. But these tribesmen are a very different people from those with whom we have been dealing in previous chapters. Primitive races for the most part, professing a debased form of Buddhism, and armed only with matchlocks, bows and slings, they present as great a contrast to Pathans or Afghans as do their jungle-clad hills to the bare rock or pine forest of the north-west. Rising from the tropical luxuriance of the valleys, and the bamboo brakes of the lower slopes, these hills soar upwards, tier upon tier, until they are crowned by the crest line of the Himalayas.

Sikkim.

Between this main range and the British district of Darjeeling lay Sikkim, and in the 80's the idea of making a road through the country to the Tibetan frontier was mooted. The proposal, however, at once roused the suspicions of the Tibetans, who sent troops to obstruct the route at a spot some miles within Sikkim territory. Such open defiance could not be permitted, and in March

1888.

1888 an expedition was despatched to turn out the Tibetans and to vindicate our treaty rights in Sikkim.

The Expeditionary Force was commanded by a gunner —Colonel T. Graham—who had seen much frontier fighting and had commanded 9/1 Northern in the Afghan and Burmese wars. It consisted of some 1,300 men, with one battery :—

9/1 Northern—Major J. Keith.

To begin with, half this force was engaged, and their advance, at first through dense bamboo jungle and eventually in deep snow, was necessarily slow. But when

March.

the stockades blocking the route were reached their defenders offered little resistance, and were turned out without even having to call upon the artillery.

The expulsion of the Tibetans from Sikkim territory having been effected, the force went into camp at Gnathong, where they were sharply attacked. But the Tibet-

May.

ans were beaten off, and after a couple of guns had been

mounted on the stockade,[1] the battery was withdrawn to Darjeeling for the "rains". In August the enemy gathered again, the battery hurried back, and the force was brought up to a strength of 1,600 men.

At daybreak on the 24th September the enemy were seen to have taken up a position on the hills facing Gnathong, and to have built during the night a stone wall, three to four feet in height and nearly three miles long. At last the guns[2] got their chance, making excellent practice on the wall, and, from the summit of the Tuke La, against its defenders in retreat. Graham (now Brigadier-General) pressed on, and in the afternoon entered the valley leading to the Jelap La. Here the guns were quickly in action against the wall with which the pass was defended, but the reply was feeble, and when the infantry advanced to the assault the Tibetans fled, leaving one brass 6-pr. S.B. field gun and a score of jingals in our hands. This concluded the fighting.

Immediately east of Sikkim lies Bhutan, its south-westerly corner only thirty to forty miles from Darjeeling. Lying in the heart of the Eastern Himalayas, it consists of a mountainous tract gradually descending from ranges of 20,000 feet or more to broad cultivated valleys, the surrounding slopes clothed with the trees of a temperate climate. Between this central zone and the plains of India there intervenes a belt of broken ranges separated by deep precipitous gorges, the whole covered with tropical vegetation.

In 1864, owing to the gross incivility with which a British Mission was treated, it was decided to take

CHAPTER
VIII.

Sikkim.
1888.
August.
24th Sept.

Bhutan.

1864.

---

[1] These were 7-prs. (of 200 lb.) which had been sent up from India. The Pioneer detachments were put through their gun-drill by 9/1 Northern before it was temporarily withdrawn.

[2] The continuous mist which prevailed at the high altitudes made the observation of the bursts of shrapnel very difficult, and the battery found it advisable to increase the proportion of common in their ammunition boxes.

CHAPTER VIII.

Bhutan.

1864.

November.

December.

1865.

30th Jan.

4th-5th Feb.

coercive measures, and a "Duar Field Force" was assembled under Brigadier-General Mulcaster. The artillery consisted of :—

5/25 R.A.—Captain J. R. Olliver.

6/25 R.A.—Captain F. C. Griffin.

The Eurasian (Christian) Battery—Captain J. E. Cordner.

The columns met with but feeble resistance from the enemy, but their advance was slow, sometimes not more than three or four miles in a day, owing to the dense and tall jungle, through which paths had frequently to be cut. A line of posts was, however, established on the outer range, and some of the troops had commenced their withdrawal, when there came a rude awakening.

On the 30th January 1865 the post at Dewangiri was attacked, and though the garrison of 800 men, with two 12-pr. howitzers, beat back the Bhutanis, the latter established themselves in a stockade commanding the post and cut off the water supply. On the night of the 4th-5th February the garrison attempted to withdraw, but the attempt ended in disaster. The elephants had not been left with the howitzers, and the infantry carriers abandoned them and the sick and wounded in a panic.

"Lieut^{nt}. C. F. Cockburn with his few gunners tried to carry away his guns and got left behind, and ultimately threw his guns over a precipice and formed his party into a rear-guard, picked up the wounded, charged and beat off the Bhutanis and behaved well".[1]

This was only one of a series of similar attacks, and on the news reaching Calcutta immediate steps were taken for the relief of the posts still holding out, and for the recapture of Dewangiri. Brigadier-General Tombs,[2] who had so greatly distinguished himself in the Mutiny, relieved General Mulcaster, and the reinforcements included 3/25 (Captain H. M. Smith).

---

[1] Letter from Lieut.-Colonel Adye, D.A.G., R.A.

[2] The Tombs Memorial Scholarship awarded to the senior cadet receiving a commission in the Royal Artillery was established in his memory.

The advance was made in two columns. The right, commanded by Tombs, with which were 3/25 and the Eurasian Battery, reached Dewangiri on the 3rd April, but the Bhutanis put up a determined resistance, some hundred and fifty barricading themselves in a blockhouse and fighting to the last. Two mortars of 6/25 and two of the Eurasian howitzers were brought into action at six hundred yards from the blockhouse, and, after an hour's bombardment, the howitzers moved in to four hundred yards. Under their fire the infantry stormed the position.

This brought the operations to a close, and with the return of the two captured howitzers peace was restored.

In the various expeditions against the hill tribes east of Bhutan guns were rarely required, and the operations were not of a nature to demand description here. But it may be recorded that No. 1 Indian M.B. was engaged in the Aka Expedition of 1883, No. 4 in the Daphla Expedition of 1874.

Before considering the Eastern Frontier we must turn our attention to the great mysterious land of Tibet, lying behind the series of tribes on the northern border which we have been considering. Isolated between three of the loftiest mountain ranges in the world, it remained a *terra incognita* until the last quarter of the XIXth century. In the 70's the Government of India took up again the question of a trade agreement, which had been allowed to lapse since the mission of Warren Hastings in the previous century. The making of a road through Sikkim was proposed, but the idea was strenuously opposed by the Tibetans, and we have seen how they attempted to block our attempts even outside their own borders in Sikkim, and how this was dealt with. Some years later the Government became aware that Russian subjects—Buddhists from Siberia—had been received in Lhasa and, as in the case of the Russian Mission to Kabul in 1878, this necessitated definite action. In order to put relations with the Tibetan Government on a proper

CHAPTER
VIII.

Bhutan.
1865.
3rd April.

Akas,
Daphlas,
Abors.

1883.
1874.

Tibet.

1888

CHAPTER
VIII.

Tibet.

1903.

1904.

March.

April.

June.

footing it was proposed to send a Mission under Colonel
F. Younghusband. A conference was arranged, but the
Tibetans proved quite unapproachable, and the Government decided that the Mission should proceed to Gyantse
with, or without, their consent.

In the autumn of 1903, therefore, a force of three
thousand men was assembled under Brig.-General J. R.
Macdonald. The artillery consisted only of a section of
No. 7 British M.B. (Major R. W. Fuller) with 10-pr. B.L.
guns, and the two little 7-prs. of the 8th Gurkhas, known
as "Bubble" and "Squeak". The winter was spent
improving the communications and it was not till March
1904 that the real advance began. The Tibetans offered
a determined resistance at Lamdang gorge, a winding
ravine overlooked by rocky ridges, but the shrapnel
cleared these, while a flanking movement turned their
right, and they were soon in full retreat. Gyantse was
reached next day—11th April—and the Mission established
in a suitable position, after which the main body returned
to Chumbi.

Gyantse lies on the lofty table-land of Tibet, some
13,000 feet above the sea. Its "Jong" or fort is a strong
place, perched on a rock rising 600 feet above the plain.
It was found deserted, but it was quite unsuitable for
occupation, and a hamlet—Changla—about a mile distant
was chosen for the accommodation of the Mission.

Here the Mission was left, but not for long in peace.
The Tibetans were soon in the field again, and the escort
—with which were the two little guns of the Gurkhas—
had to fight hard. General Macdonald advanced again,
and this time with a section of No. 30 Indian M.B. (Lieut.
Marindin) as well as No. 7 British. Before he could
reach Gyantse, however, he had to storm a strong
monastery whose forty-foot walls blocked the way, and
even when the whole force was concentrated at Changla
the enemy clung to the monasteries and villages round,
and had to be driven out at the point of the bayonet.

In these affairs the accurate fire of the mountain guns was of the greatest value in enabling the infantry to close with little loss.

The final fight was at Gyantse itself, where the Jong, monastery, and town on its isolated hill had been made into a veritable fortress. Before dawn on the 6th July a lodgment had been effected in the town, and at daybreak the guns opened fire against the Jong from two positions. It was not until 3.30 p.m. that the 10-prs. succeeded in making a practicable breach in its strong walls, but the way opened, the infantry were soon masters of the Jong, and during the night the Tibetans evacuated the Monastery.

A few days later the force started on its final achievement, the march to Lhasa the mysterious. There was little further opposition from the Tibetans, but at the action of Karo Sar on the 16th July the guns were in action at over 16,000 feet. On the 7th of August the Holy City was occupied, and at the end of September the Mission withdrew, leaving, as Resident in Gyantse, Captain W. F. O'Connor, R.A.

We may now turn to the Eastern Frontier. From the valley of the Brahmaputra to the mouths of the Irrawaddy, Bengal is bounded by a tangle of hills separating it from Burma. Working southwards from the Brahmaputra along these hills the first tribal territory to be encountered is that of the Nagas. In only one of the punitive measures carried out in their hills did guns bear a part, and then only a couple of the smallest size, so that they would have found no place in Regimental History had it not been that this occasion affords a good example of the improvisation of an efficient unit under peculiar difficulties.

In 1880 an expedition had penetrated some distance into the hills when the general telegraphed for the addition of artillery. Lieut. A. Mansel and three bombardiers of the battery in garrison at Fort William (16/9) were promptly despatched; two 7-prs. of 150-lb. and a hundred 9-pr. rockets were drawn from the Ordnance; and at the frontier they were met by thirty men of the 44th Gurkhas, detailed as gunners, with eighteen elephants to carry the outfit. Next morning they were off, but the elephants were small and unaccustomed to the work, the route was execrable, trees had to be felled, rocks blown up, bridges repaired, and the marches averaged twenty miles. It is therefore not to wondered at if some of the ammunition had to be left behind to lighten the loads. One elephant laden with rockets fell over the khud, and it is noted that the shaking the rockets received made their shooting more erratic than usual.

The enemy's position at Khonoma, for the attack of which the assistance of the artillery had been demanded, was strong. The outer wall, 4 to 5 feet thick, was far beyond the power of the low velocity 7-prs. of 150lb., so common shell with time fuzes set to give bursts just over the wall were resorted to. The next work was only half as thick and of loose stones, so double shell with percussion fuzes soon blew a hole in it. But the final assault was repulsed with heavy loss, and it was then perhaps that the artillery rendered their greatest service. The tribesmen were following up the retirement when the bugler with the guns sounded the "Lie down," and shell fire only a few feet over the infantry drove the Nagas back to their works. It was a daring expedient.

Manipur.

1891.

Tucked away behind the Naga Hills lies Manipur, and in 1891 the Government of India decided that the brother and heir of the Maharaja must be deported, and sent the Chief Commissioner of Assam with a strong escort to carry this out. So confiding, however, were all that the Chief Commissioner, the commanding officer, and other officers, entered the fort unarmed. They were at once murdered, their heads stuck on the palace walls, and the guns turned on to the Residency.

The sorry story of the withdrawal of the escort is no

concern of ours, so we can turn at once to the expedition
despatched to avenge the massacre. It was organized in
three columns, of which the first two each included a
half of No. 8 British M.B., while four guns of No. 4 were
with the third. This last column, which was commanded
by Brigadier-General T. Graham, R.A., was the only one
to encounter serious opposition. The Manipuris were
found to be holding an entrenchment, and there was
desperate hand-to-hand fighting even after this had been
bombarded by the guns and the infantry had forced an
entry to the work. The three columns met at Manipur
at the end of April, and the military operations came to
an end.

The Lushai Hills, further south than those we have
been considering, consist of a mass of mountains,
averaging some three to four thousand feet in height.

The first of the expeditions which calls for remark
here is that of 1871, which was caused by the murder
of a planter and the carrying off of his daughter.
Two columns, each 1,500 strong, were employed, starting
from Cacher and Chittagong, the former under Brig.-
General G. Bourchier, R.A. With each column went half
No. 3 Indian M.B.[1], and Lord Roberts, who was staff
officer with the Cachar column, gives a vivid description
of the effect of artillery fire on Lushais:—"Whilst
Captain Blackwood was preparing the fuzes I advanced
with the infantry. The first shell burst a little beyond
the village, the second was lodged in its very centre, for
a time completely paralysing the Lushais. On recovering from the shock, they took to their heels and scampered
off in every direction, the last man leaving the village
just as we entered it."[2]

Soon after this the girl who had been carried off was

---

[1] Lieut.-Colonel J. Hills.

[2] *Forty-one Years in India.* The Captain Blackwood mentioned
was killed at Maiwand when commanding E/B. Lord Roberts' well-
known Arab charger "Vonolel" was so named after a legendary hero of
the Lushais.

CHAPTER
VIII.

Lushai.

1888-90.

given up, none the worse for her alarming experience, and at the end of February 1872 the Lushais made their submission.

There was another expedition in 1888-9 in which a section of No. 2 Indian M.B. again took part, but there was no fighting. This was followed in 1889-90 by a more formidable undertaking known as the "Chin-Lushai Expedition," in which columns from Burma operated against the Chins, as told in Chapter IX, while one from Chittagong dealt with the Lushais. It was, however, unaccompanied by artillery and so has no place here.

COMMENTS.

Organiza-
tion.

The effect of the various re-organizations of the Royal Artillery on the designation of batteries has been explained in the Introduction to this volume, but attention may be drawn here to the examples afforded by the expeditions recorded in this chapter. Thus in 1864 we find all the batteries in Bhutan belonging to the 25th Brigade,[1] one of those formed from the Bengal Artillery at the amalgamation. The title of the battery in the Naga Hills—"16/9"—shows that the brigade system was still in operation. But in Sikkim in 1888 we have "9/1 Northern" indicating the introduction of Territorial Divisions. And finally in the Manipur Expedition of 1891 we get the first mention of British Mountain Batteries as a separate branch with their own numbers.

The "Eurasian" Battery requires some explanation. It was raised in Bengal immediately after the Mutiny under the title of "The Eurasian and Native Christian Company of Artillery," and as such rendered good

---

[1] It may be noted that 5/25 retained its mountain equipment after the Bhutan expedition, and was thus the first British mountain battery maintained as such on the peace establishment.

service in the Khasia and Jaintra Hills in 1862-3. In 1865 it was re-organized as a mountain battery under the title of "The Eurasian Battery of Artillery," and took part in the Bhutan Expedition, after which it was disbanded. We have seen what a fine effort the gunners made to save their guns when the infantry deserted them, but they were not physically up to the heavy work of mounting and dismounting the 12-pr. howitzers, let alone of carrying them.

The "Gurkha" guns mentioned on several occasions also need explanation. These were a special feature of the equipment of the 42nd, 43rd and 44th Assam Light Infantry, which became the 6th, 7th and 8th Gurkhas during the period covered by this chapter.

In Armament, as in Organization, the expeditions recorded in this chapter show many changes. In the earliest—Bhutan—the four batteries engaged were all differently armed—6/25 had $5\frac{1}{2}''$ and $8''$ mortars, the weapons of smooth-bore heavy batteries. The Eurasian Battery had the $4\,^2/_5''$ (12-pr.) howitzers which with 3-pr. guns were the regular equipment of mountain batteries in the smooth-bore era.

3/25 had these same 3-pr. bronze guns, but bored out and rifled.[1]   5/25 had 6-pr. R.B.L. guns, which had been introduced as the mountain gun of the "Armstrong" series. They proved their worth by their accuracy and range, but they were found to be too heavy for mule-back, and so were never adopted as the regular mountain equipment.

The guns with the Assam battalions were the first 7-pr. R.M.L. They weighed 150lb., had a very short range, and were soon superseded by the 7-pr. of 200lb. This gave place in turn to the $2\,5''$, and finally the 10-pr. B.L. with which the British battery in Tibet was armed.

Chapter VIII.

Organization.

Armament.

---

[1] These were the first rifled mountain guns. Some further particulars regarding their fate will be found on page 178.

In accordance with the ideas of the period these were at first provided only with shrapnel, but for the stone walls which the Tibetans used so extensively as defences shrapnel were of little use. Fortunately the question of a common shell was still under consideration, and some hundreds of these had been manufactured for experimental purposes. The expedition provided an admirable opportunity for practical trial, and 300 were sent up in May 1904, and proved so satisfactory that the supply was continued. They made the final assault of the Jong at Gyantse practicable.

In the jungle country of the Northern and Eastern Borders, where the ground was soft and the grazing good, the elephants suffered from none of the disabilities which led to their unpopularity in Afghanistan, and they were extensively used by the artillery.

Not much is to be learnt as regards tactics from the operations on the Northern and Eastern Frontiers. This could scarcely be expected in view of the prehistoric armament of the tribes and the nature of their country. But one incident deserves a place in the record owing to the vivid picture it presents of reversion to the method of the Tudor period.

It was in the attack of a fort in Bhutan. Perched high, and awkward of access, long range bombardment by the Armstrong 6-prs. and 8″ mortars failed to effect a breach, so a couple of $5\frac{1}{2}″$ mortars were carried up to within three hundred yards. Here the made-up cartridges were found to give too great a range, so a cask of powder was brought up and opened near the mortars so that more appropriate charges might be weighed out. And then by an unfortunate mischance a shell burst at the muzzle. This exploded the powder in the cask, and the battery commander,[1] two subalterns, and four gunners were killed

---

[1] Major F. C. Griffin. He had distinguished himself in the Ambela Expedition and been given brevet promotion, but the news had not reached Bhutan when he was killed.

on the spot, while an officer of the Engineers died afterwards of his injuries, and several gunners were more or
less badly wounded.

It is the country over which the gunners had to take
their guns rather than the enemy they had to fight that
makes these campaigns worthy of a place in Regimental
History. We have seen columns blowing up rocks,
felling trees and building bridges to make a way for the
elephants with the guns, or cutting their way through
tall jungle of bamboo and elephant grass, knee-deep in
the marshy ground under a burning sun. But if the
enemy were seldom seen, their arrows daily found a mark,
or the boulders they dislodged came crashing down into
the camp at night.

In Tibet on the other hand there was the long march
to Lhasa and back—four hundred miles over four passes
ranging from fourteen to seventeen thousand feet. A
lofty table-land of bitter frosts and cutting winds, and
the mountain guns in action at an altitude far above that
of the summit of Mont Blanc.

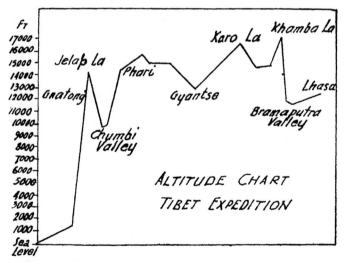

It remains only to note a point of great regimental
interest. The appointment of Brig.-General Tombs to the

command in Bhutan would appear to be the first occasion in the XIXth century for an officer of the Royal Artillery to be entrusted with a force of all arms in war. Officers of the artillery in the service of the Hon. East India Company had held such commands, and with great distinction, but in the British service they had been rigidly excluded.

MEDALS & CLASPS.

The following clasps to the Indian General Service Medals were given for expeditions on the Northern and Eastern Frontiers :—"Bhootan", "Naga, 1879-80", "Sikkim 1888", "Looshai", "Lushai 1889-92", "Manipur", "N.E. Frontier 1891." A special medal was given for Tibet with clasps for "Tibet 1903-4" and "Gyantse."

LIST OF UNITS.

(With Designation in 1914.)

6/25,	R.A.	—Reduced.	
3/25,	do.	— 16,	R.G.A.
16/9,	do.	— 54,	do.
5/25,	do.	—107,	do.
9/1 Northern,	do.	— 3, M.B. do.	
4, Mtn. Battery, do.	— 4,	do.	do.
7, do.	do.	do.	— 7, do. do.
8, do.	do.	do.	— 8, do. do.
1, Indian Mtn. Battery—21, Kohat	M.B.	(F.F.)
2, do.	do.	do.	—22, Derajat do. do.
3, do.	do.	do.	—23, Peshawar do. do.
4, do.	do.	do.	—24, Hazara do. do.
30, do.	do.	do.	—30, Mountain Battery.

CHAPTER IX.

BURMA.

Introduction—The Flotilla—The Voyage to Mandalay—The Advance to Bhamo—Annexation and Pacification—The Frontier Tribes—

COMMENTS.

The Flotilla—Armament—Employment—Medal & Clasps—List of Units.

MAP 7.

THE country known under the general name of Burma comprises a vast territory lying on the east of the Bay of Bengal, and extending from Tibet and China in the north to Siam in the south. It includes Burma proper, or Ava, occupying the middle reaches of the Irrawaddy; Pegu or Lower Burma comprising the lower valley of that river; and the coastal tracts of Arakan and Tenasserim facing the Bay of Bengal. But, as a result of many hostile acts and most insulting treatment of remonstrances, the King of Ava had been shorn of Assam and Manipur, Arakan and Tenasserim, after the First Burmese War in 1825; of Pegu after the Second Burmese War in 1852. At the time we are now considering the Kingdom was, therefore, reduced to a purely inland territory, bounded on the west by the watershed of the Irrawaddy, and on the east by the independent Shan States.

British interests were represented by a Resident at the capital—Mandalay—but relations were unsatisfactory, and with the accession of King Thibaw in 1878 the situation became intolerable. Thibaw was found to be negotiating with France, and the British Government was at last convinced that action was necessary. In October, 1885, an ultimatum was despatched, and, at the same time, orders were issued for the assembly of a force of 10,000 men under Major-General N. D. Prendergast.

CHAPTER IX.

Introduction.

1878.

1885. October.

The force consisted of a naval brigade, three infantry brigades, three mountain batteries, and a siege train. The following is the detail of the artillery :—

C.R.A. —Colonel W. Carey.

Adjutant—Captain M. W. Saunders.

Mountain Batteries.

9/1 Cinque Ports, British M.B.

No. 4 Indian M.B.

No. 1 (Bombay) Indian M.B.

Siege Train.

Lieut.-Colonel W. T. Budgen.

Captain R. D. Anderson.

Q/1 Field Battery (dismounted)

3/1 Scottish, Garrison Battery

4/1 North Irish, Garrison Battery

Garrison of Lower Burma.

6/1 Southern, Garrison Battery

8/1 London, Elephant Battery

While the troops were on their way from India, Colonel Carey at Rangoon lost no time in making preparations for the river voyage. He had lately paid a visit to Mandalay and made himself acquainted with the river defences, so he realised that the siege train might be required to deal with several forts, and many batteries on the banks, as well as having to undertake the siege of a walled city after landing. Every gun must, therefore, be mounted so as to be able to fight on board ship, while being at the same time ready for landing.

The armament of the siege train consisted of :—

4 40-pr. R.M.L. Guns.

12 6·3″ R.M.L. Howitzers.

18 5½″ Mortars.

Four dredging or "mud" barges were fitted, each to carry either two 6·3″ howitzers, or one howitzer and a 40-pr. gun : the remaining six howitzers and two guns were mounted on a "coal" barge, the guns in the bow, the

howitzers on the broadside. The mortars were also
carried in this barge.

The fitting of these barges involved the construction of
gun-decks[1] and magazines, and the erection of iron shields,
but the work was carried out so expeditiously by the
Bombay and Burma Trading Company that all was ready
for the batteries when they arrived. The barges had also
been tested by the garrison battery at Rangoon with both
6·3″ howitzers and 40-pr. guns, at ranges up to 4,000
yards, and the practice had shown that in smooth water
the shooting could be as good as on shore. A drill had
also been worked out for running the guns on shore
without the use of sheers, and the necessary stores had
been provided.

The mountain batteries were carried in "flats," their
guns in action in the bows, but no strengthening of the
decks were required for these. For the British, arrangements were made to allow of sub-divisions filing on shore
simultaneously, the mules saddled up. The Indian
batteries were not bringing their mules.

The naval brigade consisted of the river steamer
"Irrawaddy" carrying two 20-pr. R.B.L. guns, several
steam launches with bow-guns, and two cargo boats, each
with a 64-pr. R.M.L. These guns were all on naval slides,
and the sailors would not consent to use on board their
25-pr. R.M.L. guns on travelling carriages, but carried
them as cargo to be employed only on shore.

As far as the frontier station of Thayetmyo the voyage
was made without incident. On the 14th November
General Prendergast received orders to advance on
Mandalay, and on the 15th the flotilla of twenty vessels
crossed the frontier, led by the naval brigade. Next day
some entrenchments on the river bank were carried without serious resistance by a landing party which included
9/1, and on the 17th a very strong modern fort with a

[1] The decks were covered with four inches of laterite soil so as to
give a firm platform and check recoil.

CHAPTER
IX.

1885.
The Voyage
to Mandalay.

large garrison was captured in the same way, the troops (again including 9/1) being landed at some distance so as to effect a surprise.[1] In the attack of the corresponding work on the opposite bank the landing party were unprovided with artillery, and they suffered considerable loss in consequence before the capture was effected. Leaving a garrison, which included a section of No. 1 (Bombay) M.B., the flotilla steamed on, meeting with no further

22nd Nov.

resistance until the 22nd when shots were exchanged by the naval brigade with a battery on a high bluff, the siege train joining in with a few rounds from a 40-pr. On the withdrawal of the defenders their guns were pitched over the cliff, and a small garrison installed which included another section of No. 1 (Bombay). On the afternoon of

24th Nov.

the 24th the large town of Mingyan with batteries all along the river front was approached. It was too late to land troops, the passage had to be forced.

The Burmese batteries had a large command, and the guns fired through embrasures so that they were well protected. Moreover any projectile which did not hit the parapet went harmlessly into space, so that the guns of the flotilla fought at a great disadvantage. But their moral effect, and especially that of the broadsides from the coal barge, must have been great, for the Burmese forces had disappeared before operations were resumed in the morning. The naval brigade had brought into action the two 20-prs. on board R.I.M.S. "Irrawaddy," and the two 64-prs. on board the cargo boats, in addition to the smaller pieces in their launches. The siege train had the coal barge with her two 40-prs. and six 6˙3″ howitzers, and also one of the mud barges with two 6˙3″ howitzers, and fired altogether 131 common shell and 10 case shot. They could fairly claim a full share in the action which practically decided the campaign.

The last section of No. 1 M.B., was left with the usual garrison, and the flotilla proceeded on its course un-

[1] The route had been reconnoitred by Colonel Carey during his visit to Mandalay.

molested, the further forts surrendering without a shot.
On the 28th November, Mandalay was occupied, and next
day King Thibaw was deported. Within three weeks of
the declaration of war the King and his Capital were in our
hands.

Throughout December columns were scouring the **The Advance to Bhamo.**
country in the vicinity of Mandalay, rounding up bands
of disbanded soldiers. Owing to the Indian mountain
batteries having come without their mules, and coolie
transport proving quite useless, the British battery was
much overworked providing guns for these columns.
When, in addition, their relief mules were impressed for
infantry transport the C.R.A. had to make it clear that, if
this course were persisted in, the battery would soon be
out of action.

The only important military operation before the end **December**
of the year was the occupation of Bhamo, the force for
which included No. 4 Indian M.B. and a section of 4/1
with two 25-prs. (handed over by the naval brigade) and
two 6·3" howitzers. There was no opposition, but getting
the siege guns up to the town proved an awkward job owing
to want of water for the barge. This necessitated putting
them on shore on the river bank a mile or two below the
town, and making a road through the jungle—with a
stream to cross on the way. The mountain gunners, the
sappers, and the infantry all lent a hand, and with the
assistance of the gyn in bad places, and much jungle grass
for roadway, the journey was safely accomplished.

On the 1st January, 1886, the annexation of the **Annexation and Pacification.**
dominions of the King of Ava to the British Empire was
proclaimed : the war had been brought to a triumphant **1886**
conclusion—on paper at any rate. An immense number **1st Jan**
of guns had been captured, ranging from 1-prs. to 60-prs.
Some were of very ancient date, some of European make,
(including 32-prs. on travelling platforms) and there was a
battery of 9-prs. overlaid with gold.

CHAPTER
IX.

1886.
Annexation
and
Pacification.

The new territory incorporated into the British Empire on New Year's Day had an area of 140,000 square miles and a population of three and a half millions. A considerable part of the vast expanse was inpenetrable jungle or swamp, and nowhere did roads or bridges exist. Among the inhabitants there was a tradition of "dacoity" which had been allowed to rise unchecked under Thibaw's lax administration. And now there had been turned loose, to swell these robber gangs, armed men by the thousand from the army who, by some strange oversight, had not been disarmed when they were disbanded.

The work of pacification meant the hunting down of the dacoits by flying columns. It was found that a column of more than about two hundred men was unwieldly, but that infantry were at a great disadvantage when working alone. In addition to guns, some cavalry or mounted infantry therefore came to be generally included.[1]

The nature of the operations demanded marching, by day and night, through jungle and swamp, over waterless waste or upland, sweltering in the mud of the rice fields or freezing on the mountain ridges. The climate in cold weather was fairly healthy, but sections were hardly in before they were called out again, and there was continuous patrolling, in which the gunners and drivers of the mountain batteries, mounted on the battery mules, took their share. The incessant toil and exposure took heavy toll on all, although every effort was made to spare them as much as possible in the hot weather when the climate in many parts was deadly. Events were, however, often too strong, and a hundred actions were fought during the four summer months of 1886.

Sustained effort and dogged persistence had their inevitable result. By the end of the winter of 1886-7 the

---

[1] Here is an example, chosen at random, with by chance a gunner commander :—

Captain O. S. Smyth, R.A.
9/1 Cinque Ports, British M.B.—Two guns.
Royal Welch Fusiliers   —A hundred rifles.
12th Madras Infantry   —Forty rifles.
2nd Madras Lancers   —Thirteen lances.

principal leaders of the dacoits had either been accounted
for or were fugitives, their bands hunted down and
broken up.

There still remained, however, the elements of disturbance presented by the wild tribes on the borders—
east, west and north—Shans, Chins, Kachins.

On the east, the Shan States between Burma and
Siam ,accepted British authority without opposition (on
the whole) although there were many local risings. This
was especially so in the northern states bordering on the
Chinese Province of Yunnan, necessitating visits by
columns on several occasions, in which Major H. T. S.
Yates represented the Regiment among column commanders.

On the opposite side of Burma the most important
operations were those known as the "Chin-Lushai Expedition." In the last chapter mention was made of the
tangle of hills that divides the watersheds of Burma and
Assam, and the conquest of Burma now allowed of this
being approached from both sides. In 1889-90 a column
from Chittagong, directed against the Lushais, co-operated
with columns from Burma against the Chins, and between
them explored much unknown country besides carrying
through a road from India to Burma.

In 1890-91 a rebellion broke out in Wuntho, an
autonomous state on the right bank of the Irrawaddy
above Mandalay. Considerable forces took the field under
Brig.-General G. B. Wolseley, the largest column being
commanded by Captain A. E. A. Smith,ˏ R.A. The
rebellious "Tsawbwa" was hunted up to the Jade Mines
in the far north, and that little-known country opened
up, but the fugitive escaped into the "back-of-beyond"
where pursuit was out of the question.

In the forest-clad hills of the north, on both sides of
the Irrawaddy, dwelt the Kachins, a strange and superior
folk unconnected with the Chins. Large tracts of their
country had never been visited, and the hills along the

CHAPTER
IX.

The Frontier
Tribes.

1891-92.

1891.

December.

Chinese Frontier served as an *Alsatia* in which armed bands could collect with impunity. It was clear that something more than occasional visits by punitive columns was necessary in order to bring them under control, and protect the numerous trade routes between China and the Irrawaddy. Four columns were despatched in the winter of 1891-92 to establish police posts, and we may well follow the fortunes of one of these as an example of the dealings with the frontier tribes.

The "Irrawaddy Column" started from Bhamo, and after following the right bank as far as Myitkyina, crossed the river—nearly a mile wide—and turned east for Sadon, a village lying at the junction of two of the trade-routes from China. Its strength was about four hundred men with two guns of No. 6 Indian M.B. under Lieut. Boyd. Sadon was occupied on the 30th December, 1891, after a brush with the enemy, in which one of the guns was brought into action at a turn in the track only a hundred yards from the stockade which blocked the way. January was spent exploring the country, and visiting the surrounding villages, while the post was being built on the hill above. All seemed friendly, so on the 5th February the column left to inquire into matters elsewhere and to enter some unexplored country, leaving a hundred and fifty men under Lieut. Harrison, R.E., to hold the fort.

Scarcely had the column got out of sight before the attitude of the Kachins changed. Crowds flocked into Sadon, and on the 7th the fort was fiercely attacked. The great difficulty—as so often in such posts—was water. This had to be fetched from springs outside, which meant fighting, and such sallies cost casualties that the little garrison could ill afford. A valuable reinforcement was, however, soon received. Lieut. G. F. MacMunn on his way to join No. 6 Indian M.B. had picked up a dozen Gurkha mounted infantrymen at Myitkyina and fought his way through in spite of stiff opposition—followed up to within half-a-mile of the fort. There was plenty of work for a gunner to do, in addition to the value of another

British officer in such a position. The supply of hand-
grenades had been exhausted, and, after assisting the
sapper to make bombs out of jam-tins, he set about pro- The Frontier
viding the fort with a gun. Of bamboo, bound round with Tribes.
wire and cord, length two feet, calibre three inches, charge
one and a half ounces of gun-powder, it fired its bamboo
shells into the Kachins to their great confusion.[1]

On the 20th February the post was relieved, after
being under constant rifle fire, day and night, since the
7th.

It was to be some years yet before trouble on the border, 1892-93.
fomented by the Chinese, was stamped out. But this
account of the work of the artillery in the Burmese War
may well close with the year 1892-93.

COMMENTS.

The salient fact in connection with the part played by The Flotilla.
the artillery in the advance to Mandalay is the shipping of
the batteries not as passengers but as fighting units,
ready at any moment to serve their guns on board or
to land them for their normal rôle on shore. Colonel
Carey's investigation of the Burmese defences had shown
him the necessity for this novel precedure, and his foresight and energy in getting the barges converted for the
purpose, and all eventualities provided for, stands out as
the first step towards the success of the undertaking. But
the part the batteries played must not be overlooked.
Called from different parts of India, and put to such
unfamiliar work, they showed the resource and adaptability
which we may safely claim as a characteristic of the arm.

---

[1] An interesting sequel was his next meeting with the Kachins in
1920 when the Burma Rifles (nearly all Kachins) distinguished themselves at the storming of the Bazian pass in Kurdistan. General
MacMunn was able on that occasion to decorate a Kachin Naick with
the D.C.M. and to give him, for it, the ribbon of the D.S.O. he had
won at Sadon.

CHAPTER
IX.

CHAPTER
IX.

The Flotilla.

The work of the siege train as floating batteries was shown at Mingyan; the mountain battery landed wherever required to assist in the attack of the river defences from the landward; and the disembarkation of the 25-prs. on the river bank and their march across country to Bhamo was no easy task. And it was not all smooth sailing for "Mother Carey and her chickens"—as the barges were nicknamed by the sailors. The steamers proved incapable of towing more than one gun barge against the stronger current met with in the higher reaches of the river, and re-arrangements had to be made. Worse still, one of the barges was found to be sinking: her 6'3'' howitzer took charge and went overboard, and the barge followed suit, settling down in 24 feet of water with all her stores. An unsuspected defect in design had allowed of water accumulating in the hold unknown to anyone, and all had to be altered.

Armament.

During the pacification of the country the only guns employed were 7-prs., but three patterns of these were in the field—those of 400, 200, and 150lb. The British mountain battery had the first-named, the 2'5'', which had been introduced during the Afghan War and became so famous as the "screw-guns". The Indian mountain batteries and the garrison batteries had the 200lb. guns which had superseded the first pattern—of 150lb.—but these still lingered on in possession of the Assam battalions.[1]

The 2'5'' had shown the value of their range and accuracy on the bare hill-sides of the North-West, but had little chance of displaying these qualities in a jungle clearing, where the double shell of the lighter guns were so effective against stockades. Moreover, each gun required five mules instead of three—a serious matter where single file was the usual order of march.

Enough has been said of the grave error of sending the Indian mountain batteries without their mules. Before

---

[1] These local battalions in Assam had each a couple of 7-prs. as explained in Chapter VIII.

the arrival of these an attempt was made to use the $5\frac{1}{2}''$ mortars on Burmese ponies, but it proved a failure. A more successful experiment was the organization of a machine-gun battery by 1/1 Eastern with four $0.45''$ Gardners obtained from the navy carried mountain battery fashion. It formed part of the expedition to the far-famed Ruby Mines.

A word must be said about the elephant battery at Tounghoo—the only one of its kind. The country had proved too rough in the dry season, and too soft in the rains, for pony draught, so, in consequence of the good work of the elephants in Lushai in 1871, it was decided to adopt them for the carriage of the guns at the frontier station.

As regards the tactical employment of the artillery there is little to be said. Unless the enemy's position was very difficult of access, or serious resistance was antici- pated, the support of the attack by artillery fire was a doubtful advantage when the object was to pin the enemy to the ground and inflict punishment. In consequence the number of guns allotted to a column was practically limited to a section. Captains and majors got their chance in command of columns or posts, and when there was no call for their guns some gallant young subalterns, not content to remain inactive, proved themselves resolute leaders of mounted infantry. It might well be called "The Subalterns' War," and it is good to know that their services were not overlooked. The Distinguished Service Order was instituted in 1886 to reward their gallantry, and the writer well remembers the excitement at Woolwich when two of its first recipients appeared on Church Parade wearing the coveted decoration.

Any attempt to record here a tithe of the incidents which reflect so much honour on those engaged would lead us far beyond the bounds of Regimental History. All that space permits here is the bald record of the batteries which took part in the campaign: it is in the histories of those batteries that the achievements of their sections are to be found. A complete list of all those engaged is given below.

Medal & Clasps.

For the operations recorded in this chapter the Indian Medal 1854 was given with the following clasps :—
"Burma 1885-87", "Burma 1887-89", "Burma 1889-92", "Chin Hills 1892-93", "Kachin Hills 1892-93."

List of Units.
(With Designation in 1914.)

Q/1	—Major J. R. S. O. Hewitt	—38, R.F.A.
4/1 Lancashire	—Major J. A. S. Colquhoun	— 1, R.G.A.
4/1 North Irish	—Major A. Broadfoot	— 7, do.
3/1 Scottish	—Major R. H. S. Baker	—22, do.
6/1 Southern	—Major F. M. Robinson	—87, do.
8/1 London	—Major W. H. F. Sorell	—90, do.
5/1 Southern	—Major F. M. E. Vibart	—91, do.
7/1 Northern	—Major G. T. Carré	—Reduced as 92 do.
2/1 Cinque Ports	—Major J. Fowler	—No. 2 M.B., R.G.A.
9/1 Northern	—Major T. Graham	—No. 3 do. do.
9/1 Cinque Ports	—Major W. Aitken	—No. 5 do. do.
3/1 South Irish	—Major G. B. N. Martin	—No. 6 do. do.
1/1 Eastern	—Major C. J. Deshon	—No. 9 do. do.

Indian Mountain Batteries.

No. 4 Punjab	—Captain C. P. Triscott	—24 Hazara M.B. (F.F.)
No. 1 Bombay	—Captain A. Keene	—25, Bombay M.B.
No. 2 do.	—Captain F. E. Sinclair	—26, Jacob's M.B.
No. 1 Bengal	—Captain C. P. Triscott	—27, Mountain Battery.
No. 2 do.	—Captain A. E. A. Smith	—28, Mountain Battery.

Note.—It was during the Burmese war that British mountain batteries made their first appearance as a separate branch with their own enumeration, and the Indian batteries were numbered in one series. In this the batteries of the Frontier Force took numbers 1 to 4, the Bombay batteries became 5 and 6, and the newly raised Bengal batteries carried on from 7.

PART II.

CHINA.

CHAPTER X.

THE EXPEDITION OF 1860.

The Expeditionary Force—The Advance to the Peiho—The Taku Forts—The Advance to Peking.

COMMENTS.

Organization—Armament—Employment—Gunnery—Medal & Clasps—List of Units.

As a sequel to previous hostilities, with which we are not concerned, a draft treaty with China was agreed upon, and in 1859 the British Minister was on his way up the Peiho River to obtain its ratification when the vessels were fired upon by the Taku Forts at its mouth. A naval attempt to reduce the forts miscarried, and the British and French Governments agreed upon the despatch of a military expedition in the following year.

Some of the troops composing the British contingent were drawn from England and South Africa, but the great bulk came from India. The total strength was 14,000 organized in a cavalry brigade, two divisions, and some additional units, with Lieut.-General Sir Hope Grant[1] as Commander-in-Chief and Major-Generals Sir J. Mitchell and Sir R. Napier as Divisional Commanders. The artillery was as follows :—

CHAPTER X.

1859.

The Expedition-ary Force.

1860.

---

[1] Major R. Biddulph, R.A., was Military Secretary to the C.-in-C.

G.O.C., R.A. —Brig.-General E. W. Crofton.

A.A.G. —Captain R. J. Hay.

D.A.Q.M.G. —Captain L. Brabazon.

Cavalry Brigade — 2/4 Field Battery

1st Division —
$$\begin{cases} 1/4 & \text{do.} & \text{do.} \\ 4/13 & \text{do.} & \text{do.} \\ 4/2 & \text{Rocket Detachment} \end{cases}$$

2nd Division —
$$\begin{cases} 3/13 & \text{Field Battery} \\ 7/14 & \text{do.} & \text{do.} \\ 4/2 & \text{Rocket Detachment} \end{cases}$$

Siege Train —
$$\begin{cases} 4/2 & \text{Siege Battery} \\ 6/12 & \text{do.} & \text{do.} \\ 8/14 & \text{do.} & \text{do.} \\ A/5 & \text{Madras Artillery} \\ 1/5 & \text{(Supplemental) do.} \end{cases}$$

Mountain Train — Madras Artillery.

The above organization was, however, modified to the extent that the 6-pr. battery (7/14) worked with the cavalry on account of its lighter equipment, while 2/4 took its place with the 2nd division, thus giving each division a 12-pr. battery.

June, July. This force was collected during June and July on the Talienwan Peninsula[1], and on the arrival of the French the combined force sailed for the Chinese coast. The place selected for landing was Peitang, nine miles north of the Peiho River, and the disembarkation was effected 1st Aug. without opposition on the 1st August[2] and following days.

The whole country between Peking and the sea is a vast alluvial plain—intersected by cart-tracks raised

---

[1] Better known in later years as the site of Port Arthur.

[2] The return of the force at Peitang on the 1st August gives Royal Artillery—1621, Madras Artillery—184. Six guns were left at the depôt at Talienwan.

several feet above the level of the land, and irrigation
channels sunk a few feet below it. In the spring the
ground is hard and firm, but in the summer it is liable
to inundation with the rains of June and July, while in
the autumn both view and movement are restricted by
the tall crops of kowliang or millet. Through this plain
the Peiho River, the best means of approach to Peking,
pursues a devious course, eventually entering the Gulf of
Chihli at Taku.

The advance from Peitang was commenced on the 12th
August, the 1st division along a causeway, the 2nd
division across country on their right, and the cavalry still
further out on that flank. Both the latter found themselves in difficulties almost at once, as vividly told in Sir
Hope Grant's diary :—"I accompanied Sir R. Napier in
order to satisfy myself that he had succeeded in crossing
the marsh; and here he encountered great difficulty in
dragging the artillery along. The horses got bogged, the
guns sunk up to their axletrees, and the wagons stuck fast.
At last we were compelled to leave the wagon-bodies
behind, and to content ourselves with the gun and wagon
limbers . . . . at one time I really thought we should be
obliged to give up the attempt. But . . . . by means of
drag-ropes and persistence the artillery were hauled over
the two miles of mud and sound ground was reached."
Further out the half-battery of 7/14 had been unable to
keep up with the cavalry. Left behind with an escort of
Indian cavalry, it was in some danger from Tartar
horsemen who swooped down upon it. So unexpected
was the attack that Stirling had only time to fire a couple
of rounds, but a charge by the escort saved the situation.

After negotiating the marshland, the whole force
approached Sinho in line of battle. The Tartar cavalry
made a gallant effort to stem the advance, but were scattered
by the fire of the fifteen Armstrong guns (1/4, 2/4, and half
7/14). The 12th August, 1860, should be a memorable
date in the history of the Royal Artillery—the first occasion

CHAPTER
X.

1860.

The
Advance to
the Peiho.

14th Aug.
on which they used rifled guns.[1] The Chinese were driven out of the entrenchments covering Sinho without difficulty, and two days later Tangku, a walled village, was carried by storm. The Chinese gunners stood stoutly to their guns, but the fire of six batteries (including two French) in line at 900 yards was too much for them. "I am happy to say," wrote General Grant in his despatch, recording the operations of the 12th and 14th August, "that our losses in these two engagements were very slight owing to the enemy being completely paralysed by the superior fire of our artillery."

The allied force had now reached the bank of the Peiho. They were upstream of the Taku forts that had defied the naval attack of the previous year, and still closed the mouth against the entrance of our vessels. Their capture must obviously be undertaken before an advance on Peking could be commenced.

The forts were very strong places, two on each side of the mouth of the river. They were, of course, primarily intended to withstand an attack from the sea, but even on the landward side they were defended by formidable walls and wide ditches, so the Siege Train was ordered up. Causeways and bridges were prepared, batteries constructed, and armed with one 8″ and two 32-pr. guns, two 8″ howitzers, and two 8″ mortars—all smooth-bore.

The batteries[2] opened fire at daybreak on the 21st and were vigorously answered by the Chinese who had mounted, amongst others, two English 32-prs. recovered from the gun boats which they had sunk during the abortive naval attack of 1859.

The Chinese garrison offered a stubborn resistance to

---

[1] Lieut.-Colonel John Pawson, who enlisted as a trumpeter in 1849 and only died in 1933 at the age of 96 always claimed to have fired the first Armstrong gun. He was a very popular Riding-Master at Woolwich in the 80's and 90's.

[2] Of the siege train supplemented by the two 12-pr. field batteries and 3/13.

the fire of the guns at only six to eight hundred yards, but when their main magazine was blown up by a lucky shell their fire slackened. The field batteries pushed forward to within five hundred yards, the infantry advanced to the assault, and the two howitzers of 3/13 went with them to breach the barrier at the gate. "Whilst the storming parties were struggling across the ditches", wrote Lieut.Colonel Wolseley, "our Armstrong guns were making admirable practice a few feet over our heads, actually knocking the wall about so that portions of it fell upon our men's heads."

CHAPTER X.
1860.
The Taku Forts.

The obstacles were soon overcome, the infantry forced their way into the interior, the 9-prs. of 3/13 were rushed up to shell the fugitives, and Captain Bedingfeld taking a dozen gunners of 6/12, laden with 32-pr. shell and cartridges, into the fort, turned its guns on to the neighbouring work.

The capture of the one fort proved decisive, the other three surrendering, or being evacuated, without a struggle, leaving nearly 400 guns, many of large calibre, in the hands of the allies. But the elements were not so kind. Hardly had the forts fallen than a terrific thunderstorm burst over the plain, converting it into a swamp. The returning army was soon in confusion, the infantry wading up to their knees in mud. "We passed Milward's battery *hors de combat*; Desborough's howitzers had to be left behind; just as we were nearing Tank Ku one of Barry's guns got so deep in the mud that it could not be moved on. By degrees the remainder of the battery got as far as the town but there the streets were in such a state that its further progress for the night was stopped".

The mouth of the Peiho river was now open, a flotilla entered, and on the 23rd Tientsin was occupied. From here the advance on Peking was commenced on the 10—12 September, but on the 18th the diplomatic party who had ridden on in advance of the troops were treacherously seized and carried off. General Grant immediately

The Advance to Peking.
23rd Aug.— 18th Sept.

attacked, the French hurried up, and the action ended with the precipitate retreat of the Chinese, leaving 75 guns in the hands of the allies. The batteries engaged were those with the cavalry and the 1st division—half 7/14, 1/4, and 4/13. Sir Hope Grant reported "The French soon got on the left flank of the Chinese—I ordered the 9-pr. battery to open fire which diverted the enemy's attention from the French flank attack. Sir J. Mitchell was sent to the left with . . . and Stirling's 6-pr. battery to act against the enemy's right. I proceeded to the front with the Armstrong guns."

The army was now traversing reasonably sound ground, and the batteries as a rule experienced little of the inconvenience caused by such heavy going as they had met with in the low-lying flats between Peitang and the Peiho. On the 21st the advance of the allies was again disputed at Palikao,[1] but here the brunt of the fighting fell to the French. With the prospect of having to lay siege to Peking, the 2nd division and a siege battery were ordered up, the guns of the latter by lighter on the river, and it was not until the 3rd November that the advance was resumed. Skirting round the east of Peking the troops were in position facing the north wall of the Tartar city on the 9th. An ultimatum was despatched intimating that unless the gate in this wall was opened by noon on the 13th the city would be bombarded. Batteries for four 8″ guns and three 8″ mortars were ostentatiously constructed within close range of the gate, and the field artillery stood ready to keep down any fire from the walls. It was a bluff, for the siege battery had nothing like sufficient ammunition to breach the massive walls, but it served its purpose. Within a few minutes of the expiry of the time limit the gates were thrown open. Next day some of the prisoners were returned, but some had been tortured to death, and in expiation the Summer

---

[1] General Montauban, the French Commander-in-Chief, was given the title of Count Palikao, which became very well-known at the time of the Franco-German War, ten years later.

Palace was destroyed. The inlaid bronze gun and carriage taken from it by 1/4 are now in the Institution.

COMMENTS.

The China Expedition of 1860 marks an important
epoch in Regimental History, for in the preceding year
far-reaching changes in organization and armament had
been decided upon. In organization the "Brigade
System," which was to be the bed-rock of regimental
administration for the next thirty years, had substituted
"brigades and batteries" for "battalions and companies".
But it remained still the custom to refer generally to units,
even in official reports and despatches, by the name of
their commander.

Brigades consisted of from seven to ten batteries, and
were either "field" or "garrison," although the Regimental Order was careful to provide that they were liable
to be changed at any time. How this provision effected
some of the batteries who had worked so well in the field
in China is shown in the following extract from a letter
from Major-General Sir Desmond O'Callaghan—"I was
appointed in January, 1862, to a very smart field battery,
4/13, which, from the fact that the detachments left their
guns and sword in hand charged the Chinese infantry, got
the name of 'Desborough's Horse.' . . . . I did not join
it till it arrived at Woolwich, when it, and all the other
batteries of the 13th Brigade, as they arrived from India,
were dismounted, and, to everybody's grief, made into
garrison artillery."

The "Mountain Train" was formed of officers and men of the Madras Artillery, the drivers being drawn from a native horse battery.

Armament.

Even more radical was the change in armament—that from smooth-bore to rifled guns. In January, 1859, the Armstrong design was definitely accepted, and before the end of 1860 a large number of field batteries were provided with the new armament, and two of these (1/4 and 2/4) were specially sent out to China. The equipment of a battery of 6-prs. was also sent out, and 7/14 was armed with these. The other field batteries (3/13 and 4/13) had the usual smooth-bore armament of four 9-prs and two 24-pr. howitzers. The mountain train had 12-pr. howitzers and $5\frac{1}{2}''$ mortars.

From the point of view of regimental history, and probably also from that of military history generally, the most interesting feature of the China Expedition of 1860 was the first appearance of the "Armstrong" guns. Almost all those who have left accounts of the campaign, from Sir Hope Grant and the future Lord Wolseley downwards, speak enthusiastically of their performance. One undoubted drawback was, however, disclosed—the liability of the lead coating of the shells to strip in flight. When infantry were working in front of the guns this was calculated to cause annoyance—at least. And there were defects in the breech-closing system which were to lead very shortly to a reversion to muzzle-loading. But these were matters of detail, quite irrelevant to the superiority of rifled guns over smooth-bores, regarding which there was no question. There ensued, however, much public controversy regarding the merits of the Armstrong guns, and Captain Hay, who had been the chief artillery staff officer, finding that his statements were distorted by opposition writers, wrote to Sir W. Armstrong to assure him that they rendered most valuable service, and were always in serviceable condition, though it would have been surprising if slight alterations had not suggested themselves considering that they were tried for the first time and most jealously watched.

To turn now to the use made of the artillery. It is clear that care was taken to utilize the various natures in accordance with their capabilities, of which the allotting of 7/14 to the cavalry as soon as it received the lighter equipment is only one instance. The employment of the rifled guns with their accuracy of fire and segment shell against the Chinese on the walls is another: the use of the howitzers of the smooth-bore batteries is a third. And the massing of guns and concentration of fire were well exemplified by the way in which the Armstrongs of both divisions were brought together. Nor was there any hesitation in bringing up the heavy pieces of the siege train when there was serious work on hand.

To revert to the Armstrong guns. It is interesting to note how their longer range was giving the artillery their first introduction to the problems which were to loom so large for their successors. There was no doubt of the accuracy of the guns, but there were no range-finders, and "ranging" was an unknown process. On more than one occasion, when the enemy's position was examined, fire which had appeared effective from the firing point was found to have gone far afield. An astute observer noted —"The distance was about 1,700 yards, and it would almost seem as if length of range was incompatible with due precision of fire" Observation of fire had yet to be cultivated.

MEDAL & CLASPS.

A replica of the 1842 medal, with clasps for "Taku Forts 1860" and "Pekin 1860", was granted to those who took part.

LIST OF UNITS.

(With Designation in 1914.)

Battery.	Commander.	1914.
4/13, R.A.	—Major J. Desborough	— 11, R.F.A.
1/4, do.	—Lt.-Col. W. W. Barry	— 12, do.
7/14, do.	—Capt. W. Stirling[1]	— 43, do.
8/14, do.	—Major J. F. Pennycuick	— 44, do.
2/4, do.	—Capt. T. W. Milward	— 62, do.
3/13, do.	—Capt. C. M. Govan	— 64, R.G.A.
6/12, do.	—Capt. P. Bedingfeld	— 77, do.
4/2, do.	—Major G. Rotton	—102, do.
A/5 Madras Artillery	—Capt. J. McK. Macintyre—	
1/5 Supplemental do.	—Capt. W. J. Mann	—
Mountain Train do.	—Capt. H. E. Hicks	—

[1] In the embarkation return the battery is shown as "Captain Mowbray's," but he was sick in India and did not accompany it to China where it was always known as "Stirling's".

CHAPTER XI.

THE BOXER REBELLION.

The Boxer Movement—Tientsin—The Advance to Peking—The Relief
of the Legations—The Arrival of the Allied Forces.

COMMENTS.

Organization—Armament—Medal & Clasps—List of Units.

THE origin of the "Boxer" movement that led to the
Allied Expedition to China in 1900, does not concern the
History of the Royal Artillery. It is, therefore, sufficient
to say that it first became noticeable in 1899, and quickly
developed into a wide-spread anti-foreign movement.
There were murders of missionaries and Chinese converts,
followed, in June 1900, by massacres in Peking, the
killing of the German Ambassador, and the attack on the
Legations.

An International Expedition was decided upon. But,
before this force could be organized and despatched,
the relief of the Legations, with which all communication
had been cut off, called for an effort with what forces were
immediately available. The first to answer the call was
Admiral Sir Edward Seymour, who quickly organized a
naval brigade from the allied war-ships collected at Taku,
and set off for Peking. The resistance encountered was,
however, so stubborn that he was obliged to fall back to
Tientsin.

Here the departure of the naval brigade had been the
signal for a fierce attack upon the foreign concessions,
and it was fortunate that the brigade returned when it
did, and that reinforcements were on their way. The
first to arrive brought with them Nos. 2 and 4 companies

Margin notes:

CHAPTER XI.

1899.
The Boxer Movement.

1900.

Tientsin.

CHAPTER
XI.

1900.

Tientsin.

22nd June.

13th July.

of the Hong Kong-Singapore Battalion, R.G.A., under Major G. W. F. St. John, with Captain E. G. Waymouth as adjutant.

No. 2 Company was equipped with four maxims carried on ponies, No. 4 with four 2·5″ mountain guns equipped for carriers.

The batteries arrived on the 22nd June and found the Allies holding out with difficulty against overwhelming odds. No. 4 was soon in action engaging the numerous Chinese guns, and assisting in repulsing attacks on the barricades with case shot. In July the Allies found themselves strong enough to take the offensive, and in the attack on the Chinese city on the 13th No. 4 supported the infantry with shrapnel directed against the defenders on the walls, while the maxims of No. 2 did good work against the enemy whenever they massed. The Chinese fought hard, and it was not until the next day that the capture of the city was completed.

In engaging the Chinese guns the black powder charges of the 2·5″ guns had been found a great disadvantage in giving away the position of the guns. For the advance on Peking the battery therefore obtained two 12-pr. Q.F. guns from the Navy, and mounted them upon the carriages of Krupp guns captured from the Chinese. With teams of four Japanese ponies, and occasional help from the horses of the field battery, the Hong Kong— Singapore battery was well up in the race for Peking. This is, however, anticipating events.

During July reinforcements had been arriving at Tientsin, and on the 5th August the advance to Peking began, the artillery of the British force consisting of :—

The Naval Brigade with four 12-pr. Q.F. guns.

12th Battery, R.F.A.[1] with six 15-pr. B.L. guns.

No. 2 Battery, H.K.-S., R.G.A., with four maxims.

No. 4 do. do. do. with two 12-pr. Q.F. guns.

---

[1] This battery had been in the 1860 Expedition as 1/4 R.A.

The Chinese were in position at Peitsang, but were driven out by the Americans, British, and Japanese, the latter coming in for the chief share in the fighting. The British guns were all in action, occupying successive positions right up to the village, but the country was covered with crops of kowliang, 12 to 14 feet high, which made the work very difficult. Lines of fire were laid out by section commanders from the tops of limbers, and battery commanders improvised "look-outs"[1] with ladders guyed up with ropes. Next day the Chinese were in position again, holding a railway embankment and entrenchments near the village of Yangtsun. These were carried without a check, the maxims of No. 2 H.K.-S. doing useful work in preventing reinforcements reaching the village when it was about to be assaulted by the British and American infantry.

There was no further opposition from the enemy, but the heat was intense and the "going" sometimes very heavy—so much so that a day or two later six horses of the field battery died from exhaustion in crossing a patch of soft sand. Peking was reached on the 14th August, and the various contingents effected their entrance into the Chinese city by different gates, but not without heavy fighting in some cases.

From the Chinese city it was necessary to enter the Tartar city in order to arrive at the Legations, and there was a good deal of firing from the walls of the latter. A gun of the 12th, run round the corner of a street at a range of only 350 yards, cleared the snipers away, and entrance was then effected through a sluice gate. A couple of guns of the 12th breached the barricade beyond, and then followed on into the Legations, the other four guns of that battery, and the 12-prs. of No. 4 H.K.-S., going to the assistance of the Russians at the south gate of the Chinese city. A gun of No. 4 also gave a hand to

CHAPTER
XI.

1900.
The
Advance to
Peking.
5th Aug.

14th Aug.

The Relief
of the
Legations.

---

[1] These look-outs had been much the fashion at practice camps during the 90's and the Okehampton scrap-book has some amusing skits upon them.

CHAPTER
XI.
1900.
The Relief
of the
Legations.

The
Arrival of
the Allied
Forces.

September.

the French in relieving the Peitang monastery. There was a good deal of fighting in one place or another before the whole was in the hands of the allies, and, outside the walls, there were constant reconnaissances with occasional skirmishes in which the 12th had some successful shoots.

The crisis was over but troops kept pouring in. The force which relieved the Legations had comprised Americans, British, French, Japanese and Russians: the Germans now arrived with Field-Marshal Count von Waldersee for whom the German Emperor had secured the supreme command. As British representative on his staff Colonel J. M. Grierson came from Headquarters in South Africa, and Brig.-General H. Pipon arrived as G.O.C., R.A., with Captain T. L. Coxhead as Brigade-Major. The British reinforcements included "B", R.H.A., from India, the Siege Train[1] from South Africa under Colonel Perrott, and pom-pom sections from home. For the siege train India sent 800 splendid bullocks, and for the pompoms all the field batteries in that country made contributions of horses. But the pompoms were landed at Shanghai and most of the Siege Train at Kowloon, only one siege battery going as far as Wei-hai-wei. "B" reached Peking, but the fighting was all over, though there were various expeditions in which, it and the 12th took part. There were also formal inspections of the different contingents by the Generalissimo, in one of which the gallop past of the R.H.A. appears to have created a sensation among the foreign spectators.

It was too late to withdraw the troops before the winter owing to the closing up of the gulf of Pechili by ice, so all had to settle down into winter quarters. Next spring when the others had embarked the 12th was kept back

---

[1] They brought with them from South Africa the "Skoda" howitzers—see page 420—so that these eventually reached the destination for which they had been nominally consigned when shipped at Trieste. But after their long voyage to Cape Town and their trek up country to bombard the modern defences of Pretoria, their voyage again from South Africa to China to breach the ancient walls of Peking, it was surely the irony of fate that it should have yielded without giving them a chance to show their power.

owing to some recrudescence of trouble. It was not until November 1901 that with the departure of this battery the withdrawal of the British Contingent was completed.

COMMENTS.

It will have been noticed that the 12th Field Battery appears under a new designation—12th Battery, Royal Field Artillery. The separation of the Regiment into Mounted and Dismounted Branches had taken place in the previous year, and the field batteries had at last established their position as mounted troops. A change indeed from the state of affairs described in the last chapter.

Another indication of progress in the preparation of the Mounted Branch for war is the fact that the horse and field batteries from India were accompanied by their ammunition columns—R/2 and R/7 respectively.

In the Dismounted Branch the greater attention devoted to the manning of fortresses overseas is shown by the appearance of the Hong Kong—Singapore Battalion. This Battalion, and the Ceylon—Mauritius Battalion subsequently merged in it, were formed in 1891 as additions to the strength of the garrison artillery at the stations named. The gunners were drawn from the fighting races of India—Punjabi Mahomedans and Sikhs—the officers from the Royal Artillery. They were thus on a footing analogous to that of the Indian mountain batteries, and as was to be expected the men gave satisfaction as steady painstaking gunners and good layers, always well turned out and well conducted. The Boxer Rebellion gave them their first opportunity of proving their worth on service, and they made the most of it. In the heavy fighting at Tientsin their value was recognized by the Japanese as well as the British command.

The Hong Kong—Singapore battery was heavily handicapped by the black-powder charges of its 2'5" guns, just as were the British mountain batteries in South Africa at the same time. The borrowing of more up-to-date weapons from the Navy, and the adaptation of Krupp carriages so as to transform them into field guns, show enterprise and resource.

The maxim battery formed by No. 2 Company may be compared with several instances of a similar nature in Afghanistan, on the North-West Frontier, in Burma and Zululand; and we shall find batteries of the Egyptian Artillery also provided with maxims in addition to their more legitimate weapons. All through the last quarter of the XIXth century there was an inclination in many quarters to hand over the manning of machine guns—especially those in fortresses—to the artillery, and this received considerable encouragement at the School of Gunnery. Fortunately wiser counsels prevailed.

MEDAL & CLASPS.

The China medal of 1860 was given for the Boxer Rebellion with clasps for the "Defence" and "Relief of Pekin".

LIST OF UNITS.

(With Designation in 1914.)

B, R.H.A.	—Major C. F. Blane	—B, R.H.A.
12, R.F.A.	—Major F. E. Johnson	—12, R.F.A.
No. 2, H.K.-S. do.	—Captain W. St. C. Bland	—2, H.K.-S., do.
No. 4, do. do.	—Captain C. C. K. Duff	—4, do. do.

PART III.

NEW ZEALAND & CANADA.

CHAPTER XII.

THE MAORI WARS.

Taranaki, 1860-1—Taranaki, 1863—Waikato 1863—Rangiriri—Orakau
—Tauranga, 1864.

COMMENTS.

Armament—Gunnery—Employment—Medal—List of Units.

ALL through the sixties of the XIXth century there were wars between the colonists and the native inhabitants of New Zealand in which British troops were called upon to support the local forces. The operations that ensued, though not perhaps of great military importance, afford several examples of the employment of artillery under unusual circumstances, and of gallant actions on the part of artillerymen which are worthy of a place in regimental history.

CHAPTER. XII.

The first operations in which regular troops took part were those in the Taranaki District on the west coast in the neighbourhood of New Plymouth. Major-General T. S. Pratt, who was commanding the forces in Australia, was ordered to New Zealand, and took with him 3/12 R.A. (Captain H. Strover). This was a garrison battery and it took the field with a miscellaneous collection of smooth-bore pieces—a 9-pr. gun, a 24-pr. howitzer, and some Coëhorn mortars. At the same time a field battery (C/4[1]—Captain H. Mercer) was ordered out from home, equipped with the new Armstrong field gun, the 12-pr. R.B.L. On board ship were also two 8" and two 10" mortars. They arrived at Auckland in March 1861, after a voyage of 99 days from Woolwich, but no sooner had they landed than they were sent off again by sea to the scene of hostilities, taking with

Taranaki 1860-61.

August.

1861. March.

---

[1] The battery came out as "3/4", its designation was changed to "C/4" during the campaign.

them the mortars as well as their own guns. Landing at the mouth of the Waitara river bullock teams were provided, and the guns were soon in action. The Armstrongs were charged with keeping down the fire from the rifle-pits directed against the sap-heads. The mortars were handed over to 3/12 and were chiefly employed in harassing fire at night. The Coëhorns were taken into the advanced parallel where they proved particularly useful, but Lieut. E. C. Macnaghten was shot dead while laying one of them. Three days after the arrival of C/4 the enemy surrendered. Back at Auckland C/4 found that "walers" had arrived for them, and were kept busy breaking them in, and taking a turn at road-making and carrying supplies for the troops engaged in the work. After making themselves useful in this way there came drill and practice.

1862 passed peaceably, but early in 1863 disturbances broke out again, and one of the first steps of Lieut.-General D. Cameron (who had succeeded General Pratt) was to order Captain Mercer to mount a hundred of his men and train them as cavalry. They were armed with a sword and either carbine or revolver, and placed under the command of Lieut. A. J. Rait of the battery. With this squadron and four of his guns Captain Mercer proceeded by sea as before to New Plymouth, where the landing of horses in an open roadstead proved a ticklish job. For the next couple of months the mounted men were continuously employed as orderlies by day and patrols by night, and it was not until June that the guns were brought into action. Provided with bullock teams they made a night march of fifteen miles over difficult country and assisted effectively at the storming of a pa on the Katikara river. As in 1861 the trouble then died down, and the battery returned to Auckland. The "cavalry" was now reduced to a troop composed of the drivers only, the rest of the horses, and all the harness, being handed over to the transport department, who undertook to provide teams of horses or bullocks for the guns when required.

There was not long to wait. In July the Waikato war began, and lasted nearly a year. The campaign opened in the valley of the Lower Waikato river, forty miles from Auckland, and the bulk of the troops, including C/4 and 3/12, took up a position on Koheroa heights overlooking the creeks. The artillery had now to turn their hand to building and arming fortifications, which meant ferrying the guns across the water and hauling them up the cliffs, finding such sheers and other skidding as was required by cutting down the nearest convenient tree. But when all was ready for the attack the Maoris slipped away to a position further up the river.

This new position was a very strong one, barring the road where it passed between the river and a lake, the defences consisting of a double ditch and a nine foot parapet. To reach this the guns—two 12-prs. of C/4 and a 6-pr. naval field gun drawn by blue-jackets—had a severe march of a dozen miles through swamps and over narrow ridges where the bullocks had to be taken out and the 12-prs. man-handled. Eventually they came into action at a range of 600 yards against the centre of the entrenchment, and after an hour and a half's bombardment the infantry advanced to the assault. All went well until the artillery fire had to be stopped owing to the proximity of the storming party to the works : men then fell fast and the attack was only partially successful. The General called on Captain Mercer[1] to adventure an onset. On the march he had been warned of the General's intention should the infantry fail, and had served out revolvers, to which the gunners had become accustomed when acting as cavalry. He now led them to the assault, falling mortally wounded at their head. It was impossible to get at the Maoris owing to the height of the perpendicular parapet, but some $5\frac{1}{2}''$ common shell were obtained from the gun-boats, and Serjeant McKay and Gunner Green of C/4 distinguished themselves by throwing these over the parapet at great

Chapter. XII.

Waikato. 1863.

Rangiriri

20th Nov.

---

[1] The township of "Mercer" on the banks of the Waikato river commemorates his deeds.

risk to themselves since the fuzes (which dated from the Peninsular War) had to be lighted before the shells were thrown. By this time, however, the enfilade fire of the gun-boats, and the appearance in their rear of infantry who had been landed above their position, were having their effect. Realizing that their retreat was cut off the Maoris surrendered.

In his despatch General Cameron paid generous tribute to the gallant attempt of Captain Mercer—"The Royal Artillery displayed great daring and intrepidity in their assault on the central redoubt, Serjeant-Major Hamilton and other non-commissioned officers and men standing on the top of the parapet and discharging their revolvers into the work."[1]

In January 1864 I/4 (Lieut.-Colonel G. Barstow) arrived with six 6-prs. as well as its 12-pr. equipment. There were thus three batteries in New Zealand, and Lieut.-Colonel E. A. Williams arrived to take command, appointing Lieut. Pickard adjutant. In addition to the two batteries of 12-pr. Armstrongs and the six 6-prs. brought out by I/4, the following miscellaneous collection of smooth-bores was in the hands of the artillery.

2—10″ mortars.	2—4 $2/_5$″ mortars.
2—8″ do.	2—32-pr. howitzers.
2—5 $1/_2$″ do.	2—24-pr. do.

These were for the most part disposed in defensive posts, and manned by detachments from the three batteries, many of the men having never before had anything to do with smooth-bores.

Space does not permit of particularising all the incidents of the fighting that ensued, but mention may be made of the following. Attempts having been made to take a pa at Orakau by assault without success, it was decided to proceed by sap. A 6-pr. of I/4 was taken up to the head of the sap to breach the parapet, Serjeant McKay[2] repeated

---

[1] Lieut. A. F. Pickard and Asst. Surgeon W. Temple, R.A., received the 𝒱𝒞.

[2] Serjeant McKay was publicly thanked by the General and mentioned in despatches.

his bombing exploit, but this time with proper hand-grenades, and Rait's drivers got their first opportunity of shock action when the defenders broke out of the pa. The ground was very rough and covered with high fern but they appear to have made the most of their chance.[1]

The operations recounted above all took place on the west coast, or in the valleys running down to that coast. But the upper waters of the Waikato, which had been reached early in 1864, were little more than twenty miles from the east coast at Tauranga Harbour, and General Cameron therefore decided to take his troops round by sea and land them there. Any advance inland was, however, barred by formidable defences, flanked by morasses, known as the "Gate-Pa." Against these he determined to make full use of his considerable, though mixed, collection of guns, and, lest their appearance should alarm the Maoris, and cause them to evacuate their defences, the artillery were called upon to get them into position under cover of darkness. The operation was rendered particularly difficult by the scarcity of transport, but, somehow or other, it was done, although a great portion of the ammunition was brought up the two and a half miles from Tauranga in wheel-barrows. Soon after daybreak on the 24th April the following batteries, manned by detachments from the three batteries and the Navy opened fire :—

Right — $\begin{cases} 4—5^1/_2'' \text{ mortars} \\ 2—4^2/_5'' \quad \text{do.} \end{cases}$ Range 300 yards.

Centre — $\begin{cases} 1—7'' \text{ Armstrong gun} \\ 2—40\text{-pr. do.} \quad \text{do.} \end{cases}$ Range 700 yards (Royal Navy).

Left — $\begin{cases} 2—8'' \text{ mortars} \\ 2—24\text{-pr. howitzers} \end{cases}$ Range 800 yards.   Range 600 yards.

With 2—6-pr. Armstrongs in reserve.

---

[1] In August 1929 there was an interesting little ceremony at Orakau when one of the surviving veterans of the defence gave an account of the fighting from the Maori side to a gathering of New Zealand officers.

M

About noon it was discovered that the swamp on the right was passable, and a 6-pr. was taken across piecemeal, and also a $4\,^2/_5''$ mortar. These took the main redoubt in enfilade, an assault was launched, and effected an entrance by the breach. But the interior of the work was honeycombed with rifle-pits, and an unaccountable panic seized the storming party. Dusk was approaching, and it was decided not to renew the attack, but to bring up two of the Coëhorn mortars to shell the pa during the night. Whether due to this fire or not, daybreak on the 30th found the place deserted. Beyond a skirmish or two it was the end of the war so far as the Regiment was concerned.

COMMENTS.

The artillery *matériel* used in New Zealand was the most extraordinary mixture that was probably ever taken into the field. We read of $5\tfrac{1}{2}''$ mortars cast in Sydney in 1846 for the New Zealand wars of that period under the direction of Captain Mann, an ex-Bombay Horse Artilleryman, then Governor of a convict establishment. They were only half the length of the service pattern, so that the shell when loaded projected beyond the muzzle, but they made good practice up to 400 yards. The 24-pr. howitzers had also been a long time in the country and were of different patterns, so that the wheels of one would not fit the other, and in the constant removals by water transport, for which everything had to be taken to pieces, this was very inconvenient. The different varieties of ammunition and stores, which were entirely new to most of the men, were very confusing, and what are we to think of their being called upon to use wooden time fuzes dated 1806-7?

The Armstrong guns were at the opposite extreme— the very latest development in gun-making. Their range and accuracy came as a revelation to all and their high velocity, giving no time to take cover disconcerted the

Maoris who christened them the "quick-shells"[1]. But while the gunners realized how far ahead of the smooth-bore guns they were in these respects, they were also convinced that for breaching parapets and similar purposes they could not replace the 24-pr. howitzers which were included in the armament of a smooth-bore field battery. It was to be more than thirty years before the value of field howitzers was to be again recognized by the creation of howitzer batteries.

The 6-pr. Armstrongs brought out by I/4 in addition to their regular 12-pr. equipment had, as previously mentioned, been intended for mountain artillery, but found too heavy except when elephants were available as in Bhutan. They came therefore to be used as a light field gun in countries where powerful draught horses were not available.

With the Armstrong guns had come the "Fuze, time, Armstrong, E"—the first metal time fuze, and an enormous improvement on the wooden "Boxer" in use previously. But their "concussion" fuzes allowed the shell to penetrate some feet before bursting—a serious defect since their charges were not strong enough to prevent the explosion being smothered.

Every man carried a rifle, and was taught to use it, an anticipation of events even more striking than the sound judgment expressed on the need for howitzers. Revolvers were also freely issued to both gunners and drivers, and the mounted men carried swords. Here again we have an intelligent anticipation, for they soon found out that they could be carried on the saddle much more comfortably than on the person.

Even more in advance of the age is the meticulous care taken to ensure the fullest possible effect from the fire. At gun-drill on the barrack square, and when in action on the drill-field, no shell was brought up to a gun without being

Chapter. XII.
Armament.

Gunnery.

, [1] Anticipating the "Whizz-bangs" of the British soldiers half-a-century later.

properly fuzed. And when supporting an attack a supply of shell with the fuzes set was prepared in advance, so that the fire might be as rapid as possible during the culminating period of the assault. During the comparative peace of 1862 Captain Mercer held a practice camp at which sections of Maori stockades of different types were erected from descriptions and plans supplied by those who had taken part in previous campaigns in the country. These were used to try out the most effective methods of attack, and during the subsequent fighting advantage was taken of every capture of a pa to examine the nature of the defences and the behaviour of the shell and fuzes.[1]

But the difficulties of applying their gunnery knowledge to the obsolete armament with which they were provided must have been immense, especially in the case of the junior officers in charge of the innumerable detachments into which the batteries were so often split up. We have an illuminating sidelight upon this aspect of the campaign in the evidence of Lieut. C. Greer before the Warde Committee on Officers' Education a few years later. When urging a more thorough training for young officers in the practical duties of their profession than time would admit at the Academy, Lieut. Greer instanced his own experience in New Zealand. He had been only six weeks at duty when he was sent to New South Wales, and after serving there for seven months was ordered on active service in New Zealand. Shortly after arrival, on the night previous to the attack on the Gate Pa at Tauranga, he found himself in command of a battery of two 8″ mortars. Just before the hour for the attack to commence he received an order to fire a light-ball in order to enable the troops to concentrate their fire on the enemy. "I was the only officer. It was most important as, owing to the night being very dark, the troops could not see the enemy's position. I had never seen one fired, and to my horror my

---

[1] The Maoris displayed wonderful ingenuity in the construction of their pas. Round shot often rebounded from the elasticity of their palisade interlaced with vines, and the huge bundles of bracken bound with flax.

non-commissioned officer was equally ignorant. We had only three light-balls. The first failed, the second, by some fortunate chance, burned brightly and answered its purpose in every way. Had I failed altogether, although through no fault of mine, I should have received the censure of those whose approbation I earnestly desired. A few months after, whilst preparing for an expedition to attack a Maori pa, my commanding officer ordered me to take charge of a rocket tube. I was obliged to admit that I had never seen a rocket fired.''

We must admire the foresight which enabled the gunners in New Zealand to anticipate many of the developments in armament and training which were not generally adopted for a quarter of a century or more. But, perhaps after all, their chief claim to have their services recorded in regimental history lies in the adaptability and resourcefulness they displayed. They served every nature of "piece" from rifled breech-loading guns to Coëhorn mortars; they moved them wherever required, by sea or road or river, with horse or bullock teams, manhandling them on mountain tracks, or carrying them piecemeal through morasses. And, not content with such gunners' work, they reconnoitred, patrolled, and charged as cavalry; they used their shell as hand-grenades, and stormed stockades as infantry; they built redoubts and bridges for the engineers; and they did not disdain the carrying of supplies for the commissariat.

MEDAL.

A medal for New Zealand was granted in 1869. There were no clasps, but different wars were distinguished by dates on the medal. For those here recorded the exact years between 1860 and 1866 during which the recipient served are struck in relief on the reverse.

LIST OF UNITS.

(With Designation in 1914.)

C/4, R.A.	...	...	38, R.F.A.
I/4, do.	...	...	76, do.
3/12, do.	...	...	70, R.G.A.

CHAPTER XIII.

THE RED RIVER EXPEDITION.

Introduction—Lake Superior to Lake Winnipeg—The Red River.

COMMENTS.

The Guns—The Route—Medal & Clasp—List of Units.

CHAPTER XIII.

Introduction.

1869.

November.

1870.
February.

ALTHOUGH, owing to the collapse of the Rebellion on the arrival of the troops, there was no fighting in the Red River Expedition, a short account may well find a place in Regimental History, if only on account of its repercussions in the Nile Expedition fourteen years later.

As regards the cause of the Expedition, it need only be said that in 1869 the proposed transfer to the newly established Dominion of Canada of the vast territory of the Hudson's Bay Company, in what was then known as Rupert's Land, caused very general "alarm and despondency" among the different communities settled there. This came to a head in a rising of the half-breed French Canadians under Louis Riel, who established his headquarters in Fort Garry, at the junction of the Assinaboine and Red Rivers, where Winnipeg now stands. There he terrorised the inhabitants, and urgent appeals for rescue poured in from English-speaking and French-speaking alike —each afraid of the other, and both afraid of the Indians. A storm of indignation swept through the British districts of Canada, and the British Government consented to co-operate in a military expedition.

Colonel Wolseley, then serving as D.Q.M.G. in Canada, was appointed to the command, with Lieut.-Colonel W. J. Bolton, R.A., as Chief Staff Officer. His force consisted of one battalion of British Infantry, two of Canadian Militia,

and a detachment of H/4 R.A. with four 7-pr. guns under
Lieut. J. Alleyne.

Very little was known of the country to be traversed, except that it was a wilderness of rock and water and stunted trees, in which no supplies were obtainable. As in the case of the Abyssinian Expedition of a couple of years earlier, the papers were filled with melancholy prognostications of the fate of the troops. If not scalped by Indians they would surely be devoured by mosquitoes.

The little army was conveyed in Lake steamers to Thunder Bay on the north shore of Lake Superior, where it landed at the beginning of June 1870, and a fort was constructed to defend the base against a threatened raid of Fenians from the United States. But the road onward, which was to have been made practicable before the expedition arrived, was found to be impossible. For six weeks the troops struggled with the task, and then, in defiance of all local opinion, Wolseley insisted that a river route must be sought out, and followed. So boats were built on the lines of naval "whalers," each carrying 8 to 9 soldiers and 2 or 3 French-Canadian "voyageurs," with two months' provisions for its crew.

The first thing that had to be done was to carry the boats over the divide between the watersheds of the St. Lawrence River and Hudson's Bay, and to help in this the Royal Artillery then stationed in Canada sent up eighty horses. Lake Shebandowan on the summit was reached at the end of July, and in 125 boats the Expedition started on its adventurous voyage by the old canoe route viâ the Lake of the Woods and Rat Portage to the Winnipeg River, and down that to Lake Winnipeg.

On arrival there alarming news of the situation at Fort Garry was received, and it was decided to push ahead with only the British battalion and the guns. Church parade over, on Sunday, the 21st August, this advanced guard set sail to cross the lake, steering for the mouth of

CHAPTER
XIII.

The Red
River.
1870.

the Red River. Up that river they pursued their way to be welcomed at the "Stone Fort," twenty miles below Fort Garry, by the populace cheering on the banks, the church bells ringing. Riel and his garrison in Fort Garry were reported to be determined to fight, but an advance across country was impossible, for the land had been reduced to a marsh by torrential rains. There was nothing for it but to put the troops ashore within half-a-mile of the fort and limber up the guns behind the local carts. All was silent, though the guns in the fort were soon plainly visible in the embrasures, and the whizz of a round shot was expected every moment. A staff officer galloping ahead found the gates open and the place empty. It was a sad disappointment, but the gunners fired a Royal Salute from the rebel guns as the Union Jack was run up.

COMMENTS.

The Guns.

The 7-prs. taken on the Expedition were sent out from England for the purpose, and have rather an interesting history. When it was decided to adopt rifled guns generally in place of smooth-bores the Indian Government asked to be supplied with mountain guns of this nature. Six bronze 3-pr. smooth-bores were accordingly bored out to 3″ and rifled, and sent out for trial in the Bhutan Expedition of 1864. These were, however, found too heavy for mule-back, so the weight was reduced to 200lb. by shortening the bore, and turning the exterior smooth. The pattern was never adopted generally for the service, being superseded by the steel 7-prs. introduced for the Abyssinian Expedition, but six of the lightened guns were sent to Canada for the Red River Expedition, and four of these are now in the possession of the Royal Canadian Mounted Police.

In the forty-seven portages which were necessitated by the rapids the guns were carried by two men, lashed to a pole. Later, as the men got hardened, individual gunners could carry a gun on a pad on their shoulder for a mile or more.

The Abyssinian Expedition advanced four hundred miles through an inhabited country where supplies were procurable. The Red River Expedition had to advance more than six hundred through a wilderness, in which no supplies were obtainable and every pound had to be transported for miles on the backs of the soldiers themselves. But in 1870 all eyes were directed to the Rhine, and the Red River drew little attention. Its most interesting feature is the impression it created on the mind of its Commander. In April, 1884, when the Gordon Relief Expedition was under discussion, Lord Wolseley wrote to Lord Hartington (Secretary of State for War): "To those who do not know what was done by the men of the Red River Expedition the possibility of reaching Berber in boats may well be doubted. The cataracts of the Nile are only what are called rapids in North America. They are without doubt very serious and very difficult rapids. During our 600 miles of journey to the Red River we had not only to ascend and descend very great rapids, but several great waterfalls—one actually a few feet higher than Niagara!" And he had sized up the young subaltern who had brought his guns through these difficulties. In succeeding chapters we shall see Alleyne rising to become one of Wolseley's most trusted subordinates. His death when commanding the Royal Artillery at Aldershot was a real loss to the Regiment.

MEDAL & CLASP.

In 1899 the Canadian Government issued a medal for the Fenian Raids of 1866 and 1870 and the Red River Rebellion of 1870. It has a clasp for "Red River".

LIST OF UNITS.
(With Designation in 1914.)

H/4, R.A. ... ... 19, R.F.A.

PART IV.

NORTH & EAST AFRICA.

CHAPTER XIV.

THE ABYSSINIAN EXPEDITION.

Introduction—The March to Magdala—The Arogi Plateau—The Capture of Magdala.

COMMENTS.

Organization—Armament—Difficulties of the Way—Medal—List of Units.

MAP 8.

THE Abyssinian Expedition was due to a generous national impulse for the liberation of some sixty European men and women—mostly British subjects—held in cruel captivity by a barbarous tyrant. Its motive was entirely disinterested, and it was entered upon with no hope of glory, but, on the contrary, with much foreboding. The country to be invaded was shrouded in mystery, and such reports as were available dwelt on the difficulties and dangers to be anticipated—the deadly climate, the impassable ravines, the ferocious inhabitants, the fabulous fortress in the heart of the mountains where King Theodore kept his prisoners. If they were to be rescued this must be the goal of the expedition, and yet not only was it impregnable by nature, but it was defended by a powerful artillery for the manufacture of which many of the Europeans had been imported.[1]

An expedition having been decided upon, the provision of the troops was left to India, while from home were sent the latest patterns of weapons for both the artillery and infantry, and much in the way of stores and transport. The command was entrusted to Lieut.-General Sir Robert Napier, the Commander-in-Chief of the Bombay Army,

CHAPTER XIV.

Introduction.

1867.

---

[1] Theodore had even paid the Regiment the compliment of applying for the services of an Instructor in Gunnery.

from which the greater part of the troops were drawn. The total strength was approximately ten thousand fighting men, with the following artillery :—

G.O.C., R.A.   —Brigadier-General J. G. Petrie.

Brigade Major   —Captain H. le G. Geary.

Lieut.-Colonels   —{ Lieut.-Colonel T. W. Milward.
                  { ,,   ,,   Hill Wallace.

G/14 Field Battery—Captain A. H. Murray.

¹3/21 Mountain do. —Lieut.-Colonel L. W. Penn.

¹5/21   do.   do. —Captain G. Twiss.

5/25   do.   do. —Major A. H. Bogle.

No. 1 (Bombay) do. —Captain P. D. Marrett.

There was also a Naval Rocket Brigade under the command of Captain Fellowes, R.N.

The following officers of the Royal Artillery served on the Staff :—

A.Q.M.G.   —Major F. S. Roberts.

D.A.Q.M.G.   —Captain B. H. Pottinger.

A.D.C.   —Captain G. Arbuthnot.

Major Roberts was 2nd Captain of 5/25, but on appointment as A.Q.M.G. he was succeeded in the battery by Major J. Hills.

The force disembarked at Zula on Annesley Bay, a little to the south of Massowah. The first artillery unit to be landed was No. 1 (Bombay) which formed part of the advanced brigade, sent ahead in October 1867 to take possession and make necessary preparations. The main body followed in December and January, and gradually moved up to an advanced base at Senafe on the first plateau 7,000 feet above the sea.² Early in February the advance began in earnest, A/21 being with the

---

¹ 3/21 and 5/21 were generally known as A and B batteries in the Expedition.

² The horses and drivers of G/14 were sent up to Senafe at once, the guns and gunners remained below. Eventually the guns and three of the 1st line wagons were taken up by bullocks.

advanced guard. For the protection of the communications the following were left at the most important posts :—

At Senafe —No. 1 (Bombay).

At Adigrat —Two guns G/14.

At Antalo —5/25 (with the exception of the two mortars).

Antalo was half-way, and from there the advance was confined to the 1st division, the guarding of the line given to the 2nd. And when Lat was reached, a hundred miles further on, the 1st division was re-organized for the dash on Magdala in three brigades, which were to follow at a day's march distance.

With the 1st brigade went A/21 and the rocket brigade.

    do.   2nd   do.   do. B/21.

    do.   3rd   do.   do. G/14 and the mortars of 5/25.

It was made clear that the real business of the third brigade was to make the road practicable for the elephants carrying the guns and mortars. Baggage and transport were cut down ruthlessly, servants and syces sent back, and tentage reduced to a minimum.[1] At last, on the 9th April, the leading troops reached the far edge of the Talanta plateau, 10,000 feet above the sea; and there in front of them, only ten miles distant, lay Magdala, the goal of all their efforts—with the Abyssinian army encamped beside it. On the further side of the great gorge through which flowed the Bechilo River, rose a rugged mass of mountains, above which towered three summits—Fahla, Selassie, and Magdala. The *massif* was in fact triangular, its apex, Magdala, at the further extremity, while its base, flanked by Fahla and Selassie, faced the British advance. Between these latter heights the top of Magdala showed over the saddle which connected them.

---

[1] The 1st Divisional Orders issued at Lat detail one tent for "Colonel Petrie, Captain Geary, Lt.-Colonel Milward, Captain Nolan, and the Officers of A/21."

CHAPTER
XIV.
1868.

The Arogi
Plateau.

10th Apr.

The valley of the Bechilo was some five miles across, and nearly 4,000 feet deep, but, fortunately for the artillery King Theodore, in his determination to get his guns to his capital, had expended vast labour in constructing a road which led down into the river bed and then up the Arogi ravine to the saddle between Fahla and Selassie. By this road, on Good Friday, the 10th April, Colonel Milward was directed to take the artillery and baggage of the 1st brigade, while the infantry followed a path to the right which would enable them to make good the ground at the top before the guns emerged from the ravine. By a misunderstanding the guns arrived first, and the Abyssinians, seeing a long train of mules apparently unguarded, swept down—five thousand strong—on what they thought was only a baggage train. They had a rude awakening, for in a moment the sailors at the head of the column had their rocket tubes in action, and the gunners of A/21 were not slow with their 7-prs. The two companies of escort and the baggage guards joined in, and the fight was hot until the main body of the infantry, who had followed the path on the right, came to their assistance. Then was heard for the first time the continuous roll of breech-loading rifles. Added to the terror inspired by such strange weapons as the rockets and the shells of the rifled guns, this was more than the enemy could face, and they broke and fled, leaving nearly 500 dead on the field.[1]

The Abyssinian guns on Fahla and Selassie had not been idle. Directed by Theodore himself they had kept up a steady fire, but, deceived by the difference in level, their shot all went over the heads of the British a thousand feet below. The sailors, more skilful, reached the ridge where Theodore was standing with a salvo of rockets, and nearly killed him. Next day he opened negotiations, and Easter Sunday saw the prisoners all safe in the British camp. But he refused to surrender his capital, and Sir

---

[1] The ammunition expended was :—
A/21—58 common, 55 shrapnel, 2 case, 15 rockets.
Naval Brigade—270 rockets.

Robert was determined on its capture and that of the King himself.

By Easter Monday the rest of the artillery had come up, and attack was ordered—the mountain batteries and rocket brigade under Lieut.-Colonel Milward to accompany and support the infantry, the heavy artillery (G/14 and the mortars) under Lieut.-Colonel Wallace to cover the advance. It soon became apparent that Fahla and Selassie, the two great mountain bastions which flanked the approach to Magdala, were unoccupied, and three of the guns of B/21 (which was with the advanced guard) were man-handled on to the summit of Selassie. Magdala was now in full view, at the end of a flat shoulder, a mile long, a quarter of a mile wide, that connected it with Selassie. It rose a sheer wall of rock crowned by a double line of defences, the scarp broken only where a steep and narrow path led up to the gate. Orders were at once despatched for the bringing up of the guns, and while waiting their arrival the advanced guard took part in a curious episode. On the shoulder between Magdala and Selassie stood a score of abandoned Abyssinian guns, and there now appeared the King in person leading an attempt to get them back into the fortress, from which they were only some 500 yards distant. But they were under the fire of the sniders, and when the King had given up the attempt, a few gunners who were with the advanced guard turned the guns on to their owners.

Meanwhile Milward had brought his two batteries and the rockets into action at the foot of Selassie, only 1,300 yards from the gates of Magdala, and the heavy artillery, following the King's road, had found a position further back on Fahla. It was impossible for them to get any nearer owing to the roughness of the ground, but the range was too long for the mortars, and Major Hills soon stopped firing. All being in position, Colonel Milward was directed to concentrate his fire upon the gate, Colonel Wallace to shell the town immediately beyond the gate where the barracks and arsenal were situated, avoiding as far as possible the further portion occupied by the civilian inhabitants. After a couple of hours of more or less continuous bombardment, the infantry advanced to the assault, the mountain guns firing over their heads, and the famous fortress was soon in our hands. The body of the King, shot by his own hand, was found lying in the path a hundred yards within the gate.

The ammunition expended by the artillery was :—

		common,	double,	shrapnel,	rockets.[1]
G/14—106 rounds		70	—	74	—
A/21—109 do.		94	15	—	10
B/21— 91 do.		91	—	—	11
5/26— 11 do.		11	—	—	—

The captives safe, King Theodore dead, his famous fortress given to the flames, his guns destroyed, the task of the expedition was fully accomplished, and the troops set out on their wearisome march to the sea. Major Roberts took home the despatches, and in *Forty-one years in India* he gives an amusing account of his difficulties in delivering them to the Duke of Cambridge who was giving a dinner-party for the Prince and Princess of Wales.

Looting had been strictly forbidden, but Sir Robert Napier presented to each unit present at the capture of Magdala some souvenir from the articles of interest which had been found in the town. That which fell to G/14 was the Greek Cross which is such an ornament to the Woolwich Mess. It was presented to the Mess by the Officers of the Battery.

COMMENTS.

At the period of the Abyssinian Expedition the organization known as the "Brigade System" was in force. Thus we see the brigadier taken from the

---

[1] Each sub-division of A/21 and B/21 had a mule load of rockets.

21st brigade, to which two of the batteries belonged, CHAPTER
and his brigade major from the 14th which provided XIV.
another. But there was no idea of the brigade as a Organiza-
tactical formation, although the "brigading"[1] of the tion.
mountain batteries under Lt.-Col. Milward, and of the
heavy artillery under Lt.-Col. Wallace on Easter Monday,
shows some indication of movement in that direction.
Nor were mountain batteries permanent formations in
either the British or the Indian Army. In both cases
garrison batteries were temporarily transformed by the
attachment of what was called a "mountain train," as
explained in the Introduction to this volume. 5/25,
which had taken part in the Bhutan Expedition (Chapter
VIII), as a mountain battery, had, however, retained its
mountain equipment and establishment, and can therefore
claim to be the first battery to keep the status of
"Mountain" for more than the duration of a campaign.
3/21 and 5/21 only received their equipment and mules
at the base at Zula, where they appear to have drawn on
the British infantry for drivers.

Elephants had been used for the transport of mountain
guns since the Nepalese War of 1814, and they had proved
admirable climbers. When, therefore, it was decided that
field guns would be required in Abyssinia it was natural
to fall back upon them, for it was anticipated that in many
places it would be impossible to make a road wide enough
to take the guns in draught. And when it was decided
that a detachment of 5/25 should take up two 8″ mortars
it was obvious that they must also depend upon elephants
for transport. The number required for a gun or
mortar, with carriage or bed, etc., was four. The loads
were hauled (or parbuckled) up skids on to the cradle on
the animal's back, where they rested in special cradles.

The great change from smooth-bore to rifled guns had Armament.
been accepted by the Royal Artillery less than ten years
before the Abyssinian expedition took place, and thus the

---

[1] They were designated "Divisions" in Orders.

armament of the Regiment was in a transition state. It was decided therefore to send out rifled guns for G/14, 3/21 and 5/21 from home, to be taken over by them at the base. For the field battery these were the 12-pr. Armstrong R.B.L. which had done good work in China and New Zealand. But already dissatisfaction with the breech-loading system was growing, and in 1866 a Committee under Sir R. Dacres had reported that the balance of advantage was in favour of M.L. field guns. Still more must this be the case with mountain guns, and moreover the 6-pr. Armstrong introduced for that service had proved too heavy. It was determined to introduce a rifled muzzle-loader in its place, and so appeared the 7-pr. R.M.L. which filled so many columns in the English press at the time. In popular parlance they were generally referred to in the expedition as the "steel guns"—and, facetiously, as the "steel pens." They were provided with shafts, but these were clumsy affairs, and were never used owing to the excessive roughness of the ground. They were abolished shortly afterwards and no doubt the experience was responsible for the dislike for such additions which persisted in the mountain artillery for many years.

The 8″ mortars weighed eight and a quarter cwt., their beds seven and a half, and each had a travelling bed weighing a hundredweight and a half. All this fitted into a cradle on the elephant's back, which again rested on a pad, so that the load of a "mortar elephant" was over fifteen cwt., and of a "bed elephant" hardly less. Skids, handspikes, and implement boxes made a good load for a third, while a fourth carried powder. The shell were carried by mules—four each.

The captured guns[1] were all found in serviceable condition except one 56-pr. which burst when first fired on Good Friday. On this occasion the guns were fired by the Abyssinian gunners, but the European artisans weighed

---

[1] Nearly 40 in number, including 56 and 18 prs., 24 and 12 pr. howitzers, 20″, 13″ and 10″ mortars.

out all the charges. In the letter sent to Sir Robert Napier
by Theodore he ascribed his defeat on this day to the
worthlessness of his artillery.

The Royal Artillery suffered no loss from the enemy's
fire, and there was little sickness. The climate was
generally agreeable, although when the higher levels were
reached the range of temperature was great, and the
bitter winds were trying to men who had been soaked to
the skin by the daily thunderstorm. The real difficulties
and dangers were all connected with the road. This
crossed a succession of ridges and ravines, and even when
an apparently flat table-land was encountered it was
usually found to be seamed with gorges, from a thousand
to nearly four thousand feet deep. The track wound along
the faces of precipices where a false step was death, the
surface sometimes smooth slippery rock, sometimes rolling
stones. These were bad enough for the mountain batteries; for the field battery they were ten-fold worse.
Wheeled traffic was unknown in the country, and, even
when the pioneers had widened the road sufficiently to
take the wheels, they could not obviate the steep descents
down which the guns had to be lowered with tackles, or
the short sharp zig-zags where they had to be unlimbered
and passed round by hand. The horses stood it wonderfully, and as far as Adigrat the guns with their teams
of eight greys were the admiration of all.[1] But the time
came when nothing could be done to make the road wide
enough for the wheels, and then the horses had to give
place to the elephants, though it was no plain sailing
even with these. Frightened by the thunderstorms, or
the slippery surface of the rocky road, they often threw
their loads, and refused to stir. The work was arduous
in the extreme, and this account may well be closed with
the words of the Historian of the British Army :—

---

[1] The wagons had been left at Senafé.

"The most difficult and dangerous expedition against an uncivilized enemy which the British ever carried through to complete success without a mistake or a hitch. . . . . If any be singled out for special praise it must be . . . . and the handful of gunners who took the Armstrong battery to Magdala and back. . . . The Royal Regiment has never done finer service than this."

MEDAL.

A very distinctive medal was given, but no clasps.

LIST OF UNITS.

(With Designation in 1914.)

G/14 R.A.	...	...	...	33, R.G.A.	
3/21 do.	...	...	...	90, do.	
5/21 do.	...	...	...	88, do.	
5/25 do.	...	...	...	57, do.	
No. 1 Bombay Mountain Battery				25, Mountain Battery.	

Lieut. General Sir William H. Goodenough, K.C.B.

CHAPTER XV.

THE EGYPTIAN EXPEDITION OF 1882.

Introduction—The Expeditionary Force— The Occupation of the Canal—The Action of Magfar—The Actions of Kassassin—The Battle of Tel el Kebir—Alexandria.

COMMENTS.

Organization—Armament—Employment—Gunnery—Medals & Clasps Regimental Commands and Staff Appointments—Army Commands and Staff Appointments—Ammunition expenditure.

MAP 9.

THE long series of Egyptian Expeditions which culminated at Omdurman in 1898 took their rise in 1882.   The intervention of European nations in the affairs of Egypt, which followed the opening of the Suez Canal in 1869, and continued through the 70's, excited an ever increasing irritation among the Mahomedan inhabitants, which found its exponent in a major of the Egyptian Army, Ahmed Arabi.   Under his leadership a military revolt broke out; the executive power was seized; the Khedive reduced to little more than a cipher; and the smouldering anti-Christian feeling throughout the country fanned into flames.   In June 1882 there were serious riots in Alexandria, and there was grave anxiety as to the safety of the large European population.   The fleet was despatched to their protection, but its safety in harbour was seriously threatened by the strengthening of the shore batteries, and attempts to block the entrance. Admiral Sir Beauchamp Seymour could do nothing while there were Europeans in the town at the mercy of the mob, but as soon as the last of the fugitives had embarked, he demanded the handing over of the fortifications under threat of bombardment.   The forts were not given up, and on the 11th July the famous "Bombardment of Alexandria" was duly carried out.

CHAPTER
XV.

Introduction.

1882.
June.

CHAPTER
XV.

1882.

The Expe-
ditionary
Force.

We must now turn to the steps which were being taken at home. As soon as the possibility of having to send an Expeditionary Force to Egypt had been accepted the command was entrusted to Sir Garnet Wolseley, and the plan he decided upon was to land at Ismailia, and from there advance on Cairo across the desert.

The Expeditionary Force was to consist, broadly speaking, of an army corps of two divisions and a cavalry brigade from England, with an Indian Contingent of two brigades—one cavalry and one infantry. The allotment of artillery to these formations was two field batteries for each division, and a corps artillery of one horse and two field batteries, with another horse battery for the cavalry brigade. The Indian Contingent brought one field and one mountain battery. There was also a siege train of four garrison batteries.

The embarkation of this force commenced on the 1st August, and transports began to arrive at Alexandria on the 15th. With the first came Sir Garnet Wolseley, and he lost no time in arranging with Sir Beauchamp Seymour a plan to deceive the enemy as to the real destination of the force. This took the form of an attack on Aboukir, and the fleet, accompanied by the transports, sailed on the 19th, and anchored in Aboukir Bay that afternoon. But as soon as it was dark they steamed away to Port Said which was reached soon after sunrise—to find the Canal safely secured throughout its entire length, and the Indian Contingent arriving at Suez.

Sir Garnet's first object was to seize the sluices and locks of the fresh water canal, upon which the army would be entirely dependent for water both at Ismailia and during the advance. But the difficulties of the disembarkation were great, for the ships had to anchor half a mile from the shore, and men, horses, and guns had to be brought to the one small pier in barges. Thus it was
not until late on the 23rd that the Household Cavalry and one section of N/A were ready to march.

The enemy were reported to be damming the canal at
Magfar, and there they were despatched first thing in the
morning of the 24th.   Though only a dozen miles it was
a trying march for horses only just landed, for there were
some miles of "soft" desert, and the only practicable
route for wheeled vehicles was along the sandy embankment of the railway.   The Egyptian outposts at Magfar
were driven in without difficulty, and they were then seen
to be in force at Tel el Mahuta a few miles further on.
Ordering up any troops now available at Ismailia, Sir
Garnet took up a position with his little force—now
increased by a battalion of infantry and a detachment of
Marine Artillery—so as to invite attack.

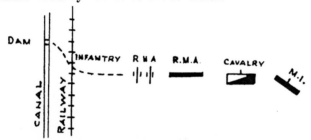

He had not long to wait.   By a little after nine
o'clock the enemy had advanced to within range, and
opened fire with four guns, soon increased to six.   The
cavalry and mounted infantry, much exposed in the open,
were ordered back, and the two guns of N/A that had,
up to this, been directed to reserve their fire, engaged the
Egyptian battery and drew their fire.   This, though
accurate, was fortunately ineffective, for the time shrapnel
burst too high and the percussion shell buried themselves
in the sand before bursting.   But the enemy were working
round our right, and about noon brought a Krupp battery
into action far out on that flank which enfiladed the
position.   By this time, however, a party of sailors with
a couple of Gatlings[1] had come up; a new line was formed

---

[1] The horse artillery had to send back pairs of horses with breast
harness to drag them through the sand for the last two miles.

CHAPTER
XV.

1882.

The
Action of
Magfar.
24th Aug.

behind some low-lying sand-hills at right angles to the
original position; and one of the guns was moved to the
shoulder of the hill, from where it was able to reply to
the flanking battery. But the labour of running up the
13-pr. guns through the heavy sand in the extreme heat
was a great strain on the gunners, to which one man after
another succumbed. The drivers insisted upon taking
their share, and the marine artillerymen lent a hand. For
five hours the section under Lieut. S. C. Hickman sustained the duel against twelve Egyptian guns. Had it
ceased fire the enemy would in all probability, have
launched an attack which the British were barely strong
enough to meet.

At dawn on the 26th August, Kassassin with its important lock on the canal was occupied by a small advanced
force, which included a section of N/A. The canal was
safe as long as this lock was in our hands, but all round
to the north and west the enemy were holding an amphitheatre of hills at a distance of from 2,000 to 3,000 yards,
and about 4.30 p.m. there were signs of hostile activity.
By that time the section of N/A had been reinforced by a
section of G/B, but neither had more than the ammunition
in the limbers, for during the rapid advance it had been
impossible to drag the wagons through the heavy sand—
all the horses were wanted for the guns. In consequence
of this shortage of ammunition the fire of the guns
slackened and then ceased, and it was thought better by
the officer commanding them for all four to go back to
Mahsama to fill up. Meanwhile the infantry maintained
the fight with the support only of the captured Krupps
which had been mounted on a railway truck by the marine
artillery. Orders had been sent back to the cavalry
brigade at Mahsama to turn out again to their assistance,
but there were accidents and misunderstandings, so that
when the cavalry brigade advanced it was under a false
impression of the actual state of affairs at Kassassin—
confirmed by the sight of the guns apparently retiring.

The sun had set, but the moon was shining brightly, and
General Drury-Lowe took his brigade round in a wide
sweep to turn the enemy's left. Suddenly, in the desert
haze, they found themselves under artillery and rifle fire
from the enemy's flank, thrown back to meet the attack.
The four guns of N/A came into action, and the Household
Cavalry delivered the "Midnight Charge" which loomed
so large in contemporary accounts.

On the 8th September the final orders for the concentration of the whole army at Kassassin were issued.
Next day the enemy made another attempt to interrupt
the proceedings by a combined attack, but by this time
more than ample troops were available to deal with it.
With the British and Indian cavalry brigades were N/A and
G/B; with the infantry were A/1 and D/1, the mountain
battery, and two 25-prs. manned by 5/1 Scottish. The
field guns were first placed in gun-pits previously made,
but when the order for a general advance was given
they moved forward, and their fire drove the enemy in
disorder before the infantry could close with them.

For the final success of the campaign it was essential
that the Egyptian main body, holding the elaborately
prepared position at Tel el Kebir, should not be merely
manœuvred out of it, but crushed by actual fighting.
The ground lent itself to movement in the dark, and the
climate also made this desirable. Wolseley decided upon
an attack at dawn after an approach march by night.

Any general description of the elaborate arrangements
made to secure the success of this operation would be
beyond our purpose, but the use made of the artillery
merits careful attention. In order to minimize the confusion inevitable in a night march, Sir Garnet Wolseley
decided to separate his two divisions, and to concentrate
between them the whole of the field artillery of the army
corps—seven batteries. Moreover such a mass of artillery
—42 guns in line—would be able at once to deliver an
overpowering fire to cover the rally of either division, or

CHAPTER
XV.

1882.

The
Actions of
Kassassin.

9th Sept.

The
Battle of
Tel el Kebir.

CHAPTER
XV.

1882.

The
Battle of
Tel el Kebir.

12th Sept.

to break down the resistance of any part of the entrenchment which the infantry failed to carry in their first rush.
The horse artillery under Lieut.-Colonel Nairne were with
the cavalry, now formed into a division.

The night was very dark, and during the march the
force insensibly assumed an irregular echelon, roughly
speaking as shown :—

2nd Division.

Artillery.

1st Division.

Cavalry Division.

The 2nd division thus struck the entrenchments first,
and when a blaze of fire burst from the whole line of
parapet they were within charging distance, while the 1st
division had still some eight or nine hundred yards to
cover.   Even in the 2nd division the fire was so severe,
and the defenders held so resolutely to their ground, that,
while some groups were able to establish a footing within
the works, others were driven back from the parapet.
As the supports came up these were borne forward again,
and about the same time the 1st division carried the whole
line of works opposed to them.   On the right the cavalry
came under the fire of the fort on the extreme flank of
the line of entrenchments, from which they were about
two thousand yards distant when the first shot was fired.
The horse artillery galloped forward, and engaged and
silenced this work, as well as a field battery in the open,
and the cavalry began to swing round on to the left rear
of the enemy.

Having thus sketched in outline the opening phase of
the battle, we can consider in some detail the action of
the artillery under General Goodenough.   The formation
adopted for the march can best be indicated by the

following diagram—the batteries were in line, the wagons covering their guns :—

H/1	J/3	C/3	N/2	I/2	D/1	A/1
Lt.-Col. T. van Straubenzie.			Lt.-Col. F. C. Elton.		Lt.-Col. B. F. Schreiber.	

CHAPTER
XV.

1882.

The
Battle of
Tel el Kebir.

13th Sept.

As soon as the alarm was given General Goodenough halted and dismounted until the light improved sufficiently, when the advance was resumed in echelon of brigades from the centre.   The general himself passed through the entrenchments at a gap close to the south of a work, which had just been carried by the infantry.   The centre battery, N/2, followed, though it had some difficulty in crossing the ditch, and one gun came to grief.[1]   As soon as it had gained the interior of the position three guns were brought into action against the interior line of parapet from which the 2nd division were suffering very severely, while the other two enfiladed the line to the north.   About the same time Colonel Schreiber's brigade came into action outside the entrenchment against the enemy still holding a part of the line opposite the 1st division, just as the divisional commander despatched a staff officer to ask for the help of the guns.   As soon as the resistance here gave way, General Goodenough stopped the fire of the brigade for fear of damage to the cavalry beyond, and directed it against the formed masses of the enemy further back.   Threatened by the cavalry, and plied simultaneously by the horse artillery on their left,

---

[1] It is from this incident that the battery derives its sub-title of the "Broken Wheel" Battery.   The following lively account is from a letter written by Lord Denbigh (then Lord Feilding and a subaltern in the battery) to his father.   "All of a sudden the smoke lifted like a curtain and we found ourselves close to a long line of entrenchments . . . . We at once went on and Major Brancker found an angle in the line just in front where the ditch was not so deep, so the right gun galloped straight at it.   It went with a bump into the ditch, and stuck fast on the face of the parapet, with most of the horses over; but a lot of the 42nd rushed to our help, and we lifted and shoved the gun over; but found one of the wheels smashed to pieces".

Chapter
XV.

1882.

The
Battle of
Tel el Kebir.

13th Sept.

the two guns of N/2 on their right, and Schreiber's brigade
in front, these gave way under the renewed attack of the
1st division, and dissolved into a crowd of fugitives.

To return to N/2. The shrapnel fire from the three
guns at a range of only a thousand yards had proved very
effective in silencing the guns and infantry along the
interior parapet, and after this had been occupied by the
2nd division the battery moved in half-batteries along
the outer face of the parapet a—c, coming into action
whenever any bodies of the enemy showed any dis-
position to stand. In this work it was shortly joined by
Colonel van Straubenzie's brigade, which had in the
meantime passed the front parapet. Finally, just as the
sun rose, N/2 came into action, surrounded by general
officers and their staffs, against trains escaping from Tel
el Kebir station. It had just brought one to a standstill
when its fire was unfortunately masked by the cavalry.

On the south of the canal the advance of the Indian
contingent and the naval brigade had met with some
opposition before dawn, but the mountain battery from
the canal bank fired at the flashes of the enemy's guns
and the infantry cleared the trenches.

The victory was complete, but it was necessary to save
Cairo from destruction, and to break the connection
between the various portions of the Egyptian army dispersed throughout the Delta. For the latter the Indian
contingent, rejoined by its field battery (H/1), pushed
on without delay to the railway junction at Zagazig, side
by side with the fugitive troops, and captured ten engines
and over a hundred railway carriages. The cavalry
division, directed upon Cairo, disconcerted Arabi's efforts
to gather fresh forces by its rapid movement, and on
the next day secured the surrender of the Citadel, and of
Arabi himself. The only point of special interest from an
artillery point of view in this movement was the way in
which the horse artillery guns were delayed by the
narrowness of the bridges over the branch canals,
intended only for pack transport.

General Sir John M. Adye, G.C.B.

The above account may well be closed by the following copy of a letter from the Chief of the Staff, Sir John Adye, written to another gunner, Lieut.-Colonel Noble, within a few days of the battle.

CHAPTER
XV.

1882.
The
Battle of
Tel el Kebir.

13th Sept.

Abdin Palace,

Cairo, 19th Sept., 82.

. . . . "We have not seen any English papers since our battle at Tel el Kebir on the 13th—I hope the people of England will think it was well done, and the march on Cairo quickly carried out. The battle was a complete surprise to the enemy as we hoped and intended. We left Kassassin in the night of the 12th halted and hid down in the desert and then went straight at their works which we had previously reconnoitred. You can imagine it was a difficult march across the desert, guided solely by the stars—not a feature of ground to help us. Our men were all as steady as possible. At about $\frac{1}{2}$ past 4 a.m., on the 13th, Wolseley and myself and the Staff halted just before dawn, and the Highland Brigade swept past us to the attack—with Graham's Brigade on their right. A few minutes before 5 a.m. there were some dropping shots and then a tremendous fire of musketry broke out all along the line, and the enemy's guns in their entrenchments opened and fired wildly in every direction, the Krupp's shells flying and bursting about but not doing much harm. The Highlanders did not even load at first, but went in with a cheer just at daylight & then the severe fighting began inside. The enemy's entrenchments were $3\frac{1}{2}$ miles long, with occasional works flanking and armed with field guns. The heavy musketry fire lasted about 25 minutes and then the grand Battery of 42 Field Guns of Goodenough came up and the enemy fled in utter confusion, our cavalry pursuing for miles—The scene inside their works was wonderful—Guns captured, amm-n, tents, supplies &c, heaps of prisoners wounded and

Chapter
XV.

1882.

The
Battle of
Tel el Kebir.
13th Sept.
unwounded all mixed up, & our troops cheering till they could cheer no longer. We took 58 guns, thousands of muskets, & any amount of camps and food. I rode along the entrenchments in the afternoon, & was surprised at the great number of dead and dying Egyptians. The Infantry ammunition of the enemy had all been laid out ready & there was a box with 1,000 rounds at every 3 or 4 yds. all along their works.

Directly the action was over the Cavalry & Horse Artillery went on South West to Belbais, & Macpherson's Brigade never stopped till they reached Zagazig. I stayed the night with Wolseley at Tel el Kebir, and slept in one of Arabi's grand reception marquees & enclose you his visiting card which you can put up in the Mess-room if you like.

Our fellows have done first rate all thro'—the Cavalry are delighted with Borrodaile & his 13-prs. made magnificent practice, his shells leaving a heap of dead at each place in the desert where his guns were in action. The Batteries have had very hard work both in marching and fighting, & have done right well as they always do. . . ."

COMMENTS.

The organization of the Army Corps which went to Egypt in 1882 is of particular interest, for it is the first example of such a formation in the British army. Although the proportion of artillery fell considerably short of that accepted as the result of the Franco-German War of 1870-1, its command and organization were in full accord with the principles then adopted, which remained in force until after the South African War of twenty years later. Thus a general officer, with suitable staff, commanded the whole of the artillery, and "divisions"[1], as

---

[1] The term "Brigade-Division" did not come into use until 1885. In the orders for Tel el Kebir the main body of the artillery under General Goodenough is termed the "Artillery Brigade".

the lieut.-colonel's command was then termed, had taken CHAPTER
XV.
the place of single batteries throughout. It is particu-
larly significant that the single field battery which formed Organiza-
tion.
part of the Indian contingent was absorbed into one of the
"divisions" from home.

Another mark of progress was the number of artillery
officers holding army commands and staff appointments.
On this point Lord Wolseley's words are worth quoting.

"In the Crimea there was not a single officer belonging
either to the artillery or engineers employed in command
of a division or brigade. But how different was the case
in the Egyptian Expedition. On the Headquarter Staff
there were 25 combatant officers, and out of these 12 were
either engineers or artillerymen. On my own personal
staff, out of 4 A.D.C's. one was an engineer and another
an artilleryman. The Chief of Staff was Sir John Adye,
who was second in command, and no general in the field
was ever more ably and loyally served than I was by
Sir John Adye".

All the batteries allotted to the force were on the
highest peace establishment, so the calling up of those
reservists only who had quitted the colours within the
previous eighteen months provided amply sufficient men
to meet all requirements. The situation as regards horses
was very different. In addition to bringing batteries up
to war establishment, the artillery were called upon to
find nearly a thousand horses for the regimental transport
of the other arms, and similar purposes. In consequence
many of the batteries left at home were reduced almost to
skeletons.

The most serious blot on the artillery organization was,
however, the absence of any provision for the formation
of ammunition columns. Once again the easy expedient
of converting a field battery into an "ammunition reserve"
was resorted to.

The 9-pr. and 16-pr. R.M.L. guns brought into the   Armament.

o

service on the reversion to muzzle-loaders in the 70's were already out of date in 1882, and in the attempt to bolster up the muzzle-loading system a 13-pr. R.M.L. had been introduced in 1880 for both horse and field artillery. In range and accuracy it was far in advance of its predecessors, but it had a heavier charge which, with a rigid carriage, gave very violent recoil. This, as we have seen, threw a severe strain on the detachments, and the equipment was practically condemned by the campaign.

The great complaint of both horse and field batteries, whether armed with 9, 13, or 16 prs., was the impossibility of keeping the ammunition wagons with the guns. The sand of Egypt proved every whit as fatal in this respect as the rocks of Afghanistan. At Kassassin as at Ali Masjid the horse artillery had to cease firing at a critical moment owing to the absence of their wagons. After the action all available horses from the batteries that had been landed had to be sent into the desert to collect the wagons left behind on the 25th August, while the guns at the front were supplied with ammunition by boat on the canal.

For the Siege Train a 40-pr. experimental B.L. equipment had been under trial at Shoeburyness in 1882, but the old armament of 40-pr. and 25-pr. R.M.L. guns and 8″ mortars was taken to Egypt.

The Watkin field range-finder had only been introduced into the service in 1881, and how much its failure in Egypt was due to want of familiarity with the instrument, and how much to the mirage, would be hard to say. There is no doubt that it got a bad reputation in Egypt from which it never really recovered.

The Expedition to Egypt in 1882 provided not only the first example of the organization of the artillery of an army corps in accordance with the lessons of the Franco-German War, but also the first example of its employment in the field in accordance with those lessons. The massing of the divisional and corps artillery under one command at Tel el Kebir shows a full appreciation of Prince Kraft's teaching.

But the old tradition was still strong. In spite of the "doctrine" which had been evolved at the famous manœuvres of the 70's, general officers were still apt to look upon their batteries as tied to the apron-strings of the other arms. General E. O. Hay, who was adjutant of the corps artillery, relates how, when sent to ask for orders at Kassassin, all that he could extract from the general in command was that they should "conform to the infantry."

It was not until the manual of 1881 that the bracket system of ranging made its appearance, so it is not to be surprised at if it was little practised in the campaign. I am again indebted to General Hay for an account of how he rode to a battery of horse artillery in action on the 9th September, and found the captain almost in despair because, as he said, "they have got our range and we haven't got their's." When asked where the major was, he replied "On the flank observing the fire".

At that time majors did not direct the fire of their batteries. Each gun was used by its No. 1 as a personal weapon under the direction of the section commander. The No. 1 layed the gun except when the section commander took it upon himself to do so,[1] there were no such persons as "layers" until many years later. There can be little doubt that the Egyptian "gunnery" was better than ours. Lord Wolseley wrote to Colonel (afterwards Sir Robert) Biddulph, R.A., after the first engagements: "Their infantry are contemptible, except behind entrenchments; their cavalry as bad, but their artillery is very good".

---

[1] In this volume will be found recorded several instances of officers being shot while laying their guns.

MEDALS & CLASPS.

The Medal given for the campaign had two clasps, "Alexandria 11th July" (for the Bombardment) and "Tel el Kebir". The Khedive gave a Bronze Star, and also made a plentiful distribution of the orders of the Osmanie and Medjidie.

REGIMENTAL COMMANDS AND STAFF APPOINTMENTS.

G.O.C.R.A.	—Brig.-General W. H. Goodenough.		
Brigade-Major	—Major A. G. Yeatman-Biggs.		
A.D.C.	—Captain G. B. N. Martin.		
Cavalry Division	Commander	—Lieut.-Colonel C. E. Nairne.	1914.
	Adjutant	—Captain E. O. Hay.	
	N/A	—Major G. W. Borrodaile	Z, R.H.A.
	G/B	—Major W. M. B. Walton	M, do.
1st Division	Commander	—Lieut.-Colonel B. F. Schreiber.	
	Adjutant	—Captain F. N. Innes.	
	A/1	—Major P. T. H. Taylor	9, R.F.A.
	D/1	—Major T. J. Jones	5, do.
2nd Division	Commander	—Lieut.-Colonel F. C. Elton.	
	Adjutant	—Lieut. H. V. Cowan.	
	I/2	—Major W. Ward	30, R.F.A.
	N/2	—Major W. G. Brancker	71, do.
Corps Artillery	C/3	—Major E. R. Cottingham	35, do.
	★ J/3	—Major L. F. Perry	72, do.
	F/1 (Ammunition)	—Major W. S. Hebbert	26, do.
Indian Contingent	Commander	—Lieut.-Colonel T. van Straubenzie.	
	Adjutant	—Captain R. H. S. Baker.	
	H/1	—Major C. Crosthwaite	29, do.
	7/1 Northern (Mountain)	—Major J. F. Free	Reduced.

Regimental Commands and Staff Appointments—*Continued.*

Commander	—Lieut.-Colonel W. Newman.	
Adjutant	—Captain H. Knight.	
Commander	—Lieut.-Colonel M. Elliot.	
Adjutant	—Captain H. G. Newcome.	
4/1 London	—Major G. A. Noyes	81, R.G.A.
5/1 do.	—Major W. H. Graham	80, do.
5/1 Scottish	—Major G. B. Macdonell	66, do.
6/1 do.	—Major F. T. Lloyd	57, do.

Siege Train

N.B.—Lieut.-Colonel Nairne and Captain Hay came out with the Corps Artillery but were transferred to the horse artillery when the cavalry division was formed from the two cavalry brigades.

ARMY COMMANDS AND STAFF APPOINTMENTS.

Chief of the Staff (Second-in-Command)	—General Sir J. M. Adye.
G.O.C. 2nd Division	—Lieut.-General Sir E. B. Hamley.
D.A.A. & Q.M.G's.	—Major J. F. Maurice, Major J. Alleyne, Captain E. R. Elles, Lieut. J. M. Grierson.
A.D.C's.	—Lieut. A. G. Creagh. Lieut. J. Adye. Captain F. G. Slade.

AMMUNITION EXPENDITURE.

Division.	Battery.	Armament.	Magfar	Mahsama	Kassassin	Kassassin	Tel el Kebir	Total.
Cavalry ...	N/A	13-pr.	280	196	117	138	88	819
	G/B	do.	—	24	37	30	36	127
1st ...	A/1	16-pr.	—	30	—	40	84	154
	D/1	do.	—	—	—	30	63	93
2nd ...	-I /2	do.	—	—	—	—	30	30
	N/2	do.	—	—	—	—	68	68
Corps	C/3	13-pr.	—	—	—	—	33	33
Artillery ...	J /3	do.	—	—	—	—	7	7
Indian ...	H/1	9-pr.	—	—	—	—	24	24
	7/1	2·5"	—	—	—	40	42	82
		Total	280	250	154	278	475	1437

CHAPTER XVI.

THE GORDON RELIEF EXPEDITION.

Introduction—The Desert Column—Abu Klea—Metemmeh—The River
Column—The Evacuation of the Sudan.

COMMENTS.

Organization—Armament—Medal & Clasps—Army Commands and
Staff Appointments—List of Units.

MAP 10.

CHAPTER
XVI.

Intro-
duction.

1883.

1884.

March.

May.

WHEN the Egyptian Army was broken up at Tel el Kebir
in September 1882 the Sudan was already in flames owing
to the revolt led by the Mahdi, Mahomed Ahmed. In
1883 a totally inadequate force of Egyptian troops under
Hicks Pasha, sent by the Khedive to restore order, was
destroyed near el Obeid. The British Government
would not agree to the Khedive's desire to re-conquer the
Sudan, but sent Gordon with somewhat ambiguous
instructions to withdraw the scattered Egyptian garrisons.

By that time, however, the centre of gravity had
shifted to the Eastern Sudan, and it was found necessary
to despatch an expedition to Suakin, as will be told in the
next chapter. It will be sufficient to say here that after
defeating Osman Digna at Tamai in March the force was
withdrawn. With this withdrawal the position at
Khartoum became daily more alarming, and it was completely cut off when Berber was captured by the Arabs at
the end of May.

Into the discussions regarding the necessity of an
expedition and the form it should take, it is unnecessary
for us to enter. For the purpose of this account it is
sufficient to say that the plan decided upon was for the
main force to follow the river route in boats while a camel
corps was held in readiness for a dash across the desert
in the event of an urgent call from Gordon. It will be
convenient to consider the latter first.

For the Camel Corps Lord Wolseley called for a thousand volunteers to be selected from regiments at home, the artillery to be supplied by 1/1 Southern (Major W. Hunter). This was a garrison battery with mountain equipment stationed at Cairo, and it was ordered to exchange its mules for camels.

In September 1884 the battery commenced its move up the Nile by train and steamer and barge as far as Wady Halfa, and thence by march route, not reaching the "Front" at Korti until Boxing Day. There it found that Wolseley had decided to send the Desert Column from Korti to Metemmeh on the Nile, where it was hoped to meet the steamers from Khartoum. Under Sir Herbert Stewart, it started from Korti, some 1,800 fighting men with 2,200 camels, and 300 Egyptian camel-drivers, etc.[1]

The going was good and the air fresh, though the sun was hot, and the column marched on a front of about fifty yards, halting and starting by bugle call. On the 14th January Jakdul was left, on the 16th the hussars got into touch with the Arabs, and the column then marched on a broader front, the three camel regiments leading in mass formation. In the evening the enemy were found in position covering the wells at Abu Klea, but Sir Herbert Stewart considered it too late to attack, and the force went into zareba. Sleep that night, and breakfast

CHAPTER
XVI.

The Desert
Column.

1884.

September.

December.

1885.
January.

14th Jan.

---

[1] XIXth Hussars   135
Naval Brigade, with Gardner gun   58
Half 1/1 Southern, with three 2·5″ R.M.L. guns ...   43
Engineers   27
Heavy Camel Regiment   400
Guards do.   do.   385
Mounted Infantry do.   383
Sussex Regiment (to garrison posts)   258
Medical, Commissariat, and Transport   104

The Hussars had their horses, all the rest were mounted on camels.

Headquarters and half the battery were kept in reserve at Korti : Captain G. F. A. Norton commanded the half-battery with the Column.

in the morning, were somewhat harrassed by desultory
sniping, and about nine o'clock orders were given to form
square for an advance. The distribution of the troops in
the square is shown in the following conventional diagram.
The battery had its gun and ammunition camels, and
there were camels for hospital and water-tanks, but all the
others were left in the zareba.

In the fight that ensued we are concerned only with
the fortunes of the artillery, and a vivid account of these
has been preserved in a letter from one of the subalterns[1]

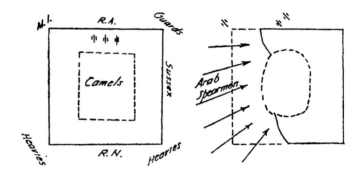

of the camel battery written only a few days later.

"Meantime preparations were being made for a fighting
square to advance and take the wells. . . . It was during
this time that Lyall,[2] when talking to me, got a bullet right
into his back, and fell in great pain. . . . The fire got
sharper as the skirmishers were withdrawn to join the
square. We took our gun camels in the square, and then
we advanced under a really nasty fire from the front, right
flank, and left flank. We moved slowly on for a mile or
so, and then we got the guns into action, and dragged
them with ropes, occasionally running them out and firing

---

[1] Lieut. N. W. H. Du Boulay.

[2] Lieut. C. N. Lyall.

at bodies of the enemy—one shell we sent back to our right rear at some horsemen . . . we learnt afterwards dispersed a large body who were making for the zareba in rear. The enemy were now all round us potting at us, but occasionally being cleared away by the skirmishers; whilst his main force was formed up on our left front in regular companies, with flags, etc. Suddenly we saw them move and come down at us, charging the left front of the square. Our three guns were got out, turned towards the enemy, and loaded with case shot. I was in charge of two, and Guthrie, who had taken Lyall's place, was in charge of one. I waited until the enemy was about 250 to 300 yards off, and then fired almost at the same time as Guthrie did. The result was an almost clean sweep of the enemy who would have come on the front face of the square. We could not well fire before we did, because there was a depression in the ground, and we should not have done much execution. Before we could load again the left flank of the square was broken, the enemy was in upon us, and we were fighting hand-to-hand. The square was broken in, but no one thought of turning or giving way; and the camels here helped us by forming a barrier between the enemy and the right flank of our square. A few minutes and the enemy turned and ran. Then we got to work with the guns again, and kept up a fire upon them for a long way. At first all was confusion; the ammunition camels were hard to find, and whilst I was running for some I came upon poor Guthrie on a stretcher. His gun was on the left, and when the square was broken, he and a few others were left alone with it, and an Arab attacked him. Guthrie knocked the man down, but the man made a cut at him from his knees. Gunner A. Smith[1] saw it, just managed to ward the blow with a handspike; and then brained the Arab; but the knife had cut Guthrie's thigh and divided the artery. Guthrie nearly bled to death, but the wound was tied up just in

---

[1] Gunner Albert Smith, ᴿᴬ.

time, and I hope will be all right eventually.[1] When all was over the square moved on to the next hillock, leaving us and the Naval Brigade, and part of the Hospital Corps, where we were. We blazed away whenever we saw any of the enemy reforming, and did really good work; dispersing them and then getting back into the square . . . . this was about 3.30 p.m. We had been at it since daybreak, and our thirst was intense, for it was awfully hot on the polished rocky ground. The 19th Hussars . . . . went on to the wells, & signalled back that there was fair water. That message just pulled us together, and we toiled on with our guns in the square and encamped at the wells, and drank, and drank, and drank.

The work our men had with the guns over all sorts of ground, right through the day, was immense. We only got in a little before dark, and had nothing to eat except biscuits that we had carried in our pockets. You can imagine how done up we all were, seeing that we had no rest the night before.

Guthrie came to us in this way; . . . he was attached to a transport company . . . and was sent on (to take camels back again); so when poor Lyall was hit, he, naturally, having no duty, and being a gunner, took his place, only to be wounded the same day.

The behaviour of the men under fire has been very good from the first advance out of the zareba before the wells (Abu Klea) to the retreat from before Metemmeh. In the first battle our guns were of course just outside the square, and we had to keep our heads clear of the men behind us—not pleasant. One gunner got a bullet right through his helmet from behind". . . . .

To the above may be added the following account of Gunner Smith's exploit given later by Serjt.-Major Watts. "I was a 'striker' by trade, and so was No. 3 on the gun, my duty being to tighten the junction-nut by hammering on the trunnion. Owing to a casualty occurring I was promoted A/Bombardier and put in charge of the 1st line

[1] Lieut. J. D. Guthrie. He died of his wound.

ammunition camels, and Gunner Smith, also a 'striker',
was put in my place.   When the face of the square began
to crumble up it was a case of 'officers to the front', and
Mr. Guthrie was stabbed, and when Gunner Smith saw
this he picked up the traversing handspike and hit the
Arab who had done it over the head.   The little locking
stud on the handspike split his head right open, and then
Smith stood astride Mr. Guthrie, and kept the Arabs off
with his handspike until the square reformed a few
minutes later.   I was standing within a few yards of the
place and saw everything".

CHAPTER
XVI.

1885.

Abu Klea.

We need not dwell on the rest of the pitiful story.
All through the night of the 18th-19th the little force
pressed on towards the Nile.   It was a terribly trying
march -through most difficult country, but at dawn their
eyes were gladdened by the sight of the palm trees in the
valley.   It was not to be reached, however, without
another sharp fight, in which the general received his
mortal wound.

Metemmeh.
19th Jan.

A zareba was formed on the ridge above the valley,
and the naval brigade, the hussars, some of the heavies,
and the battery were left there to guard the camels and
the wounded.   The remainder then moved down towards
the river in a fighting square under the eyes of those left
behind.   Fascinated by the sight they watched the enemy
closing in on the square, first from the right, then from
the left, and finally in great force from the front.   The
gunners were fortunate enough to be able to help against
the first two attacks, but it was too risky to fire over the
heads of the square at the third, which came on with great
swiftness.   The steady fire of the square was sufficient,
however, and no Arab got within a hundred yards of it.

Next morning the whole force moved down in peace to
the Nile, and all were able to quench their thirst at last.

20th Jan.

A demonstration against Metemmeh showed that an
attack would not be justified, and on the same day the
expected steamers from Khartoum came into sight.   On

CHAPTER
XVI.

1885.

Metemmeh.
24th Jan.

14th Feb.

the morning of the 24th January Sir Charles Wilson (who had succeeded to the command) started on his voyage. As everyone knows he arrived too late.

The Desert Column had lost its *raison d'être*. A great hazard had been taken in order to be ready to send immediate assistance to Gordon should his situation require it. To have kept the column in its position near Metemmeh when its object no longer existed, and the risks were even greater, would have been unjustifiable. On the 14th February the Desert Column turned its back on the Nile.

The battery camels were still in very fair order owing to the care bestowed on them, but many had to be given up, and officers and men had to march back on foot. After a threat of attack on the camp at Abu Klea had been dispersed by the cavalry and a few shells, the return to Korti was accomplished without further incident.[1]

A word must now be said as to the River Column, for, although its only artillery was an Egyptian battery,[2] several officers of the Royal Artillery, destined to rise later to positions of great prominence, found in it the opportunity for distinction.

The picture of the flotilla winding its way up the river, navigated by Canadian *voyageurs*, escorted on the banks by British cavalry, Egyptian camel corps, and Native levies, is a fascinating one, and all went well until the news of the fall of Khartoum brought orders to halt. The advance was resumed after a few days, but now the way was found blocked by the enemy at Kirbekan. Here they were in occupation of a chain of rocky hillocks, out of which they were driven on the 10th February, but, as in the case of the Desert Column, with the loss of the commander—General Earle. The command devolved upon

---

[1] Major Hunter had come up with the other half battery while the Desert Column was halted on the Nile.

[2] Commanded by Lieut. A. Crauford, R.A.

General Brackenbury, R.A., and the advance was continued until the 24th when orders arrived for its withdrawal.

With the return of the River Column all the troops were back in the Province of Dongola, and arrangements were put in hand for summer quarters there. But on the 11th May Lord Wolseley, under instructions from Home, ordered the withdrawal of the troops from the Sudan.

The operation was not a simple one, for it was certain that anarchy would follow the departure of the army, and probable that the dormant hostility of the tribes would find vent in attacks upon the retreating troops. Under Major A. E. Turner, R.A., steps were taken for the evacuation of the many thousand refugees, and the command of the rear-guard was entrusted to General Brackenbury. On the 5th July, 1885, he left Dongola bringing in with him all steamers, boats, and stores.

"The last to come in was General Brackenbury's column. The Camel Battery of the Royal Artillery came in half-batteries escorted by cavalry . . . . across the desert from Akasha to Wady Halfa."

COMMENTS.

There is little to be added in the way of comment, for only one battery—or rather half of it—was engaged, and the fine story of their fight has been vividly portrayed by those who took part in it. But some further details of the preparation of 1/1 Southern for its novel rôle deserve a place in Regimental History.

On the 27th February, 1884, the gunners (suddenly transferred from Malta) found themselves in the citadel of Cairo looking after the 40-prs. which dominated the city. It was not for long. They had brought 7-pr. (200 lb.) equipment with them from Malta, and on the 5th March they were practising as a mountain battery with borrowed mules. There followed a move to Abbasiyeh, the issue of mules of their own, and 2·5″ guns instead of the old 7-prs. Inclusion in the Gordon Relief Expedition brought an order to provide themselves with camels, and Lieut. Du Boulay went off to do the buying, with the assistance of a young veterinary surgeon, a gunner to do the branding, and an interpreter.

Each gun and carriage required five camels, with another for ammunition, a total of six. The detachments were mounted—horse artillery fashion—and they took another six. Officers, non-commissioned officers, and trumpeters accounted for twenty. Then followed a second line with more ammunition, spare parts, rations, stores, etc., as shown in the following table :—

	Weight lb.	British Personnel.	Egypt- ian.	Cam- els.
Gun-breech—Load with driver	551	—	6	6
Gun-muzzle— do. do. do.	540	—	6	6
Trail — without do.	350	—	—	6
Axle — do. do.	350	—	—	6
Wheels — do. do.	350	—	—	6
1st Line Amtn. with do.	543	—	6	6
Detachments— do. do.	323	30	6	36
Ammunition— do. do.	543	18	12	30
Tools & Spares do. do.	400	6	—	6
Line Gear & Forge	323	—	1	4
Medical	323	1	—	1
Rations	400	—	1	4
British Officers	323	7	—	7
Egyptian do.	323	—	2	2
N.C.O's. & Trumpeters	323	10	—	10
Spare		15	7	23
Total		87	47	159

The armament was, perhaps, unnecessarily up to date. The 2·5″ R.M.L. "screw" gun had, as we have seen in

previous chapters, been brought into the service in 1880,
during the 2nd Afghan War, and had gained a great
reputation there.   But its range and accuracy, invaluable
in the mountains, were of little value in the desert.
The 7-pr. which it had supplanted would have been just
as effective and much handier in action, and would have
saved a camel for each gun. Its use would also have
saved a contretemps which might have had serious con
sequences.

Before leaving Cairo a supply of star shell was indented
for from home. They reached the battery of Korti, but
when the boxes were opened, the shell were found to be
of three-inch diameter, while the calibre of the gun was
only two and a half.   When first introduced the 2˙5″ gun
had been given the name of "7-pr." following the usual
custom, although there had been two previous 7-prs.
When, then, an indent for 7-pr. star shell was received
those belonging to the earlier equipments were sent.
Someone had blundered, but the name of the equipment
was changed to "2˙5″" to avoid such accidents in future.

MEDAL & CLASPS.

For the Gordon Relief Expedition the Egyptian Medal
of 1882 was given with clasps for "The Nile 1884-85",
"Abu Klea", and "Kirbekan".

ARMY COMMANDS AND STAFF APPOINTMENTS.

D.A.G.[1]   —Colonel H. Brackenbury.
A.A.G.   —Lieut.-Colonel J. Alleyne.
D.A.A.G.   —Major F. G. Slade.
  do.   —Major A. E. Turner.
A.D.C. to C.-in-C.—Major A. G. Creagh.
  do.   do. —Lieut. J. Adye.

---

[1] And Second-in-Command River Column.

P

LINES OF COMMUNICATION.

A.A.G. —Lieut.-Colonel J. F. Maurice.
D.A.A.G.—Lieut.-Colonel F. T. Lloyd.
do. —Captain H. C. Sclater.
A.D.C. —Lieut. F. R. Wingate.

SPECIAL SERVICE.

Lieut.-Colonel F. Duncan.
Major G. B. N. Martin.
Lieut. H. M. L. Rundle.

LIST OF UNITS.

(With Designation in 1914.)

1/1 Southern, R.A. — 120 R.F.A.

CHAPTER XVII.

SUAKIN.

I. THE EXPEDITION OF 1884—The Rise of Osman Digna—El Teb—Tamai.

II. THE EXPEDITION OF 1885—The Fall of Khartoum—Hashin—McNeil's Zareba—The Withdrawal.

COMMENTS.

Organization—Armament—Medals & Clasps—List of Units.

MAPS 10 & 11.

I—THE EXPEDITION OF 1884.

IN the last chapter the sending of an expedition to the Eastern Sudan in 1884 was mentioned : the reason for this must now be explained.

After Tel el Kebir, a slave-trader named Osman Digna, indignant at the suppression of the slave-trade, raised a revolt in the Eastern Sudan, and massacred several parties of Egyptian soldiers in the country round Suakin. By the end of November 1883 Tokar was invested, Suakin attacked, and the despatch of reinforcements became imperative. A force of Gendarmarie under Valentine Baker Pasha landed at Trinkitat and advanced to the relief of Tokar, to be routed with the loss of a couple of thousand men killed and wounded, and a battery of Krupp guns captured. It was evident that there was no hope of relieving Tokar with Egyptian troops, and it was decided to send a force drawn from the British troops still available in Egypt.

For the Expeditionary Force some four thousand men were detailed, including 6/1 Scottish Division, R.A. (Major F. T. Lloyd) which had formed part of the Siege Train in 1882, and had been in garrison in Cairo since.

CHAPTER
XVII.

The Rise
of Osman
Digna.

1883.
November.

1884.
4th Feb.

El Teb

It was now ordered to form two camel batteries—"A" and "B"—to be organized as a brigade, with Major Lloyd as commander, and Lieut. Sir G. V. Thomas as adjutant.

This force, under Sir Gerald Graham, landed at Trinkitat on the 22nd February 1884, marched on the 28th, and on the next day attacked the enemy in position at El Teb.

The advance was made in square formation, and, to avoid crowding the interior with camels, the gunners themselves dragged their guns.[1] As soon as they came under the fire of the enemy's artillery a battery was pushed out to the front, and as the square advanced both batteries were in the front line firing whenever the square halted. The enemy's gunners stuck to their work well, though they burst their shrapnel too high for effect. Their chance came when the British got within case range, but then the Arabs swarmed down with sword and spear in a desperate effort to break the square and the British batteries got their turn. Round after round of case was poured into them, but though they fell in great numbers, more than one reached the batteries—to be shot with a revolver or knocked out with a rammer. One was felled by a gunner with a round of case in his fist! The fight was not yet over, however, for large numbers of Arabs had still to be shelled out of the villages, and their batteries silenced. But once this was done they were soon in full retreat, leaving four Krupp 9$^{c/m}$ and three bronze 84$^{m/m}$ guns in our hands. The ammunition expended by "A" and "B" batteries was 61 shrapnel, 31 common, 25 case.

The next day the advance to Tokar was resumed, the gunners still dragging the guns, though now assisted by some donkeys they had captured and put in the shafts. No real resistance was met, and on the 4th March the

---

[1] A useful reinforcement was received on the night before the battle by the arrival of half-a-dozen time-expired gunners on their way home from India. They had slipped ashore from the transport and tramped out to lend a hand.

force was back in Trinkitat, from whence they sailed to Suakin.

At Suakin a field battery (M/1) was found in camp, disembarked from an Indian transport homeward bound. Major E. H. Holley had borrowed four 9-pr. R.M.L. guns from the Navy, and with some old Egyptian harness, and mules from the transport, had managed to turn out a field battery. In the advance which followed this battery marched with the 2nd brigade, the two camel batteries with the 1st.

Osman Digna had collected some twelve thousand men at Tamai, and here he was attacked on the 13th March, the force formed in brigade squares. The action began with a sudden rush upon the 2nd brigade by a mass of Arabs who had been concealed in a ravine. This threw the square into some disorder, but M/1 came into action on the right rear, and opened fire with shrapnel, changing to case as the range shortened. Unsupported as the battery was, the steadiness of its fire proved of great service in preventing the enemy working round to the rear face of the square. A few minutes later it was the turn of the 1st brigade, but it stood firm, and the fire of its camel batteries materially contributed to the recovery of the other square. Later on the enemy appeared in great force on the right of the 1st brigade, and one of the guns was ordered to that face of the square, where it was soon busily engaged at a range of under fifty yards. Its case shot exhausted, shrapnel reversed had to be resorted to. The other three guns of this camel battery were pushed forward on to a spur from which they could command the ravine in which the Arabs had been hidden, and M/1 was ordered to join the 1st brigade, taking up its position first on the left rear, and later on the right rear. By a little after 10 a.m. the action was over, and the 1st brigade, with all the artillery, crossed the ravine and reached Osman Digna's camp. Next day this was burnt and the force returned to Suakin. The ammunition expended was :—

Battery.	Shrapnel	Common	Case	Total
M/1	103	34	45	182
6/1	53	34	45	132

After this success it was decided to withdraw the force from Suakin, and at the end of March both batteries embarked.

II—THE EXPEDITION OF 1885.

The fall of Khartoum. 1885.

With the fall of Khartoum in January 1885 the destruction of Osman Digna's power became a matter of vital importance if, as was naturally expected, its recapture was to be undertaken. The first idea was that the troops to be sent against Osman Digna should eventually join hands with the River Column at Berber, so that the two main lines of approach to Khartoum might be in our possession. Sir Gerald Graham, who was again to command, was, therefore, given the double task of breaking up Osman Digna's gathering and of constructing a railway from Suakin to Berber.[1] The force placed at his disposal was considerably larger than that of 1884, and included not only an Indian but an Australian Contingent[2] —the first occasion on which Colonial troops fought alongside British and Indian. The artillery was as follows :—

C.R.A.   —Lieut.-Colonel S. J. Nicholson.

Adjutant   —Captain R. Bannatine.

---

[1] This railway was the subject of much discussion at the time, but does not concern us here. Its interest regimentally is only derived from its utilization, after return to England, at Lydd, Plumstead, Shoeburyness, etc., where for many years its carriages with "Suakin-Berber Railway" still prominent on their sides were in daily use.

[2] Consisting of a staff, a field battery, a battalion, and an ambulance corps.

G/B—Horse Battery	—Major J. F. Meiklejohn.
5/1 Scottish—Mtn. Bty.	—Major J. J. Congdon.
6/1 do. —Amtn. Coln.	—Major W. O. Smith.
N.S.W.—Field Battery	—Lieut.-Colonel W. W. Spalding,

and the following artillery officers served on the staff :—

A.A. & Q.M.G.	—Major R. McG. Stewart.
D.A.A. & Q.M.G.	—Major A. J. Pearson.
do. do.	—Lieut. J. M. Grierson.
Officer i/c Signalling	—Major E. T. Browell.
A.D.C.	—Captain A. N. Rochfort.
do.	—Lieut. W. C. Anderson.

CHAPTER XVII.
1885.
The fall of Khartoum.

The first task of the expedition was to break up a gathering of Arabs at Hashin, a group of rocky hills seven miles from Suakin, whence nightly raids on our posts were despatched. The Arabs were driven off, a zareba and redoubt constructed on the hills, and the force returned to Suakin leaving a battalion to hold the post. In this affair the only artillery engaged was G/B, which covered the advance and withdrawal of the infantry in spite of the difficulties of the thick bush.

Hashin.
20th Mar.

The next step must be an advance on Tamai where Osman Digna was reported to have concentrated seven thousand men. A necessary preliminary was the establishment of an intermediate supply depôt in the desert, and in doing this there occurred, on the 22nd March, the unfortunate affair known as "McNeil's Zareba." Although there were no guns with the force, part of the ammunition column accompanied it, and, being outside the square, was exposed to the Arab charge. On the 2nd April the advance was resumed, and on the following day Tamai was reached. There was little resistance, but pursuit into the mountains was hopeless. After burning Osman Digna's camp, and destroying large quantities of ammunition, the troops returned to Suakin.

McNeil's Zareba.
22nd Mar.

CHAPTER XVII.

1885.

The Withdrawal.

1st May.

Osman Digna had not been brought to book,[1] but for political reasons the withdrawal of the Suakin Expedition, as of the Nile Expedition, was decided upon. And so the troops dispersed, their object unaccomplished.

The railway was then taken in hand, and 6/1 Scottish found another part to play. As an ammunition column it had been represented in every engagement, and at McNeil's Zareba had suffered severely. Now it was given Gardner guns, and patrolled the line by day and night with these mounted on railway trucks.

And when the force was broken up, 6/1 was fated to remain in garrison, taking over, in addition to the armoured train, the mountain battery equipment and mules of 5/1. Throughout the summer the battery remained at Suakin, continually engaged in small affairs with the Arabs, and suffering severely from the climate. It was not until May 1886 that it left.

COMMENTS.

Organization.

For the two expeditions recorded in this chapter "Improvisation" would be a better heading than "Organization," so far at least as the artillery was concerned, and the same may be said as regards armament.

Thus in 1884 we have the turning of 6/1 into two camel batteries, and some further details of this transformation may now be given. Its strength was made up to 9 officers and 116 British non-commissioned officers and men, to which were added 2 officers and 105 Egyptian artillerymen. These latter were employed as camel leaders, and on other duties unconnected with the service of the guns, which was reserved entirely for the British gunners.

---

[1] It was not until 1900 that Osman Digna was finally run to earth. He died at Wady Halfa, a political prisoner.

There were 80 camels and 19 horses, with mule carts for baggage, etc.

In the 1885 expedition the most interesting item as regards organization was the conversion of 6/1 into an ammunition column. Here is an account given at the time :—

"In the end we got all the ammunition transport required. It was a Noah's Ark. It consisted of :—

(i). Mules from Gibraltar with "Rock Scorpions" as drivers.

(ii). Mules from Malta with Maltese drivers.

(iii). Mules from India with Punjabi drivers—the best.

(iv). Camels from Africa with Arab drivers.

But we got on very well."

The guns taken over from the Egyptian artillery in 1884 by 6/1 were 84$^{m/m}$ bronze rifled guns, but Major Lloyd doubted their accuracy, and the gunners were puzzled by sights marked in Arabic, so on arrival at Trinkitat eight 7-prs. of 200 lbs. were borrowed from the Navy. To allow of these being mounted on the Egyptian carriages leather collars were fitted round the trunnions—they had to be renewed after each engagement—and the elevating screws replaced by wooden quoins.

In 1885 5/1 were armed with 2'5" R.M.L. jointed guns as had been the battery with the Desert Column. These were the service mountain artillery weapon, but, as pointed out in the last chapter, were not so suitable for the African desert as for the Indian hills.

The 9-prs. borrowed by M/1 from the Navy were an odd lot—two of 6 cwt. and two of 8 cwt.—but they were complete with field carriages and limbers, though these had no splinter-bars. Swingle-trees were taken off some old Egyptian equipments, and with the assistance of ship's carpenters and sail-makers, and some Egyptian harness, the equipment was complete. Fifteen horses for officers and non-commissioned officers and mules for

the gun-teams, were obtained from the regimental transport, and the battery was fit to take the field.

The New South Wales Battery was armed with 16-pr.
R.M.L. guns, but these were considered too heavy for
the Sudan, so they were left in Australia, and 9-prs. sent.
out from England to meet the battery at Suakin.

MEDAL & CLASPS.

The Egyptian Medal of 1882 was granted with clasps
for "Suakin 1884", "El Teb", "Tamai", "Suakin
1885" and "Tofrek", and the Bronze Star was also issued
to those who had not already received it.

LIST OF UNITS.

(With Designation in 1914.)

G/B,	R.H.A.	M, R.H.A.
M/1,	R.A.	45, R.F.A.
5/1 Scottish, do.		66, R.G.A.
6/1 do. do.		57, do.

Volunteer Field Battery, N.S.W.A.

CHAPTER XVIII.

The Re-Conquest of the Sudan.

The Recovery of Dongola—The Occupation of Berber—The Atbara—
The Final Advance—Omdurman—The Second Phase—The Death
of the Khalifa—The Eastern Sudan.

Comments.

British Organization — Egyptian Organization — British Armament —
Egyptian Armament—Employment—Medals & Clasps—Army Commands and Staff Appointments—List of Units.

Maps 10 & 12.

DURING the decade which followed the withdrawal from the Sudan, the Dervishes, under the Khalifa Abdullahi, who had succeeded the Mahdi in 1885, made more than one attempt to invade Egypt. These were checked at Ginniss in 1885, at Sarras in 1886, and Toski in 1889, without the assistance of British batteries, and do not therefore concern the Regimental History, although we may note with pride the part played by officers of the Regiment— Parsons, Rundle, Wingate and Wodehouse. But the control of the upper Nile was of vast importance to Egypt, and there was reason to fear French designs of encroachment in those regions. Sentimental as well as political reasons urged the recovery of Khartoum, and these motives for the re-conquest of the Sudan were strengthened by the decision of the Italian Government, after the disaster of Adowa, to evacuate Kassala. In March 1896 the Sirdar received the long-looked-for permission to undertake the recovery of the Dongola Province.

The Egyptian Field Force, always kept in readiness, hailed the opportunity, and on the 7th June 1896 the Dervish frontier post at Firket was captured by a surprise at dawn. The artillery under Major C. S. B. Parsons con-

CHAPTER XVIII.

The Recovery of Dongola.

1885-89.

1896.

7th June.

CHAPTER
XVIII.

The
Recovery of
Dongola.

1896.
September.

sisted of one horse and two field batteries of the Egyptian Artillery, and there was also a maxim battery formed from the machine-gun sections of two British battalions. The railway was hurried on, the gun-boats were dragged up the second cataract as the river rose, the troops marched along the banks. There were delays and difficulties— storm and flood and cholera—but by the beginning of September it was possible to resume the advance. At Hafir the gun-boats were received with a heavy fire from guns cleverly concealed where the river narrowed, and from riflemen in the palms who commanded their decks. The three Egyptian batteries came into action at 1,200 yards and silenced the guns, the infantry cleared the palm-trees with long range volleys, and the flotilla got past.

During the night the Dervishes abandoned their position. The gun-boats pushed on to Dongola and bombarded the defences; the army followed; and on its arrival the Dervishes refused the unequal combat and evacuated the town. Captain de Rougemont, who was serving in the gun-boats,[1] was given command of the "Zafir," and sent ahead to turn the Dervishes out of Merowe, and the great bend of the river at Korti and Debbeh was occupied. The Province of Dongola was restored to Egypt.

For a further advance a shorter line of communication than that afforded by the Nile was the first essential, and after much discussion it was decided to build a railway across the desert from Wady Halfa to Abu Hamed. Commenced on the 1st January 1897, it was pushed steadily forward until it approached too closely to the enemy's position at Abu Hamed. A flying column, which included No. 2 Battery, Egyptian Artillery (Captain Peake), marching along the banks of the river then surprised and

captured that place at the point of the bayonet, and by

---

[1] The gun detachments were provided by the Egyptian Artillery.

the end of October the railway arrived there. The great
bend of the Nile had been short-circuited.

Berber was occupied without opposition, but the
position was somewhat precarious, for it had been necessary to leave a considerable force to guard the Dongola
Province. When, therefore, news was received of the
move northward of the Dervish army the Sirdar immediately telegraphed for a brigade of British infantry,
and concentrated all the Egyptian troops available in an
entrenched camp situated in the angle between the Atbara
and the Nile. By the end of January 1898 the British
brigade had arrived, and in the next month the Dervishes
were on the move from Metemmeh. It was fortunate that
their advance had been so leisurely.

The force under the command of Sir Herbert Kitchener
on the Atbara consisted of some fourteen thousand men,
formed in one cavalry and four infantry brigades, of which
latter one was British. The artillery was as follows :—

C.R.A.   — Lt.-Col. C. J. Long.

16/Eastern—Maxim Battery — Captain W. C. Hunter-
                                        Blair.

No. 1 Egyptian Horse do.   — Major N. E. Young.

No. 2   do.   Field   do.   — Captain M. Peake.

No. 4   do.   do.   do.   — Major C. E. Lawrie.

No. 5   do.   do.   do.   — Captain C. H. de
                                        Rougemont.

Each of the Egyptian field batteries had a section of
galloping maxims in addition to its six guns. There was
also the maxim battery manned by British infantry which
had taken part in the recovery of Dongola, and a detachment with rockets from the gun-boats under Lieut.
Beatty, R.N.

The Dervishes from Metemmeh crossed the Nile and
marched down the right bank until almost in touch with
the Anglo-Egyptian position behind the Atbara, when they
turned along that river, and crossed it higher up. The

Sirdar moved in the same direction, keeping in touch, and struck with all his might as soon as he judged the time was ripe.

Previous to this there was constant patrolling by the cavalry and maxim sections to enliven the waiting. On the 30th March and 5th April attempts to entice the Dervishes to come out and fight resulted in some sharp encounters in which the horse battery and the maxim sections of the field batteries were closely engaged, retiring alternately to cover the withdrawal of the cavalry.

Dawn on the 8th April found the troops lying down in attack formation along a low ridge facing the Dervish zareba at a distance of about half-a-mile. The four batteries and rocket detachment were disposed on two natural positions in front of the centre and left of the infantry line : the maxim battery of 16/Eastern on the extreme left, the Egyptian maxim sections with their batteries, except that of No. 5 which was with the 1st Egyptian infantry brigade, its guns drawn by infantry detachments.

Shortly after six o'clock the batteries opened fire, and were soon busily engaged in a methodical search of the zareba. At a quarter to eight the bugles rang out, and the infantry swept forward, the batteries following up as soon as their fire was masked. The Dervishes held their fire until the infantry were within three hundred yards when a furious fusilade broke out. But nothing could check the attack. Crashing through the thorn fence and the trenches beyond, the infantry pushed right across the interior of the enclosure, bearing down all resistance.[1] At half-past eight the "Cease-Fire" sounded.

The dimensions of the enclosure were about a thousand by eight hundred yards, and 690 shrapnel, 270 common,

---

[1] The maxim section of No. 5 carrying their guns and tripods, went right through the zareba with the infantry, and incidentally rescued Mahmud, the Dervish general, from Sudanese soldiers who were on the point of killing him.

145 double, and 30 case were expended by the artillery in searching it. 13 rockets were fired.

"High Nile" was not due until July, so the troops were distributed in rest camps along the river while preparations for the final advance were made. These included the strengthening of the Anglo-Egyptian army by another brigade of British infantry, a regiment of British cavalry, a field gun and a field howitzer battery, and a portion of a siege battery—all of the Royal Artillery. In July these reinforcements were passing through Cairo and up the Nile to Wady Halfa, whence they were carried by the new railway across the desert to Atbara. There they found the movement south in full progress—infantry and guns in barges on the river, mounted troops marching along the banks. Just south of the sixth cataract an advanced base was established upon Royan Island, and communications dropped—thenceforward the army must depend upon the supplies it carried with it.

The following was the constitution of the army :—

CAVALRY.

A British regiment.
An Egyptian brigade.
An Egyptian Camel Corps.

INFANTRY.

A British division of two brigades.
An Egyptian division of four brigades.

ARTILLERY.

C.R.A.	—Colonel C. J. Long.
Staff Officer	—Captain J. W. G. Dawkins.
32nd Field Battery	—Major W. H. Williams.
37th do. do.	— do. F. B. Elmslie.
16/Eastern Siege Section	—Lieut. E. G. Waymouth.
do. Maxim Batty.	—Captain C. O. Smeaton.[1]

---

[1] Captain Hunter-Blair had been ordered home on promotion.

Chapter XVIII.

1898.

The Final Advance.

July.

EGYPTIAN ARTILLERY.

No. 1 Horse Battery   —Major N. E. Young.
No. 2 Field   do.   —Captain M. Peake.
No. 3 do.   do.   —Lieut. C. G. Stewart.
No. 4 do.   do.   —Major C. E. Lawrie.
No. 5 do.   do.   —Captain C. H. de
                            Rougemont.

On the 26th August the final advance began. It must
have been an imposing array—the masses of infantry and
artillery with the cavalry screen in front, the camel corps
on the right, the gun-boats on the left, and beyond them
again on the far bank the "irregulars." The movement
was unhurried, an average of ten miles a day, bivouacking
every night in zareba, and it was not until the 1st September that the Sirdar and his staff, riding up on to the
Jebel Surgham, got their first view of Omdurman.
And in the plain between, scarcely three miles away,
fifty thousand men were advancing directly upon them.

On the river to the left the gun-boats in line ahead
were tackling the forts, while the howitzers of the 37th
battery were landed on the opposite bank. Disembarkation was a difficult job for the current was strong, and the
attempt had to be abandoned in the case of the
40-pounders, but by half-past one the howitzers opened
fire on the Mahdi's tomb, its dome prominent above the
mud houses of Omdurman. With the third round a hit
was registered.

The army, meanwhile, had just gone into bivouac,
and the gun-teams were filing on to their picquet lines,
when the order came to take up their defensive positions
at once, for attack was imminent. All the afternoon, and
all that night, the two armies lay in battle array, not five
miles apart, with only the ridge of Surgham between.
Every possible range was taken and recorded, and the
star shell were laid out ready by the guns. But for some
reason the expected attack was not delivered. The sketch
below shows the distribution of troops in the Anglo-
Egyptian bivouac, which stretched for a mile and a half

along the river, the front forming a rough semi-circle.
From it an open sandy plain, sloping gently upwards,
stretched away to the horizon, bounded on both sides by
hills—Kerreri on the right, Surgham on the left.

CHAPTER
XVIII.

1898.

The Final
Advance.

1st Sept.

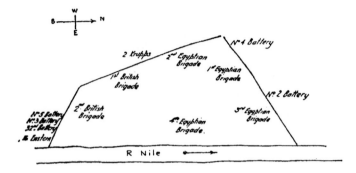

The orders for the 2nd September were that, if not
attacked, the army should advance at six o'clock, and
some of the batteries had started to take up their places
for this movement when they were hastily recalled.
The Egyptian cavalry and camel corps left the zareba
at dawn to reconnoitre desertward, with them the horse
battery and a maxim battery made up of the maxim
sections of Nos 1 and 3 under Captain G. McK. Franks.
This mounted force occupied the Kerreri Hills, but as soon
as it was clear that the whole Dervish army was advancing,
the cavalry were ordered to clear the field of fire from the
zareba, the camel corps and the maxims to return to it.
This was, however, easier said than done. On the rocky
hills the advantage lay with the Dervishes, and the
mounted troops were rather roughly handled when the left
wing of the attack swept down upon them. Eventually
the cavalry drew away into the desert to the north, and
the camel corps got back into the zareba under cover of
the fire of a gun-boat, but with heavy loss. The horse
battery was in the worst case. They had fallen back with
the cavalry from the western summit of the ridge, and had
taken up a second position, but hardly were the guns in

action again before the enemy were creeping up to close rifle range. A further withdrawal was imperative, but at a critical moment the wheelers of one gun were shot and another gun was overturned by the rocks, and both had to be abandoned.

It was on the other flank, however, that the main attack on the zareba first developed. At about a quarter past six there was a glimpse of banners and spear-heads showing over the crest of Surgham, and Wingate asked the Sirdar if he might give the order for the artillery to open fire. The 32nd commenced at 2,700 yards, quickly followed by Nos. 3 and 5. The Dervishes came on over the high ground on both sides of the peak, where they had three guns in action, but the shell fire of the three batteries took toll from the first, for the ranges were all marked on the rocks, and the banners fell fast. But as this attack slowed down another huge mass surged over the shoulder further to the left. Coming down the gentle slope at not more than 1,700 yards it offered a perfect target, and men fell in heaps with every shell. Under this fire the dense masses dissolved into ragged lines, and the attack did not get more than a couple of hundred yards down from the sky-line.

Then the fire was turned back on to those first engaged, who were coming on again with incredible valour—to the amazement and admiration of the British. Under the hail of bullets from guns, maxims, and rifles, they swerved away towards the centre and right, without ever getting within six hundred yards of the front of the British division. The batteries on the left could no longer bear upon it, so the two Egyptian batteries moved off towards the centre, where they were soon in action again. The Dervishes had found dips in the ground in which they could lie safe from the flat trajectory fire of rifles and machine guns, and pour a steady fire into the zareba, but the shrapnel searched them out, and, as they rose to escape, they came again under the rifles, and so gradually the front was cleared. Following the sound of the firing

the two batteries moved on again until they reached the CHAPTER XVIII.
extreme right, where they joined Nos. 2 and 4 in beating
off the attack from Kerreri. Eventually the threat on
this flank also faded away, and the first phase of the battle
was over.

If protracted street-fighting was to be avoided the city
of Omdurman must be occupied before the Dervishes could
get back there, and shortly after nine o'clock the
Anglo-Egyptian army was on the move. The 21st
Lancers went ahead to clear the way, and the rest
followed in echelon of brigades from the left. The 2nd
British brigade came first, then the 1st British, followed
by the 2nd and 3rd Egyptian, the 32nd and No. 3 batteries
with the two British brigades. The direction was over
the south-eastern slopes of Surgham. There was some
opposition from scattered parties of the enemy, and both
batteries had to come into action more than once to clear
out of the way the larger gatherings of Dervishes through
which the 21st had recently ridden in their well-known
charge.

With such slight interruptions the advance proceeded
according to plan. The British division had just crossed
a low ridge between Jebel Surgham and the Nile, the
2nd Egyptian brigade was ascending Surgham, the 3rd
was at its foot, when heavy gun and rifle fire was heard
from behind the crest of Surgham on the right. We must
turn now to what was happening there.

While the main body was moving on Omdurman
between the Nile and Surgham, Macdonald was ordered
to take his brigade (1st Egyptian) by the other side of that
hill, and to clear out the Dervishes still there. He was
given three batteries—Nos 2, 4 and 5—while the Camel
Corps, with which were Franks' maxims, covered his right
flank. The line of advance was almost directly on
Surgham, and it was soon evident that the Dervishes were
occupying the sandhills beyond in great force. It was
indeed the flower of the Khalifa's army, some fifteen

CHAPTER
XVIII.

1898.

The Second
Phase.

2nd Sept.

thousand men under his Black Banner, waiting to swoop down on the isolated brigade. The three batteries moved out to the front and opened fire, and at once the enemy came on, and at a tremendous pace.

The whole picture was plain to the Sirdar on Surgham, and his orders were promptly issued. The two Egyptian brigades wheeled to their right, those of the British division did the same, which brought them into line, facing west. The two batteries crossed the spurs of Surgham, and came into action in front of the infantry, just under where the Headquarter Staff had taken up its position. Under their fire, and the determined resistance of Macdonald's brigade, the Khalifa's attack weakened, and gave way. It was fortunate, for a new and greater danger now threatened. The left wing of the original attack, which had cleared the cavalry off Kerreri in the early morning, and had followed them in their retreat northwards, now reappeared, coming over the hills to the right rear of Macdonald's brigade.

It was at this critical moment that Macdonald made his famous change of front. Turning to Franks, whose maxims were on the right of the infantry line, he told him to "get his guns round," and used them as the pivot on which to build up his new line, bringing up batteries and battalions right and left as they could be withdrawn from their places in the old line. The manœuvre was carried out with admirable precision in spite of the desperate situation. It was the most critical moment of the day, for the new attack was coming on with desperate determination, the cavalry in chain mail charging with banners flying. The shells ploughed lanes through their ranks, but they were within two hundred yards before they caught the full force of the rifle fire. Even when the dead were piled in a solid line the survivors struggled on with dauntless courage, and individuals actually reached the line. It was touch and go for ammunition was running short, but the Lincolns came up at the double with fixed bayonets, and the Egyptian cavalry,[1] reappearing over the Kerreri Hills as the Dervishes broke, scattered the fragments.

In the meanwhile the 32nd and No. 3 batteries had taken heavy toll of the remnant of the Khalifa's attack as it streamed across their front, and the 32nd only just missed bagging the Khalifa himself. Colonel Wingate pointed him out to Major Williams as he was crossing the front of the battery at the head of his bodyguard of forty or fifty men. The range was only 2,100 yards, the second round burst in front, but just as the third went off the Dervishes broke into a trot, so the rear of the party instead of the leader got the benefit. Before another round could be fired they were out of sight.

The battle was over, and the whole Anglo-Egyptian Army advanced in one long line, with bands playing, driving the Dervishes before them into the desert. By half-past eleven the Sirdar resumed his interrupted march to Omdurman, and rode into the city with the leading Egyptian brigade and the 32nd Battery[2], whose trumpeter sounded the "Cease Fire" from a house-top near the Mahdi's tomb.

There is no need to carry the story further, but the work of the 37th battery must not be omitted. From the opposite bank of the Nile, where they had been landed on the previous afternoon, this battery had been methodically carrying out its allotted task of battering the Mahdi's tomb to pieces and breaching the great stone wall which encircled the Khalifa's headquarters. Three breaches in the walls opened ways of entrance to the troops, and some forts and barracks were also cleared. The gun-boats assisted in the latter part, but their chief duty was to drive the Dervish army from the river.

CHAPTER
XVIII.

1898.

The Second
Phase.

2nd Sept.

---

[1] The horse battery came back with the cavalry, its guns recovered.

[2] An unfortunate accident occurred owing to a section of the battery, that had been left outside the walls, opening fire on the Khalifa's house, when the Sirdar arrived there. It had been posted by Kitchener himself with orders to shell the Tomb if the Khalifa reached it, and his black banner carried behind the Sirdar naturally gave that impression. No blame was attached to the battery.

CHAPTER
XVIII.

1898.

The Second
Phase.

2nd Sept.

The decision not to land their 40-pounders was, of course, a bitter disappointment to 16/Eastern, but, like good gunners, they found other ways of rendering service. All hands were landed and worked hard at unloading the reserve ammunition from the barges and packing it on to camels to keep the batteries supplied. Not content with this their barge was converted into a hospital, and first aid was rendered to a hundred and fifty wounded.

The ammunition expended by the artillery in the day was :—

32nd Field Battery	420
37th do. do.	410
The Egyptian Batteries (except No. 4) average	360
No. 4 Egyptian Battery	913

The three batteries with Macdonald's Brigade fired 150 case.

It remains only to add that after entering Khartoum in triumph the Sirdar continued his progress up the Nile to Fashoda, taking with him No. 2 Battery. There he found the Marchand Mission, and successfully arranged the removal of the French threat to the Sudan.[1]

The Khalifa contrived to escape pursuit and took refuge with a considerable force of cavalry in the inaccessible regions of Kordofan. Here he defied all attempts to bring him to battle until November 1899, when Sir Reginald Wingate, who was to succeed Lord Kitchener as Sirdar, ran him to earth and utterly defeated him. The Khalifa himself was killed with some thirty of his principal Emirs, and his camp and thousands of his followers were captured. Thus it fell to a gunner to crown the great work of the pacification of the Sudan by shattering the last forces of Mahdism.

To avoid interrupting the narrative of the main course of the operations, nothing has been said so far of events

---

[1] Of the four present at that momentous meeting—Marchand, Mangin, Kitchener and Wingate—the latter alone survives.

in the Eastern Sudan. Although no artillery units had a share in these, the principal parts were played by two well-known officers of the Regiment, and a sketch of the operations may therefore find a place here.

The temporary occupation of Kassala, by the Italians, had been agreed upon, on the understanding that they would restore it to Egypt when required, and its return was effected in 1897. Colonel Parsons, R.A., appointed Governor, summoned to his assistance Major G. E. Benson, R.A., who happened to be travelling in Erytrea, and gave him the command of the troops taken over from the Italians and of some Arab irregulars. With these— and one English serjeant—Benson was sent to block the retreat of Osman Digna from the Atbara. There were several encounters in which more than three hundred of the enemy were killed and nearly six hundred taken prisoner, with large numbers of horses, camels, etc., and Osman Digna himself was wounded and narrowly escaped capture. Nor must mention be omitted of the fine performance of Colonel Parsons in capturing Gedaref after a stiff fight, and holding it against fierce attacks until relieved by a force under Rundle from Omdurman.

COMMENTS.

The 32nd battery and 16/Eastern had been garrisoning Egypt for some years, and Rundle, Wingate, and last, but not least, Lord Cromer (who had also been a gunner) did their utmost to insure their being included in the Expeditionary Force. But English horses were excepted, and so the order went forth that the blacks of the 32nd—many of which had come out to Egypt with the battery five years before—were to be left in Cairo, while the 37th would come out without theirs. At a conference held by Lord Kitchener to discuss such preliminary arrangements Major Williams was informed of this decision and asked what he proposed. The answer "mules" brought a round of laughter, for the idea of mules in gun-teams was then a novel one. Williams burst out: "Damn it, Sir. I'll take them up with goats if you'll give me enough of them". He got his mules, and good ones too, for the Damascus coach was being closed down, and its mules were bought, and proved a magnificent lot.

The normal team was eight, but for emergencies three more were hooked in down the centre, so that there were three in the lead, the lead-centre, and wheel-centre, a team of eleven. Breast harness was designed and made up. Officers, staff serjeants, Nos. 1, coverers, range-takers, etc., had Arab horses, and gunners rode the off mules, except the wheel. The remainder of the gunners picked up donkeys or any other conveyance they could find.

Ammunition wagons were not taken, but with each gun there were four rounds in portable magazines on the carriage in addition to those carried in the limber. Two Maltese carts formed a link between guns and twelve camels, each carrying twenty rounds.

There were also two British maxim batteries. The first was formed by the infantry before the Dongola campaign, the second was formed by the Royal Artillery, and is therefore entitled to record here. No. 16/Eastern had been stationed in Egypt for some years, and when the Sirdar telegraphed for a brigade of British infantry in 1897 Sir Francis Grenfell ordered three of the battalions in the brigade to hand over their machine guns to the company to form a maxim battery. The establishment was fixed at 2 officers, 28 N.C.Os. and men, 28 mules, and 30 Egyptian drivers. Four of the guns were on infantry travelling carriages drawn by three mules tandem, the other two were carried in pack. The battery arrived with the British brigade in time for the Atbara, and went through the whole of the rest of the campaign.

The Egyptian horse battery had teams of eight Syrian Arab horses, and ammunition wagons, which always accompanied their guns. The so-called "field" batteries were really "pack," the gun and carriage being carried on four mules, although there were also shafts which could be attached to the trail for draught. A hundred rounds a gun were carried on mules or camels with the guns, another hundred on camels in an ammunition column, with a further supply in the barges.

A British officer commanded each battery, with a specially selected Egyptian officer as second in command. The Egyptian officers were drawn as a rule from the Effendi class, the juniors selected from the cadets of the Military College. They proved capable, loyal, and reliable, and there was no single instance of an artillery officer being concerned in the troubles which occurred in the Sudan during the South African War. The men were selected from the annual contingents, chiefly for physique and eye-sight, and although illiterate made good gunners, invariably steady in action.

The sections of "galloping maxims" attached to the batteries were, perhaps, the most interesting feature of the artillery organization. These had teams of six horses and mounted detachments as in horse artillery. At first only the field batteries had them, but after the Atbara one was formed for the horse battery also.

The two British field batteries were armed respectively with the 15-pr. B.L. gun and the 5" B.L. howitzer, the new armament adopted for the field artillery in the 90's. The considerations which led to this re-armament have been discussed at some length in Volume I of this History, but the salient features may be recapitulated here. They were, in the first place, the abolition of common shell for field guns and the introduction of field howitzers to provide the destructive effect which the common shell of the guns had failed to give. Combined with these major changes was the replacement of gunpowder in cartridges and bursting charges by cordite and lyddite. Omdurman was the first occasion on which a British battery fired high explosive shell except at practice, and then under pre-cautions. Before leaving Cairo practical trials had been carried out to see whether the artillery would be equal to dealing with the defences of Omdurman where the Khalifa was reported to have built a formidable wall round his headquarters. A replica was constructed on the artillery practice ground near Abbassiyeh to test the powers of the different natures, and as might have been expected the shrapnel of the 15-prs. proved of little value for such work. With the idea that the battery might possibly be called upon to breach the wall some common shell were made in the Egyptian army workshops, and camels carrying these wandered in rear of the battery at Omdurman. But they were never used—howitzers and guns adhered to the rôles for which they had been designed. The shrapnel of the 15-prs. found an ideal target in the massed attacks of the Dervishes, the howitzer lyddite did all that was called for in blowing out the dome of the Mahdi's Tomb, and breaching the six-foot walls of the citadel.

Both the 15-pr. and the 5″ howitzer equipments were however found to be on the heavy side both for march and action under desert conditions. Full detachments of nine were provided, but the howitzers had to call on their infantry escort for assistance in running up.[1]

For range-finding the mekometer was used with success. There were none of the complaints about the mirage which had been held accountable for the failure of the Watkin range-finders in 1882.

The Egyptian horse battery was armed with $7.75^{c/m}$ Krupps, an unsatisfactory old equipment. The absence of brakes, especially, delayed the fire very materially, and this had serious consequences when the battery found itself in a tight place on the Kerreri Hills. The field batteries

---

[1] It will be remembered that the gunners of N/A had also found the running up of their 13-prs. too much for them in the desert sand in 1882.

also had Krupps—6˙5ᶜ/ᵐ in their case—during the Dongola
campaign, but, as soon as the re-conquest of the Sudan was
decided upon, the Sirdar sent Colonel Long home to select
the best pack equipment he could find for these batteries.
The one he pitched upon was the 75ᵐ/ᵐ Vickers-Maxim, a
12½-pr. with, also, an 18lb. double shell.   The fuzes were
Krupp pattern, similar to those adopted for the 13-pr. and
18-pr. on the re-armament with Q.F. guns after the South
African War and their accuracy impressed all the British
officers.   The gun was nominally a quick-firer, having
fixed ammunition and a hydraulic buffer, but the recoil was
not entirely suppressed.   With detachments of five, however, a rate of fire of twelve rounds a minute could be
reached.   The muzzle velocity was comparatively low, but
good practice could be made with time shrapnel up to
3,600 yards.   Four mules were required, the loads being
gun, jacket, wheels, and trail—averaging 300lb. saddles
included.

The two newly raised batteries were armed with this
equipment in 1898 in time to take part in the Atbara, and
their effect there was so good that the other two were
re-armed before the final advance, though No. 3 retained
also two of its old Krupps.

The maxims of the Egyptian artillery were on "galloping carriages", drawn by artillery teams.   An interesting feature was that the gun could be fired without
unlimbering.

As we have seen, the batteries at Omdurman were at
first distributed more or less throughout the defensive line,
although the chief strength was on the left where the
heaviest attack might be expected.   Later, when the weight
of attack moved towards the right, batteries were moved
from the left, first to the centre, and then to the right, to
meet it.   And in the second phase the brigade which was
given the most dangerous task was given three out of the
five batteries.

But the most interesting point in the employment of

the artillery is the use made of the field howitzers. On this, the first occasion of their use in war[1], they were employed in full accordance with the principles approved by Lord Wolseley; that is to say they were reserved for the tasks for which they had been designed—a refreshing contrast to the way in which some of the batteries sent to South Africa were so soon to be misused there.

The mutual support of the artillery and infantry is very noticeable, and so close was this that battalions took the distances for their long-range volleys from the batteries alongside them. The use of the guns to clear out the Dervishes from sheltered positions where the rifles could not reach them is also to be noted. But the artillery might well claim that it was a "gunners' day", and indeed there were many infantrymen ready to complain that the gunners had "spoilt their show".

Although the arsenal at Omdurman contained many Krupp guns and also machine guns, the Khalifa made no use of any in his attack on the British zareba, except the three which fired the opening rounds, and these effected nothing.

MEDALS & CLASPS.

For the Recovery of Dongola in 1896 the Khedive gave a medal with clasps for "Hafir" and "Firket". For 1897 the clasps were "Sudan 1897" and "Abu Hamed". For 1898 "The Atbara", "Khartoum", and "Gedaref". For 1899 "Abu Aadel" and "Gedid".

A British medal without clasps was given for the operations from Abu Hamed to Omdurman.

ARMY COMMANDS AND STAFF APPOINTMENTS.

Major-General H. M. L. Rundle—Chief of Staff.
Lieut.-Colonel F. R. Wingate —Director of Intelligence.
    do.   C. S. B. Parsons—Governor of Kassala.
Major G. E. Benson   —Brigade Major.

---

[1] Field howitzers had actually been used in the previous year at the battle of Domokos in the Turko-Greek war.

LIST OF UNITS.

(With Designation in 1914.)

32nd Field	Battery, R.A.	...	32, R.F.A.
37th do.	do. do.	...	37, do.
16/Eastern, Garrison do.	do.	...	94, R.G.A.

CHAPTER XIX.

SOMALILAND.

Introduction—The Expedition of 1901—The Expedition of 1902—·The
Expedition of 1903—The Expedition of 1904.

COMMENTS.

Organization and Armament—Medal & Clasps.

CHAPTER
XIX.

Introduc-
tion.

SOMALILAND is, roughly speaking, a triangle, occupying
the north-eastern corner of Africa between the Gulf of
Aden and the Indian Ocean.   It is inhabited by nomads,
the only permanent settlements being those on the coast.
At the time in question it was divided into spheres of
influence, that of the British being on the northern side
opposite Aden, the Italian the eastern side facing the
Indian Ocean, and the French the north-eastern side with
the port of Jibuti.   The hinterlands of all owed allegiance
to Abyssinia.

In 1898 the British sphere, which had previously been
administered under the Governor of Aden, was transferred
to the home authorities, and very shortly afterwards the
Haji Mahomed Abdullah, widely known as the ''Mad
Mullah,'' began to give trouble.   Preoccupation in South
Africa prevented any action against him until 1900, but
at the end of that year the raising of a local levy for the
protection of the country was sanctioned.

The Expe-
dition of
1902.

With this force Lieut.-Colonel Swayne attacked the
Mullah in the following year.   The scene of action was
the fertile valley in the southern part of the Protectorate
known as the ''Nogal''; and after being defeated here the
Mullah was driven southwards across the waterless bush
country (termed the ''Haud'') into Italian Territory.   In

this little campaign the Regiment was only represented by Captain C. M. D. Bruce, who was in command of the Camel Corps.

Next year the fight was renewed, the levy having been strengthened by a battery of 7-prs. from Aden for which Somali detachments were trained. Four of the guns were put into posts, while the other two, with camel transport, joined Colonel Swayne, who was advancing into the bush country south of the Protectorate. At Erigo his force was attacked from all sides and Captain J. N. Angus, commanding the section, was killed at his guns. The Somali detachments stuck to their work, plying the enemy with case until they were almost at the muzzles. Eventually the attackers were driven off, but the levies were so shaken that Colonel Swayne had to fall back upon the frontier fort at Bohottle, leaving the Mullah in possession of the rich oasis known as the "Mudug."

It was plain that matters could not be allowed to remain in such an unsatisfactory condition, and also that the levies could no longer be relied upon in close fighting. For the next campaign, therefore, troops were drawn from East, Central, and South Africa,[1] as well as from India. The Somali detachments of the 7-prs. were replaced by Sikhs belonging to the King's African Rifles, under Lieut. J. A. Ballard, and the section became known as the "K.A.R. Camel Battery." At the same time the contingent from India included a section of the 28th Indian Mountain Battery under Lieut. H. E. Henderson.

The command of the combined force was given to Brig.-General Manning, the Inspector-General of the King's African Rifles, with Lieut.-Colonel G. T. Forestier-Walker, R.A., as Staff Officer. The plan of campaign was for two forces, each two thousand strong, to converge on the Mullah's oasis from Berbera on the Gulf of Aden

[1] A "Burgher" contingent was included in the South African troops.

and Obbia[1] on the Indian Ocean respectively, while an Abyssinian force acted as a "stop."[2] The section of the 28th Indian M.B. went with the Obbia force, but the feeding, and more especially the watering, of their mules proved a great difficulty, and they never got to the front. The K.A.R. Battery was with the Berbera column, and proved useful in dispersing such parties of the enemy as were encountered. But there was little fighting until a detachment under Colonel Plunkett was surrounded and wiped out at Gumburru, and a flying column under Gough was attacked and suffered heavily. These two events upset the scheme for the envelopment of the Mullah's main body, and all that could be done was to roll up the line of communication from Obbia, and, as in the previous year, concentrate the whole force on the frontier at Bohottle.

Preparations for a fourth campaign were put in hand, and these included the re-equipping of the section of 28th Indian M.B.. with 7-pr. guns and camel transport. The artillery of the force thus consisted of two sections of 7-prs. with Indian *personnel* and camel transport, and one of these was included in each of the two brigades in which it was organized. The whole strength amounted to some seven thousand men, and the command was given to Major-General Sir Charles Egerton, with Lieut.-Colonel G. T. Forestier-Walker, Major H. E. Stanton, and Major F. C. Owen, all of the Regiment, as staff officers. Up to now fortune had been with the Mullah, but in December 1903 he made the fatal mistake of transferring his whole force to the "Nogal," thus giving the long-looked-for opportunity of bringing him to book. On the 10th January, 1904, Sir Charles Egerton advanced to the attack over ground nearly as open and flat as a parade

---

[1] Obbia was in Italian Territory, but permission was given for the column to land and advance through the Italian Protectorate.

[2] Colonel A. N. Rochfort, R.A., went with the Abyssinian Force as British Representative

ground. As soon as the square arrived within range of the zareba it halted while the guns moved out to the front and opened fire at 1,600 yards. This brought the Somalis swarming down from the bush in which they had been concealed, but they could not face the fire, and the nearest they got to the square was four hundred yards. Then the whole array broke up and fled, followed by the shells of the guns until the mounted troops took up the pursuit. The Mullah escaped but his power had been broken.

COMMENTS.

The history of the "Camel Battery" begins with the existence in store at Aden of six 7-pr. R.M.L. guns of 150 lb., very possibly left behind after the Abyssinian War, in which, as we have seen, these guns were used for the first time. The guns were periodically taken out for drill by one of the garrison batteries, with camels borrowed from the Transport, for which saddles were kept in store with the guns. When, therefore, Colonel Swayne asked for guns for his second expedition the battery was sent over from Aden, and Captain J. N. Angus trained Somali detachments for them. These stood to their guns well at Erigo, but were replaced, with all the other Somali levies, before the next expedition, and their place taken by Sikhs under Lieut. J. A. Ballard. He improvised a limber arrangement so as to admit of the guns being moved on their carriages, with the camels in draught, when single file was necessitated by the dense bush, so as to reduce the time required for bringing the guns into action if attacked.

The section of the Indian Mountain Battery brought its Indian establishment, but in consequence of the difficulty of finding forage and water for the mules General Egerton had its 2·5″ guns and mules replaced by 7-prs. and camels. This battery also found that camel transport was apt to cause delay in coming into action owing to the timidity of the camels, and they got over the

CHAPTER
XIX.

Organiza-
tion
and
Armament.

difficulty by retaining enough mules to pull the guns in draught, which was always employed in the vicinity of the enemy. The establishments adopted for the respective sections were as follows :—

	28th Indian M.B.	K.A.R. Battery.
British Officers	1	1
Indian do.	1	—
do. Non-commissioned officers & gunners	33	21
do. Mule Drivers	5	—
Somali Camel Drivers	22	23
Camels	22	20
Mules	5	—

MEDAL & CLASPS.

For the campaigns in Somaliland the African General Service Medal was granted, with clasps for "Somaliland 1901" and "Somaliland 1902-04".

PART V.

WEST & SOUTH AFRICA.

CHAPTER XX.

THE WEST COAST.

Introduction—The Coastal Belt—The Hinterland—Nigeria—The West African Frontier Force—The Mahomedan States—Medals and Clasps.

N.B.—For "Comments" see next chapter.

THE difficulty experienced in dealing with the large number of frontier expeditions in India is enhanced when we come to West Africa by the fact that the expeditions there were undertaken by half-a-dozen different governments with military, or semi-military, forces of the utmost diversity. The three Ashanti Expeditions will, therefore, be described in some detail, and, for the rest, all that can be attempted is to indicate the nature of the forces available, and of the difficulties they encountered. This is their due, for although no unit of the Royal Artillery took part, no column was complete without its gun or two, and many an artillery officer found opportunity for distinction in command of these, as also in service with the other arms.

A few words must first be said as to the country. The "West Coast" had been familiar to European merchants ever since its discovery by the Portuguese in the XVth century, and the names by which its different parts were commonly known bear witness to the nature of their products—Gold Coast, Ivory Coast, Slave Coast. The rainfall was prodigious—up to 200 inches in the year: the only communications were winding paths from village to village, constantly blocked by fallen trees: the only transport native carriers. In such a country and climate neither horses, mules, or donkeys could live, and all supplies for Europeans and rice for native troops and followers, as well as guns and ammuni-

Marginal notes: CHAPTER XX. Introduction. The Coastal Belt.

CHAPTER
XX.

The Coastal
Belt

tion, had to be carried on men's heads. It is not surprising that it became known, before the discovery of the
anopheles mosquito, as "The White Man's Grave".

Along this coast the British possessions were scattered
—Gambia, Sierra Leone, The Gold Coast, Lagos, The Oil
Rivers—sometimes linked together in uneasy partnership,
sometimes standing alone.

The governments of these various settlements naturally
found it necessary to raise armed forces for the maintenance of law and order in their territories, and these
usually included some gunners. In the 50's, for instance,
the "Gold Coast Artillery" figured for a few years in the
Army List, but its existence was brief and stormy. A
change came with the recruitment of Hausas and other
fighting tribes as armed constabulary, and an artillery
section showed that they had the makings of good gunners
in the Ashanti campaign of 1873-4.

The
Hinterland.

1896.

1898-9.

Beyond the settled districts of the various settlements
there extended in most cases a vast indeterminate "hinterland" of a very different character to the coastal belt.
Thus, beyond Ashanti and its gloomy forest, through
which the sun never penetrated, stretched the "Northern
Territories", a land of grassy plains and stony mountains,
inhabited by well-clothed, semi-civilized people, with
horses and herds of cattle. Behind the Crown Colony of
Sierra Leone dwelt turbulent tribes whose inroads
necessitated numerous expeditions in the 80's. In 1896
a Protectorate was established over these regions, followed
by measures to repress slavery, and the imposition of a
hut tax to cover the expense of policing the area. Such
interference with their customs, and demand on their
goods, had the natural result. The Temni and Mendi
tribes rose in revolt, sacked the mission stations, and
murdered the missionaries. The military operations that
ensued for the suppression of what became generally
known as the "Hut Tax Rebellion" occupied a considerable
time in 1898-9, during which columns traversed the whole

country, reducing the tribes to order. We may take as an example the march of the main column which covered not less than a thousand miles in three months—no mean performance through tropical jungle. The artillery under Lieut. A. F. Becke,[1] consisted of a 7-pr. and a maxim manned by the native battery from Sierra Leone, with a corporal from the British battery to assist him. The maxim was the first of that sort to reach the West Coast, and the 7-pr. had lost its sights, but a wooden one made by the subaltern (and carried in his pocket) served its purpose. Witness is borne to the seriousness of some of the fighting by the fact that at the end of one encounter there were only six rounds left for the 7-pr. and one box for the maxim (1,100 rounds).

CHAPTER
XX.

The
Hinterland.

The Niger Coast Protectorate only extended from the coast to the junction of the Benue River with the Niger. But as exploration advanced up the river it was discovered that here again there was a vast hinterland, differing both in natural features and type of inhabitant from this coastal belt. Swamp and forest gave place to a pastoral land of grassy plains and park-like glades where sheep and cattle grazed and horses flourished. This was the country of the Mahomedan Fula who had brought all Hausa-land under their sway. Unfortunately they were in the habit of raiding the pagan tribes for slaves, and, in order to enable the National African Company to deal with this propensity, it was given a charter as the "Royal Niger Company" in 1886. A force of constabulary was at once raised, and in its first expedition captured Bida, the capital of the Emir of Nupé, after a stiff fight under its walls, in which the Company's two guns (a 9-pr. and a 12-pr.) played a prominent part. But, as we shall now see, European rivals were appearing on the scene, and to deal with them something more than a Chartered Company and its constabulary, however efficient, was required.

Nigeria.

---

[1] He had won the Gold Cup with "No Name" just before going out.

CHAPTER
XX.

Nigeria.

In order to check the game of grab started by the Germans a conference was assembled at Berlin in 1884-5. This laid down the respective "Spheres of Influence" of the Powers interested, and, in the next few years, agreements regarding the frontiers of most of the British possessions were signed. But this did not prevent the French sending expedition after expedition into the unoccupied territory in order to gain a footing there. There were tense moments when *Tirailleurs Sénégalais* and Hausa Riflemen faced each other with bayonets fixed on the unsurveyed and unmarked boundaries. But the sound sense, tact, and patience of the officers on both sides averted actual bloodshed,[1] and in 1898 the rival claims were settled by a Convention signed in Paris, and the boundaries demarcated by a Joint Commission.

The West
African
Frontier
Force.

1897.

The constabulary of the Chartered Company had acquitted themselves most creditably, but both European Powers and Mahomedan Chieftains resented, and attempted to ignore, the political and military interference of a "Trading Company." Mr. Chamberlain saw that an Imperial Force was necessary, and in 1897 he entrusted the raising of it to Major Lugard. On the

1900.

1st January 1900 the Chartered Company's rule was absorbed by the Imperial Government and its constabulary disbanded.

Thus came into being the "West African Frontier Force". It was recruited from the most reliable tribes with a sound backing of selected British non-commissioned-officers. Its British officers were seconded from the Regular Army and the Militia, and among them were many artillerymen—not only in the batteries. The "Waffs" soon became a thoroughly well organized and disciplined corps, and early in the XXth century nearly all the colonial constabulary and other military forces in

---

[1] Except for the unhappy accident at Waima in 1896, when both parties were hunting the same rebellious tribe.

West Africa[1] were being merged into it. At the close of the period covered by this volume it consisted of the Northern and Southern Nigeria Regiments, the Gold Coast Regiment, the Sierra Leone Battalion, and the Gambia Company, each (except the last named) including a battery of four $75^{m/m}$ guns. There was also a battalion of Mounted Infantry in Northern Nigeria, raised and commanded by Lieut. T. A. Cubitt, R.A.

Its first Inspector-General was also a gunner— Brigadier-General G. V. Kemball.

Under the new conditions the suppression of slave-raiding was actively pursued, and Emir after Emir was brought to book. By the end of 1902 all the country lying on the banks of the Niger and Benue Rivers was settled under civil administration. There remained only the Mahomedan States which lay along the northern border, and their acceptance of British overlordship was essential to the safety of the trade routes. Efforts were made in the first place to secure the co-operation of their suzerain the Sultan of Sokoto, but without success, and an appeal to arms became unavoidable. In January 1903 a column of W.A.F.F. took the field, and met with no serious opposition until they arrived under the walls of Kano, the great Hausa capital. These were found to be fourteen miles in circumference and too thick to breach, but a gate was blown in by artillery fire, and the place stormed with little loss. After installing a Resident the expedition under Brigadier-General Kemball continued on its way to Sokoto, the headquarters of the Fula Empire. Here also the resistance was feeble, thanks largely to the effect on the enemy's masses in open country of the shrapnel fire of the battery under Captain G. C. Merrick. The Sultan having fled, the High Commissioner (who had arrived at Sokoto two days after its capture) decided that

---

[1] With the exception of the Imperial troops at Sierra Leone, the West African Regiment maintained there for garrison duty, and the civil police in the towns.

no further military operations were necessary, and broke up the force. But the deposed Sultan and some of his Emirs organized a centre of disaffection at Bormi, and a force sent to bring them to reason met with a rebuff. It was only after a desperate fight, in which the Sultan was killed, that reinforcements succeeded in breaking up the rebel gathering.

MEDALS & CLASPS.

No attempt has been made to describe, even to enumerate, the innumerable operations which were conducted by the various forces in the British settlements on the West Coast of Africa. All that has been attempted is to give an idea of the work in which so many officers, and a smaller number of non-commissioned officers,[1] of the Regiment were engaged. In conclusion a list of the medals granted, and more especially of their clasps, is appended. It tells its own tale.

A medal similar to that given for the Ashanti Expedition of 1874 was introduced in 1892 as an "East and West Africa" medal, and was given for the Yonnie, Gambia River, Tambi, and Jebu expeditions with clasps for "1887-8", "1891-2", and "1892". Further clasps to this medal were "Gambia 1894", "Brass River 1895", "Niger 1897" (for Egbon, Bida, Ilorin), "Benin 1894", "Benin 1897", "Dawkita 1897" "Sierra Leone 1898-99" (for the Hut-Tax Rebellion).

A reversion was then made to date-clasps, as follows :— "1896-8", "1896-99" (for Northern Territories of the Gold Coast and Lagos Hinterland), "1897-8", "1899", "1900"

With the XXth century a new African medal was introduced, and this soon amassed a formidable array of clasps :—"Nigeria" (for 1900-01), "S. Nigeria" (for 1901), "Gambia" (for 1901), "Aro 1901-2" (for the expedition under Colonel Montanaro, R.A.), "N. Nigeria 1902, 1903,

---

[1] Two non-commissioned-officers were promoted to Sergeant for gallant conduct at Bida in 1897.

1903-4, 1904, 1906'' (for the campaigns against Bornu and Kontagora, Kano, and Sokoto, and the Munshis), ''S. Nigeria 1902, 1902-3, 1903-4, 1904-5, 1905-6''. There followed the clasps ''West Africa 1906'' and ''West Africa 1908''.

CHAPTER XXI.

THE ASHANTI EXPEDITIONS.

Introduction—The Wolseley Expedition—The Scott Expedition—The Willcocks Expedition.

COMMENTS.

Organization—Armament—Employment—Medals & Clasps.

CHAPTER
XXI.

Introduc-
tion.

DURING the XVIIIth century the Ashantis became the paramount power on the Gold Coast, and by the middle of the XIXth the jurisdiction of the Gold Coast Government, which nominally extended to the river Prah, was bounded in reality by the range of the guns on their castle walls. Beyond this the whole country was terrorised by the fierce warriors from beyond the Prah, who raided the tribes mercilessly in order to obtain victims for their sacrifices. Emboldened by abortive attempts to bring them to order, they went so far as to carry off European missionaries as prisoners, and even to threaten the coast towns. The Government was at last convinced that steps must be taken to recover the prestige which had been lost by our inability to secure even those living under British protection. It was decided to entrust the task to Sir Garnet Wolseley, recently returned from the successful conduct of the Red River Expedition.

The
Wolseley
Expedition.
1873.
October.

Sir Garnet landed in October 1873, and found himself in the undignified position of not having sufficient force even to ensure the safety of Cape Coast Castle. "Special Service Officers" were set to work at once to train native levies, and among these were Captain A. J. Rait[1] and Lieutenant F. H. Eardley-Wilmot of the Royal Artillery. With the assistance of a couple of non-commissioned officers of the Marine Artillery, they were soon turning Hausas into passable gunners, and before the end of

---

[1] His work with the "cavalry" in New Zealand will be remembered. He retired in 1877 with the rank of Captain and a C.B.

October their "battery" of 7-prs. and rockets[1] had its *baptême de feu* and lost its subaltern—the first to fall of the little band of officers who did such wonders in shaping the levies into soldiers. By the end of November the adoption of the offensive by Wolseley had caused the withdrawal of the Ashanti force from British territory—the first phase was over.

On the 20th January, 1874, the main body of the expedition crossed the Prah at Prahsu, 70 miles from Cape Coast Castle, and took the direct road to Kumasi, the capital of Ashanti, 75 miles further on. There was no serious opposition until the 30th, when the Ashantis were found in occupation of a ridge which barred the way, just short of the village of Amoaful. Against this the British advanced in three columns, with the two 7-prs. leading the centre and the rockets with the flanking columns.

The position was covered in front by a swamp, and the whole field consisted of dense forest through which the flanking columns had to cut their way. The advanced guard became hotly engaged, and as soon as a favourable position could be found the 7-prs. were ordered into action. Nothing could exceed the skill and energy with which Captain Rait got his guns across the swamp and up the hill, and a dozen rounds along the path caused such a slaughter in the dense masses of the enemy[2] that the position was carried with a rush. The next ridge, however, found the Ashantis disputing the road again: the guns were brought into action, and after a few minutes concentration of gun and rifle fire, a charge cleared the way. It was the last serious stand, and, with the centre of resistance broken through, the flanking columns were able to make progress, the rocket sections busily engaged.

---

[1] It may be noted that in 1870 a Mr. Loggie, acting Inspector-General of Police, and "formerly in the Royal Artillery", did excellent service with his rockets, complaining only that "the most fatiguing part of the day's work was acting as whipper-in to these people."

[2] Forty dead bodies lay close together showing unmistakably the effect of shell fire. "We could not have forced our way but for Rait's guns." (Sir A. Alison).

CHAPTER
XXI.

The
Wolseley
Expedition.
1874.

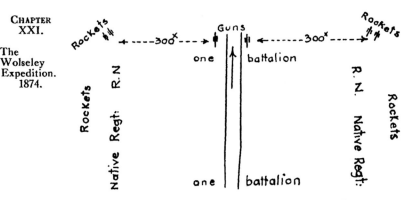

2nd Feb.   On the 2nd February the advance was continued from
Amoaful. Little opposition was encountered, but there
were serious attacks on the convoys in rear, and Wolseley
decided to cut adrift from communications for a dash on
Kumasi. Next day, however, the opposition stiffened, and
4th Feb. on the 4th the advanced guard was soon brought to a stop
and a general action developed, in which the 7-prs. were
again prominent. Each "bound" along the road was
effectively supported by the fire of the guns, but it was
only after six hours hard fighting that the little force
formed up in the main square of Kumasi, and gave three
cheers for the Queen.

The King had fled, his army dispersed and panic-
stricken, but the weather had broken, and with the rivers
rising across the road back, Sir Garnet dared not linger.
6th Feb. On the 6th the Palace was blown up, the town set on fire,
and the army marched away leaving Kumasi a heap of
smoking ruins.

The final despatch bore generous tribute to the work
of the little force of artillery. "In all the actions and
skirmishes the gun and rocket fire had been most effective.
. . . . I consider Captain Rait to be one of the best
soldiers I have ever served with." At the sale of the loot
the seven artillery officers[1] with the force were allowed to

---

[1] Captain H. Brackenbury and Lieut. J. F. Maurice were serving
on the staff as Assistant Military Secretary and Private Secretary
respectively.

buy, for its weight in gold, the very curious gold rams head
from the King's Palace now in the Regimental Mess.

The expedition under Sir Garnet Wolseley had released
the captives, but it had not accomplished its object of
ending the human sacrifices. It was not long before King
Kofi Kalkali had to be deposed : civil war ensued, and a
break up of the Kingdom. In 1888 the Government
stepped in and secured the election of King Prempeh, but
things went from bad to worse internally, while the King's
external policy threatened damage to British influence and
trade. There was also evidence of his revival of human
sacrifice. Sterner measures were obviously necessary, and
it was decided to send an expedition to Kumasi under
Colonel Sir Francis Scott.

The artillery consisted of a battery armed with six
7-prs., two maxims, and two rocket troughs, manned by
the Gold Coast Constabulary. For this twelve non-commissioned officers of the Royal Artillery were sent out
from England under Captain G. E. Benson. But since the
guns were never in action it is not necessary to follow their
doings. Suffice it to say that the expedition reached
Kumasi without opposition in January 1895, and that the
battery was left in garrison there, while the representatives
of the Royal Artillery returned to England.

King Prempeh had allowed Sir Francis Scott to enter
his capital unopposed, thinking no doubt that he would be
contented with the establishment of a Resident and would
then withdraw the troops. To the dismay of the Ashantis
the King, the Royal Family, and the principal Chiefs were
removed as political prisoners, their country was declared
a British Protectorate, a Resident was installed, and a
fort was built and garrisoned.

For four years the Government was carried on by a
native council under the Resident, but the Ashantis were
all the time nursing hopes of vengeance. With the
outbreak of the South African War their priests assured

s

CHAPTER
XXI.

The
Willcocks
Expedition.
1900.
March.

them that their chance had come. And when, in March 1900, the Governor of the Gold Coast arrived in Kumasi with only a small escort it seemed that fate was playing into their hands. The spark which fired the train was his demand for the surrender of the Golden Stool, the emblem of Ashanti sovereignty.[1]

There was a general rising which gradually assumed serious proportions, but the fort was strong and armed with four 7-prs. as well as maxims, and it was not until

May that the Governor called for help. His call was responded to from all sides, though the various detachments which got in to strengthen the garrison only did so by hard fighting. Although the attacks on the fort could make no impression on the defence, the sorties of the garrison were equally fruitless, and the close investment cut off all communication with the outer world, while the reinforcements, which had increased the garrison to over seven hundred, brought the question of food prominently to the fore. Eventually it was decided to attempt to break through the rebel lines with the bulk of the force, leaving a hundred to hold the fort. Although weakened by reduced diet during a long and trying siege, disheartened by the repulse of their sorties, and encumbered by a thousand non-combatants, including women and children, the column forced its way to the coast.

We can now turn to the Relief Expedition. As soon as the seriousness of the situation at Kumasi was realized at home orders were sent to Colonel Willcocks, commanding the West African Frontier Force, to proceed to the Gold Coast and assume command of the forces for the relief of Kumasi. He landed at Cape Coast Castle in May, and lost no time in organizing an expedition. The difficulties were great. Alarm and despondency prevailed everywhere : the coast tribes were paralysed with terror ;

---

[1] The Golden Stool was found by chance in 1922. A new war was only averted by prompt proclamation of the Government's intention to hand it back to its rightful custodians.

and food and ammunition were still at sea.[1]   What troops
were available were spread along a line of communication
nearly a hundred and fifty miles long, the first half of
which, though it boasted a road, was unbridged, while the
remainder consisted only of a forest path.   Worst of all
the advanced detachments were being forced back with
heavy loss.[2]

On the 22nd of June the West African Regiment
arrived from Sierra Leone, and were pushed up the line
to reinforce the company of the Northern Nigeria Regiment which had been skilfully holding its own at Bekwai,
only twenty miles from Kumasi.   The enemy's position
at Dompoassi was carried with the bayonet—the first
encouragement—only to be followed, unfortunately, by
a severe reverse shortly afterwards.   At the same time
came the news of the Governor having broken out of
Kumasi, which cleared the situation, though it left the
relief of the garrison in the fort still to be undertaken.
The 15th was the last day to which they could hold out,
and it was decided to abandon the posts on the lines of
communication, and to concentrate the whole force at
Bekwai.

From Bekwai, General Willcocks advanced on the 13th
July with a thousand men, two 75$^{m/m}$ guns of the
W.A.F.F., and four 7-prs. manned by a detachment of the
West India Regiment.   By taking advantage of a little-
known path, and spreading false information, the little
force arrived within three or four miles of their destination
before meeting any opposition.   Orders had been given
for all guns and maxims to be at once concentrated on the
stockades, and the two 75$^{m/m}$, three 7-prs. and three
maxims were soon in action at a range of well under a
hundred yards.   The story may well be left to General
Willcocks :—

---

[1] All supplies for Europeans, and the rice for the natives, had to
be obtained from overseas.

[2] At Dompoassi, on the 6th June, Lieut. W. E. Edwards, R.A.,
though severely wounded, continued to work his 7-pr. after the whole
gun detachment had been killed or wounded.

CHAPTER
XXI.

The
Willcocks
Expedition.
1900.
13th July.

"My object was to extend all the troops of the advanced guard and main body under cover of our shell and maxim fire, and, as soon as this was attained, to order a general bayonet charge. . . . I could see that the 'millimetres' were smashing up the great logs in the stockades, but even at that short distance could not breach them.[1] Our infantry was now practically all up, and covered a front of six or seven hundred yards, so I told Phillips[2] to keep up a heavy fire from his 12-pounders for a minute, after which I would sound the 'Cease Fire' and 'Charge'. . . . Like a wave up rose the ranks, and there was no doubt now. No Ashanti ever born could stand before that line of steel".

All was over, and, as soon as the pursuers could be collected, the advance was resumed until the sight of the fort with its flag still flying gladdened all hearts. An arranged signal of star shell from the guns took the news to Bekwai, to be passed on to the coast, and thence cabled home.

After establishing a new garrison in the fort, General Willcocks took the relief force back to Bekwai. There it was brought up to a strength of nearly three thousand native troops, with a hundred and fifty British officers, by the arrival of Sikhs and Central Africans from Somaliland and special service officers from England, and General Willcocks was able to set about its reorganization as the "Ashanti Field Force" for the systematic subjugation of the whole country. In this force Lieut.-Colonel A. F. Montanaro was given the command of the artillery, but was frequently employed in command of columns—as was Lieut.-Colonel H. E. J. Brake, R.A., who had brought the Central African Regiment round by the Cape.

The work done during the next three months was stupendous—columns traversing the country in every

---

[1] A mistake. Colonel Montanaro writes:—"The largest stockade was 150 yards long, 5 feet high, and 5 feet thick. The scene behind it was simply awful, and a notable feature was the destructive effect of the 75″ guns, the shells of which had penetrated the stockade and burst beyond, mangling many bodies.

[2] Lieutenant E. H. Phillips, R.A., in command of the 75″ guns.

direction, destroying stockades, breaking up war camps, burning rebel villages, cutting down sacrificial trees, and generally harrying all reactionary elements. They had to contend with stubborn resistance, so that there was incessant fighting. But the greatest obstacle was the interminable rain, flooding the whole country, so that half the marches consisted of floundering through quagmires, often followed by a bivouac in the mud. It was months before the last remnants of the rebels were brought to book.

CHAPTER
XXI.

The
Willcocks
Expedition.
1900.

COMMENTS.

With the exception of the two Whitworth guns got out by the Royal Niger Company for the expedition against Bida, the artillery *matériel* in use in West Africa prior to the formation of the West African Frontier Force, consisted of rockets and 7-pr. guns of 150lbs. Rockets were in very general use in colonial wars up till nearly the end of the XIXth century—although their propensity to return, like a boomerang, to the firer was occasionally embarrassing—and instruction in their use was included in the "Colonial Courses" at the School of Gunnery. The 7-pr. of 150lb. introduced for the Abyssinian Expedition of 1867-8, though deficient in both power and range, had the great advantage for jungle fighting of portability and simplicity. The 75$^{m/m}$, brought out by the W.A.F.F., was the same gun as that in use by the Egyptian Artillery at Omdurman, and by the C.I.V. Battery in South Africa, but shortened so as to reduce its weight for carriage. The consequent reduction of muzzle-velocity was immaterial for jungle fighting.

There is no doubt that the 75$^{m/m}$ was admirably adapted for the species of warfare for which it was required in West Africa, where the heavy double shell was invaluable, and its low muzzle velocity and comparatively short range were no disadvantage. It was immensely popular in the

West African Frontier Force under the quaint nickname
of "the millimetre gun."

Armament   In West Africa carriers were the only means of transport, and for the gun and carriage the 7-pr. required 11
and the $75^{m/m}$ 16, with the same number as relief in each
case, so that the actual requirements per gun were 22
and 32 respectively. But "reliefs" were a luxury rarely
enjoyed. On an emergency one carrier has been known
to take a 7-pr. gun across a log bridge on his head.

The ammunition was packed in "bearers"—each a
man's load—containing six complete rounds for the 7-pr.
and three for the $75^{m/m}$ (or two of double shell). Ammunition had, therefore, to be carefully husbanded, but
as it was rare to get a range of more than a hundred yards
little had to be expended in "ranging."

Employ-   The tactics generally, and therefore the employment
ment.   of the artillery, were dominated by the fact that columns,
(which must necessarily include large numbers of carriers)
were almost always strung out in single file along paths
on which it was often impossible to see twenty yards
ahead. In such veritable tunnels the opportunity for an
ambush must be ever-present. It might be sniping from
men concealed in the bush or in the tree-tops, or a more
serious attack from a stockade parallel to the path and
only a few feet from it, but so skilfully concealed that its
presence would never be suspected until a blaze of fire was
poured into the column. Most serious of all was the
definite defiance of a stockade directly blocking the path,
generally placed just round a sharp bend.

A typical force would consist of a fighting column of
(say) 800 men with three guns and four maxims, and a
supply column of 400 men with one gun and two maxims.
The fighting column would have a $75^{m/m}$ with the advanced
guard, and another with the main body, and a 7-pr. with
the rear-guard. The supply column would have its 7-pr.
with the rear-guard, for, as on the Indian Frontier, a
retirement always brought the natives swarming round the

tail of a column, and a force advancing through tribal country was apt to be looked upon as retiring by the natives who had been passed by. So rear-guards had always to be strong. The most important point, however, was that there should be a $75^{m/m}$ with the advanced guard to be brought into action at once should the road be blocked, for these were the only guns that were effective against solidly built stockades. Their 18-lb. double shell penetrated into the centre before bursting with immense destructive effect, and three to six rounds were all that were required to make a practicable breach.

Stockades parallel to the path could generally be enfiladed and for this the 7-prs. with the rear-guard were well-placed and effective.

MEDALS & CLASPS.

The following medals and clasps were awarded for the three expeditions :—

The Wolseley Expedition.—A medal, which subsequently became a "General Service" medal for East and West Africa, with a clasp for "Commassie."

The Scott Expedition.—A cross, without clasps.

The Willcocks Expedition.—A medal with clasp "Kumassi."

CHAPTER XXII.

THE KAFIR WARS.

Introduction—The Wars in Cape Colony—Morosi's Mountain—The
Gun War—Sekukuni—The Matabele—The Mashonas.

COMMENTS.

Organization—Armament—Employment—Medals & Clasps.

MAP 13.

CHAPTER
XXII.

Introduc-
tion

THE British had only been in possession of Dutch South
Africa for five years when they became involved in
the first of a long series of Kafir Wars. The frequency
of casual allusions to these wars, and even to the names
of the chiefs, in the light literature of the mid-Victorian
period shows clearly the impression they made upon
public opinion at home.

Although regular troops took little part in those which
come within the period covered by this volume, guns were
in much request, and the operations afford many examples
of the value of artillery, as well as some illustrations of the
difficulties and dangers incidental to the presence of a gun
or two with small forces.

The Wars
in Cape
Colony.

1877.

In Cape Colony the wars were in the main, due to the
rival efforts of the Kafirs from the east and the Europeans
from the west to expand along the coastal regions in the
occupation of the Hottentots. The eighth and last, the
only one which falls within the period covered by this
volume, followed the usual trouble between the tribes
settled within the new boundary and these still independent. This came to a head in 1877, and the "Frontier
Armed and Mounted Police"[1] had to go to the rescue of

---

[1] The Frontier Armed and Mounted Police became in 1878 the
Cape Mounted Rifles, thus reviving the name of an old corps which
had formed part of the British Army until disbanded 1870. "Fingo"
was the name given to fugitive Kafirs from Natal who had settled
under British protection.

the Fingos. About 1,500 of the latter, with 80 police, were attacked by several thousand Kafirs. The police had a 7-pr. gun, by which they set great store, but after the tenth round its trail broke. The officer in command of the police ordered it to be withdrawn, telling off an escort of 25—out of his total strength of 80—to ensure its safety from the enemy. But he failed to realize the effect its retreat would have upon the natives upon his side. At the sight of the gun in movement to the rear the Fingos broke and fled. The police fell back to their headquarters, where they were fiercely attacked, and the carriage of their other gun collapsed. The attack was beaten off, but not before it had caused considerable alarm for the safety of King William's Town.

The Imperial troops in Cape Colony consisted only of a couple of battalions and "a few artillerymen and sappers," but a naval brigade was landed, volunteers collected, and the rebellious tribesmen driven out of British territory. As soon as the danger appeared over, however, the volunteers began to drift away home—as such forces have a way of doing—and in December the Kafirs were on the war-path again. In default of regular artillerymen the bandsmen of the 88th under Lieut. Kell formed a 7-pr. battery, and by the middle of January 1878 it was possible to take the offensive. The little army advanced against the Kafirs with the guns of the 88th and the rockets of the naval brigade in the centre of the line, and the enemy were soon forced to abandon their position. But it was difficult to bring them to book in the rugged forest country, and the war dragged on until June. By that time the death of Sandile, the leading spirit among the Kafirs, and the arrival of regular troops from England put an end to hostilities.

Among the reinforcements was N/5, a battery about which there will be much to say in the next chapters. It now marched to Pieter Maritzburg with a column under Sir Evelyn Wood.

CHAPTER
XXII.

Morosi's
Mountain.
1879.

Ten years later there was trouble again, owing to the action of a Basuto clan, that, under a chief Morosi, had fortified an isolated mountain in Cape Colony and there defied the authorities. In 1879 a force of Cape Mounted Rifles and Cape Yeomanry was collected, and two thousand Basutos were called out to assist in bringing to order this recalcitrant offshoot of their own race. The artillery consisted of two 7-prs. belonging to the Cape Mounted Rifles.

March.

The first attack upon the rebel stronghold was made in March, but the little shells of the 7-prs. made no impression upon the defences, and the attack was brought to a standstill. In this emergency Serjeant Scott of the artillery troop, with some of his men, volunteered to carry up the shells and use them as hand-grenades to clear the enemy out of the *schanzen*, the fire from which was holding up the attack. Unfortunately the second shell thrown by Serjeant Scott burst prematurely and shattered his arm, and the attack was not pressed further.[1]

Next day reinforcements arrived, and these included Captain G. E. Giles[2] to command the artillery, bringing with him a 12-pr. R.B.L. (Armstrong) gun.[3] But in a second attack on the mountain both it and a 7-pr., which were covering the advance of the main body, broke down. For a second time the attempt failed.

For the next effort a $5\frac{1}{2}''$ mortar was sent from the Castle at Cape Town, and there was a three days' preliminary bombardment. The mortar broke down, but not until it had fired 367 rounds, and caused many casualties. Its continuous fire, day and night, searching out the hitherto untouched recesses of their stronghold, seems to have broken the spirit of the defenders, and the assault was completely successful, Morosi himself being killed.

---

[1] Serjeant Scott received the Victoria Cross.

[2] Late R.A. He had handed over his section of N/5 to Lieut. Slade on being posted to the Cape Mounted Rifles in April 1879.

[3] Purchased from the Orange Free State for £1,241.

The danger of the possession of modern rifles by the native tribes had become apparent, and the Cape Government decided to disarm the Basutos, who had adhered to their policy of keeping themselves supplied with the most modern fire-arms. Fully occupied elsewhere in South Africa, the Imperial Government were unable to assist, so that, here again, the force employed was entirely Colonial. The artillery troop of the Cape Mounted Rifles, under Captain Giles, furnished the artillery, its armament consisting of three 7-prs. and two $5\frac{1}{2}''$ mortars, subsequently increased by three 6·3" R.M.L. howitzers. Between October 1880 and April 1881 there were several engagements in which the mounted charges of the Basutos proved somewhat disconcerting to some of the colonial corps who had been accustomed to easy victories over the coast tribes, and the 7-prs. had to meet more than one attack with case shot. Eventually peace was patched up, but the Basutos kept their rifles.

CHAPTER XXII.

The Gun War. 1880-81.

Another Basuto tribe, further north, was giving trouble to the Transvaal in the 70's. Under a Chief Sekukuni they ravaged the Lydenburg district, and the efforts of the Boers to reduce them to order failed somewhat ingloriously. With the annexation of the Transvaal in 1877 the British Government found themselves saddled with the dispute, and so insolent was Sekukuni's attitude that action was imperative. Unfortunately the regular troops in the Transvaal consisted only of half a battalion and a detachment of artillery, and had, therefore to be supplemented with volunteers and native levies. The Royal Artillery, under Lieut. Nicholson, manned a section of 7-prs., and Lieut. Slade took over a couple of 6-pr. Armstrongs with infantry detachments.

Sekukuni. 1877.

During the autumn of 1878 there was a good deal of fighting in which the guns proved their value in supporting infantry attacks and clearing the enemy out of villages— freely acknowledged in orders. But the operations failed in effecting their purpose before the approach of the

1878.

sickly season, and then the outbreak of the Zulu War led to their being called off. The two gunner subalterns joined Sir Evelyn Wood's column at Utrect on the Transvaal border, Nicholson taking his section with him, Slade handing over his 6-prs. to the 80th Regiment.

A year later, when the Zulus had been disposed of, it was Sekukuni's turn. Operations were resumed under Sir Garnet Wolseley, and his stronghold in the Olifants

Mountains was stormed in November 1879. Sekukuni was taken prisoner and his tribe gave no further trouble.

Among the tribes which broke away from Chaka's tyranny in the early years of the XIXth century were the Matabele who settled in the country north of the Limpopo. From their chief Lobengula, Cecil Rhodes obtained the concession in which the British South Africa Company was formed in 1890. But differences

arose between the natives and the settlers, and in 1893 the Matabele rose in arms against the white men. Fortunately, Dr. Jameson, the Company's representative in Rhodesia, had seen what was coming, and taken measures accordingly. Within a month the Company's troops had occupied Buluwayo, and driven Lobengula to his death as a fugitive.

The Matabele had never been really conquered, and the memory of the war rankled. Then there came the

terrible scourge of rinderpest that destroyed their cattle, and their new rulers denied them the traditional resource of raiding the Mashonas. Resentment grew apace, and when the Jameson raid called away the greater part of the armed police, they saw their opportunity, and rose in revolt.

Many white people—farmers and miners with their wives and children—were brutally murdered, and there was consternation throughout South Africa. Lieut. Colonel Plumer formed at Mafeking (then rail-head)

a "Matabele Relief Force" of 800 mounted men, in which the Maxim Detachment of fifty was commanded by Lieut.

G. D. Wheeler, R.A. An Imperial contingent for the Relief Force was also assembling at Mafeking, and this included a section of No. 10 Mountain Battery from Natal under Lieut. R. H. F. McCulloch. Owing to the scarcity of supplies in Matabeleland, this section was, for the time being, the only portion of the Imperial contingent called upon, and it was taken over bodily by the British South Africa Company, who retained—and paid at colonial rates—one officer and eight non-commissioned officers and men. The rest of the *personnel* were provided from the local forces, with a second officer in Lieut. N. W. Fraser of the West Riding Regiment.

For the journey of nearly six hundred miles to Buluwayo, McCulloch took the guns and gunners by the so-called "coach", and Fraser marched the mules. On the 11th June the section joined Plumer and saw plenty of fighting during the remainder of that month and July, ending up with the capture of the Matabele strongholds in the Matoppo hills in August. The biggest fight took place on the 5th, when five Impis were attacked by between seven and eight hundred men and completely routed. On this occasion the artillery of the force, consisting of the two mountain guns, a 1-pr. Hotchkiss, a maxim, and a rocket trough, were detached with an escort of a hundred men to take up a position on a ridge from which they would be able to support the attack. Sufficient account had not, however, been taken of the rapidity with which the Matabele could move. Seeing their opportunity in this separation they hastened from all sides to take advantage of it, and, creeping up gullies, under cover of the bush, made a sudden and determined rush upon the guns. The carriers of the Hotchkiss dropped it and bolted, but McCulloch took one of the 2·5″ to the right, Fraser took the other to the left, the guns opened fire with case, the maxim got into action and the situation was saved. In his final despatch Sir F. Carrington mentioned "Lt. McCulloch, commanding Royal Artillery, and Lt. Fraser, W.R. Regiment, attached Royal Artillery, for great

CHAPTER
XXII.

The
Matabele.
1897.

The
Mashonas.
June.

coolness and steadiness when the enemy were in force within 50 yards in the attack on Sikombo on the 5th August. Both remained with their guns directing their fire after being wounded."

Before this, however, the revolt had spread to Mashonaland. On the 16th June the peaceful Mashonas rose without warning, and murdered the whites in all solitary stations. The Imperial contingent which we left at Mafeking was ordered up, and two companies of mounted infantry, which had been waiting at Cape Town, took ship to Beira. They had been brought from Aldershot by Lieut.-Colonel E. A. H. Alderson, and he now took on also a couple of 7-pr. guns and half a dozen gunners from the Cape garrison under Lieut. S. C. C. Townsend. The intention was that they should conduct the guns to Salisbury, and there act as instructors to men of the local forces. At Beira, however, they found a transport taking troops to Mauritius, and among them details for the 24th, 25th, and 26th Companies, Western Division, R.A. With every prospect of having to fight before reaching Salisbury, Colonel Alderson decided to take the opportunity of making up the detachments of his two guns, and so eight gunners were taken off, bringing the strength of the little party up to 1 officer, 2 serjeants, and 12 gunners.

We are not concerned with the difficulties and delays of the journey through Portuguese Territory—50 miles by river and 80 by rail. Suffice it to say that Umtali was reached at last, and Colonel Alderson assumed command of the "Mashonaland Field Force". Leaving the "Umtali Artillery" to man a 7-pr. which protected the laager, Alderson pushed on with the main body. There were several encounters with the Mashonas on the 200 miles march to Salisbury, but the want of transport was the great difficulty. The Royal Artillery had to get their guns along with teams of two mules instead of four, and to be content with donkey-wagons for their baggage. Their escort is worth a note. At Umtali a dozen business men, anxious to return to Salisbury, petitioned to be allowed to accompany the column with their own ox-wagon. Per- mission was given, provided that they would act as permanent escort to the artillery, and this duty they accepted and performed punctiliously.

On arrival the "Salisbury Artillery" was merged in the artillery troop of the Rhodesia Horse, one 7-pr. and a couple of maxims, under Captain W. F. G. Moberly, late R.A. With the Royal Artillery, this provided the guns required for the larger patrols which could now be sent out, the usual allotment being one to a patrol, and never more than two. In the engagements that ensued the guns came in for a good deal of attention from the enemy, and were repeatedly hit, but the gunners are reported to have worked them coolly and effectively, though the Mashonas' habit of going to ground in the caves with which their kopjes were honeycombed must have been somewhat disheartening to artillerymen.

COMMENTS.

In view of the small number of guns ever employed there are few matters connected with their organization to require attention. Attention may be drawn, however, to the free use of infantrymen as gunners. The 24th, the 80th, and the 88th Regiments all appear to have found detachments at different times, usually under artillery officers, and sometimes with a nucleus of artillerymen. The Artillery Troop of the Cape Mounted Rifles,[1] which provided the artillery in the majority of the campaigns included in this chapter, was raised in 1879 by Lieut. J. C. Robinson, late R.A., and was commanded in the Basuto Wars by Captain G. E. Giles also of the Regiment until posted to the C.M.R. in April 1874. The

[1] In 1881-4 it was known as the Cape Field Artillery; in 1884—1913 as the C.M.R. Artillery; and after 1913 as the 1st Battery, South African Field Artillery.

CHAPTER
XXII.

Organiza-
tion.

popularity of the artillery arm among the less regular
forces is also worthy of note—the "Cape Town Volunteer
Artillery", the "Graham's Town Artillery", the "Umtali
Artillery" the "Salisbury Artillery".

Armament.   The 7-pr. guns were mounted on low narrow-track
carriages,[1] which had the disadvantage of turning over
very easily on rough ground, but had the great advantage,
for use in the Kafir wars, of being able to negotiate narrow
bush paths which would have been quite impassable by any
vehicles with an ordinary track. They were drawn by
three mules, harnessed tandem fashion, and could also be
carried by the mules in pack when the ground was
altogether too bad for draught. They were occasionally
packed on oxen, but this was an emergency measure, not
provided for in the section establishment.

The 6·3″ howitzers obtained for the Gun War of 1880
we shall meet again twenty years later, disinterred from
store in King William's Town to render notable service in
the defence of Ladysmith.

Employ-
ment.

From the point of view of artillery history, the incident
of most general interest is the effect produced by the
retirement of the 7-pr. in the 8th Kafir War, since it
presents an almost exact parallel to the withdrawal of the
smooth-bore battery at Maiwand three years later. The
Fingos' reply, when reproached for their behaviour, puts
the case clearly enough—'When we saw the gun move
we thought it time for us to leave".

The other lesson is the danger of depending upon guns
of insufficient power when the breaching of defensive works
has to be undertaken. It is almost pathetic to note the
efforts to obtain heavier pieces after the failure of the
7-prs. The 12-pr. bought from the Boers, the mortar—
it had been cast in 1802—from the Castle at Cape Town.

---

[1] A brigade of field artillery was equipped with 2·5″ guns on
similar carriages as an experiment at home in the 90's.

We may contrast this with the elaborate trials made to see
whether the artillery to be taken to Omdurman would be
equal to the work required of it, and the similar action by
the artillery in New Zealand.

And there is the larger question of whether it is
generally advisable to include artillery in such forces as
those with which we have been dealing in this chapter.
In 1882 the Cape Government asked the opinion of
Major-General C. Gordon, then "Commandant-General"
of the Cape Forces. He replied that he considered the
7-prs. on wheeled carriages a great impediment. The
effect of their feeble shell against an enemy hidden in
rugged fastnesses and wooded kloofs was almost nugatory.
The guns always gave the enemy a point on which to concentrate their fire : they were always a cause of anxiety
to a commander : the ammunition train was a continual
embarrassment.[1] "I am altogether against the taking of
field artillery with movable aggressive columns. They
should be with the reserve, and for the defence of posts,
where they are invaluable".

This is a very different story from that of the infantry
who volunteered to carry the 7-prs. over the Shandur
Pass to Chitral in order that they might have their
support when the fighting began. "Circumstances alter
cases", and perhaps we may leave it at that. But there
can be no difference of opinion regarding the gallantry of
the gunners who used their shells as hand-grenades when
they failed in their more legitimate use. And here again
we have a parrallel in the Maori War.

MEDALS & CLASPS.

The South African General Service Medal of 1854 was
given with clasps for "1877", "1878", "1879", singly or
combined, only one clasp being given to each individual.

In 1900 the survivors of the "Gun War" received the

---

[1] At Morosi's Mountain the artillery column was a thousand yards
long. 1443 rounds were fired.

T

Cape of Good Hope General Service Medal with a clasp for
"Basutoland".

For the operations in Matabeleland and Mashonaland
in 1893 the issue of a medal by the British South Africa
Company was officially authorized. It has clasps for
"Rhodesia 1896" and "Mashonaland 1897".

List of Units.

(With Designation in 1914.)

N/5	R.A.	86, R.F.A.
10, M.B. do.		107, R.G.A.

CHAPTER XXIII.

THE ZULU WAR.

Introduction—Ekowe—Kambula Hill—Isandhlwana—The Second Phase
—Ulundi.

COMMENTS.

Organization and Armament—Medal & Clasp—List of Units.

MAP 13.

IT is time to turn to the real source of all the trouble in
South Africa described in the last chapter. Whenever and
wherever there was discontent or disorder among the
Kafirs, there in the background brooded the shadow of the
Zulus. It had been a terror to the natives in the days of
Chaka : under his grandson Cechwayo it became the dread
of the whites, for he had maintained the organization
and the discipline which was the origin of the nation's
strength under Chaka, and had revived its martial spirit
which had been allowed to decline during the reign
of his father. In 1878, after five years on the throne, he
had a well-drilled army of fifty thousand men, a standing
menace to Natal and the recently annexed Transvaal.
Sir Bartle Frere, High Commissioner for South Africa,
decided not to wait until a Zulu invasion brought fire and
sword to one or both, and called on Cechwayo to disarm.
In the event of his not complying with this ultimatum, the
Commander-in-Chief, Lord Chelmsford, was directed to
march on the Zulu capital.

The problem of invasion was a difficult one on account
of the length and shape of the frontier, and the strength
and mobility of the Zulu army. Eventually it was decided
to enter the enemy's country in three columns—the right
crossing the Tugela near its mouth, the left starting from
Utrecht on the Transvaal border, and the main body in
the centre crossing the Buffalo River at Rorke's Drift.

CHAPTER
XXIII.

Introduc-
tion.

1878.

The total strength of the three columns was rather more than seven thousand British troops, with a native contingent of nine thousand. The artillery was as follows :—

    C.R.A.—Lieut.-Colonel F. T. A. Law.

    N/5 Field Battery—Lieut.-Colonel A. Harness.

    11/7 Garrison do. —Major E. J. Tremlett.

    Section   —Lieut. F. Nicholson.

    Rocket Battery—Bt. Major F. B. Russell.

The guns were all 7-prs., some on "Kaffraria" carriages, some on the low narrow-track carriages described in the last chapter. There was also a naval brigade with two 12-pr. Armstrong guns and two rocket tubes.

The period allowed by the ultimatum expired on the 31st December, 1878, but ten days' grace were given. No reply having been received the three columns crossed the border on the 10th January, 1879. Like those in the invasion of Afghanistan six weeks before, they were quite independent, and it will therefore be most convenient to follow their fortunes separately.

The right column, under Colonel Pearson, had with it for artillery, the naval brigade and a section of 11/7 (Lieut. W. N. Lloyd). After driving off one Zulu attack on the way, the column had advanced as far as Ekowe when the news of the disaster to the centre column reached it. Pearson decided to stand fast with his infantry and artillery, sending back the mounted men. This left him with a garrison of between eleven and twelve hundred, and some three hundred non-combatants, and he set to work at once to build a redoubt round the Mission Station, clear the foreground, and make all preparations to meet an attack.

The guns were placed in blinded emplacements, and were well supplied with ammunition—150 rounds a gun. The only shortage was of case shot, and as at Kabul and in Burma the inventive genius of the Regiment was equal to the occasion. It was noticed that Morton's jam tins exactly fitted the bore, and the order went forth that as soon as empty they were to be deposited with the C.R.A. for conversion into case shot.[1]

CHAPTER XXIII.

1879.

Ekowe.

April.

The Zulus made only one attack—and that a half-hearted one—and the garrison made one sortie—in which one of the guns took part. The excellent arrangements made had secured the safety of the little post against attack, but the supplies had nearly run out by the time the scouts of the relieving force were sighted.

The left column under Colonel Evelyn Wood had for artillery 11/7 (less the section with the right column). In its place was the section brought by Lieut. Nicholson from Sekukuniland. There was also a rocket detachment under Lieut. A. J. Bigge.

Kambula Hill.

After crossing the Blood River the column advanced unopposed, but on hearing of the disaster to the centre column Colonel Wood at once fell back to a position on Kambula Hill, where he could cover both Newcastle and Utrecht, and so give confidence to both the Natal and Transvaal borders. There was heavy fighting on the Intombi River and at Inhlobane Mountain in March, and on the 29th of that month Kambula Hill was attacked in force.

January.

March.
29th Mar.

In the defence of that position Nicholson's two guns were in a redoubt, while 11/7 was in action outside. Lieut. Nicholson was mortally wounded early in the day, and his place was taken by Major H. Vaughan, Director of Transport. The Zulu attacks were pressed home with the utmost resolution, the guns pouring in shrapnel and case until they grew so hot that water had to be poured over them to allow of the breech being served.[2] Cechwayo, when a prisoner at Cape Town, stated that it was with the

See 275

---

[1] A specimen is to be seen in the R.A. Institution.

[2] With muzzle-loading guns one of the numbers had to place his thumb on the vent during the sponging out after each round, to prevent a draught which might cause any smouldering fragments of the previous cartridge to burst into flame.

greatest difficulty that his men could be persuaded to face the guns—one round of case, he said, killed ten head men of his own regiment, besides wounding others.

The centre column had for artillery N/5 and the Rocket Battery. After crossing the Buffalo River at Rorke's Drift, and clearing the Zulus out of a position on the Ingutu mountain, the column reached Isandhlwana on the 20th January and went into camp there. At dawn on the

22nd Lord Chelmsford with half the force moved off to support a reconnaissance, leaving in camp only one battalion of the 24th Regiment, one section of N/5, and the Rocket Battery. How this little force was attacked and destroyed a few hours later is well known, and requires no description here. As to how the gunners fared in the disaster we have fortunately the first hand account of the sole survivor, Lieut. H. T. Curling of N/5[1], in a letter to his mother written within a few days of the affair.

Helpmakaar, Natal,
February 2nd, 1879.

"Now things have quieted down again a little, I can tell you more about what has happened. I trust you had no false reports; I saw the first man who went into Pieter Maritzburg with the news, so I hope you may have no anxiety. On the morning of the fight the main body left at 3.30 a.m., a little before daylight, leaving me with two guns and about seventy men. About 7.30 a.m. we were turned out, as about one thousand Zulus were seen on some hills about two miles from the camp. We did not think anything of it, and I was congratulating myself on having an independent command. I had out, with my guns, only twenty men, the remainder, fifty in number, stayed in the camp. We remained formed up in front of the camp (it was about half a mile long) until 11 o'clock,

---

[1] The Rocket Battery was surprised and cut up early in the day when in action on a small knoll in front of the camp.

CHAPTER
XXIII.

1879.

Isandhl-
wana.

22nd Jan.

when the enemy, disappearing behind some hills on our left, we returned to camp. We none of us had the least idea that the Zulus contemplated attacking the camp, and having in the last war[1] often seen equally large bodies of the enemy, never dreamt that they would come on. Besides, we had about six hundred troops (regulars), two guns, about one hundred other white men, and at least one thousand armed natives. About twelve, as the men were getting their dinner, the alarm was again given, and we turned out at once. Major (Stuart) Smith[2] came back from the General's force at this time, and took command; this of course relieved me from all responsibility as to the movement of the guns. We, being mounted, moved off before the infantry, and took up a position to the left front of the camp, whence we were able to throw shells into a large mass of the enemy that remained almost stationary. The 24th Regiment came up and formed skirmishing order on both our flanks. The Zulus soon split up into a large mass of skirmishers that extended as far round the camp as we could see. One could form no idea of the numbers, but the hills were black with them. They advanced steadily in the face of the infantry and our guns, and I believe the whole of the natives who defended the rear of the camp soon bolted and left only one side of the camp defended. Very soon bullets began to whistle about our heads, and the men to fall. The Zulus still continued to advance and we began to fire case, but an order was given to retire after firing a round or two. At this time out of my small detachment one man had been killed, shot through the head, another wounded, shot through the side, and another through the wrist. Major Smith was also shot through the arm but was able to do his duty. Of course, no wounded man was attended to, there was no time or men to spare. When we got the order to retire we limbered up at once, but were hardly in time as the Zulus were on us at once, and one man was killed (stabbed) as

---

[1] The Kafir War of 1877-8 described in the last chapter.
[2] Major Stuart Smith was the captain of N/5.

CHAPTER
XXIII.

1879.

Isandhl-
wana.

22nd Jan.

he was mounting on a seat on the gun carriage. Most of the gunners were on foot, as there was not time to mount them on the guns. We trotted off to the camp, thinking to take up another position there, but found it in possession of the enemy, who were killing the men as they came out of their tents. We went right through them, and out the other side, losing nearly all our gunners in doing so, and one of the two sergeants. The road to Rorke's Drift that we hoped to retreat by was full of the enemy, so no way being open we followed a crowd of natives and camp followers, who were running down a ravine. The Zulus were all among them stabbing men as they ran. The ravine got steeper and steeper, and finally the guns stuck, and could get no further. In a moment the Zulus closed in and the drivers who now alone remained were pulled off their horses and killed. I did not see Major Smith at this moment but was with him a minute before. The guns could not be spiked, there was no time to think of anything, and we hoped to save the guns up to the last moment. As soon as the guns were taken, I galloped off and made off with the crowd. How any of us escaped I don't know; the Zulus were all among us, and I saw men falling all round. We rode for about 5 miles hotly pursued by the Zulus, when we came to a cliff overhanging a river. We had to climb down the face of this cliff and not more than half of those who started from the top got to the bottom. Many fell right down, among others Major Smith, and the Zulus caught us up here, and shot us as we climbed down. I got down safely, and came to the river, which was very deep and swift. Numbers were being swept away as they tried to cross, and others were shot from above. My horse fortunately swam straight across, although I had three or four men hanging on to his tail, stirrup-leathers, etc. After crossing the river we were in comparative safety, though many were killed afterwards, who were on foot and unable to keep up. It seems to me like a dream. I cannot realize it at all. The whole affair did not last an hour from beginning to end. Many got

away from the camp but were killed in the retreat. No
officer or men of the 24th Regiment could escape; they
were all on foot, and on the other side of the camp. I
saw two of them, who were not with their men, near the
river, but their bodies were found afterwards on our side
of the river. Of the fifty men we left in the camp eight
managed to escape on spare horses we had left in camp.
One sergeant (Edwards) only of my detachment got away.
Altogether we lost sixty-two men and twenty-four horses—
just half the battery. Those who have escaped have not
a rag left, as they came away in their shirt sleeves. We
always sleep at night in the fort, or laager (as it is called),
and in the open air. It is very unpleasant, as it rains
every night, and is very cold. We none of us have more
than one blanket each, so you can see we are having a
rough time. The first few days I was utterly done up,
but have pulled round all right now. What is going to
happen no one knows. We have made a strong intrench-
ment, and are pretty safe even should we be attacked.
The only thing we are afraid of is sickness. There are
fifty sick and wounded already, who are all jammed up at
night in the fort. The smell is terrible. Eight hundred
men cooped up in so small a place. Food fortunately is
plentiful, and we have at least three months' supply. All
spies taken are shot; we have disposed of three or four
already; Formerly, they were allowed anywhere, and our
disaster is to a great extent due to their accurate information of the General's movements. What excitement this
will cause in England and what indignation. The troops,
of course, were badly placed, and the arrangements for
defending the camp indifferent, but there should have been
enough troops; and the risk of leaving a small force to be
attacked by ten or fifteen times its number should not have
been allowed. As you have heard there were no wounded
—all the wounded were killed, and in a most horrible way.
I saw several wounded men during the retreat all crying
out for help, as they knew the terrible fate in store for
them. Smith-Dorrien, a young fellow in the 95th Regi-

CHAPTER
XXIII.

1879.

Isandhl-
wana.

22nd Jan.

ment, I saw dismount and try to help one. His horse was killed in a minute by an assegai, and he had to run for his life, only escaping by a miracle. . . . .

(Sd.) H. T. Curling.''

The Second Phase.

The first phase of the war had ended with a rebuff— the right column hemmed in, though within the enemy's country : the left column at a standstill with its back against the frontier ; and the main body forced back across the border with a loss of half its number. Before the second phase could commence the force in Natal must be strengthened, and the despatch of reinforcements had been put in hand as soon as the news of Isandhlwana reached England. The artillery on the way consisted of drafts for N/5 and 11/7 and the following batteries :—

From St. Helena — 8/7 Garrison Battery (with 7-prs.)

From Mauritius —10/7   do.   do. (with Gatlings)

From Home   —M/6 Field   do. (with 7-prs.)

    N/6   do.   do. (with 9-prs.)

    O/6   do.   do. (as Amm. Column)

These commenced to arrive early in March, and the first task to be undertaken was the relief of Ekowe. This, as we have seen, was successfully accomplished by Lord Chelmsford, one of his two brigades being under the command of Lieut.-Colonel Law, R.A.

For the main operation of destroying the Zulu power the British force was organised in two Divisions and a "Flying Column", to which the artillery was allotted as follows :—

C.R.A.—Colonel W. E. M. Reilly[1]

Brigade-Major—Captain J. R. Poole

---

[1] Colonel Reilly met with an accident and could not accompany he advance, so Lieut.-Colonel Brown took his place, and Lieut.-Colonel Harness commanded the artillery of the 2nd Division.

1st Division   Comdr.—Lieut.-Colonel F. T. A. Law

              Adjutant—Captain T. C. Cooke

                M/6—Major W. H. Sandham.

                8/7—Major H. L. Ellaby

                11/7—Lieut. W. N. Lloyd

                O/6—Major A. W. Duncan

                           (Ammunition Column)

2nd Division   Comdr.—Lieut.-Colonel J. T. B. Brown

              Adjutant—Captain J. Alleyne

                N/5—Lieut.-Colonel A. Harness

                N/6—Major F. S. le Grice

                O/6—Captain R. Alexander.

                           (Ammunition Column)

Flying Column   11/7—Major E. J. Tremlett

                10/7—Major J. F. Owen

                           (Gatling Battery)

CHAPTER XXIII.

1879.

The Second Phase.

During June the 2nd Division and Flying Column advanced steadily and methodically from the Blood River under the command of Lord Chelmsford, and on the 4th July completely defeated the main Zulu army at Ulundi.

June.

The formation adopted by the combined force was a large square with the guns disposed on all faces so as to be ready to meet the Zulu rush wherever it might come. The following sketch shows the distribution of the batteries, the only alteration made during the action being the move of a section of N/6 from the left rear to the left front.

Ulundi.

4th July.

The attack commenced at about 9 a.m. and the guns were at once brought into action—either just outside the infantry or in gaps left for them—opening fire as soon as the mounted troops cleared their front. The range was rather over 2,000 yards, the ground was clear of bush, and the fire at once took effect, but it could not prevent the great circle of Zulus drawing in to musketry range, and in some places almost closing with the defence. At the right rear corner, where the ground offered some

cover, the dead were counted in groups at less than thirty
yards from the muzzles of the section of N/6.

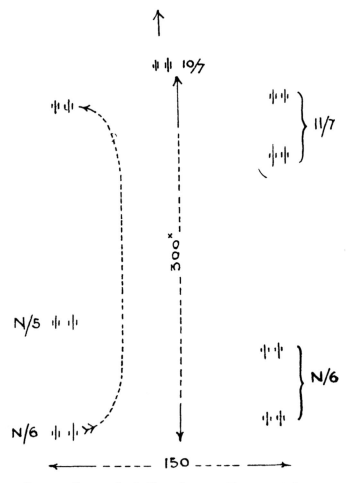

It was all over in half an hour. The attack began to
falter, and the mounted troops were launched in pursuit—
then there were signs of a rally and a section of N/6 were
called for, and soon dispersed the gathering. The
military power of the Zulus, which had been the domina-

ting force in South Africa for nearly half a century was shattered.

The action was so short that the expenditure of ammunition was small, though individual sections expended all their case, and were obliged to fall back upon reversed shrapnel.

In conclusion a word may be added regarding the sad death of the Prince Imperial. We are not concerned with that unfortunate incident, but it is satisfactory to know that the Royal Artillery were able to accord due honour to his remains. When his body was brought into camp it was carried to the gun-carriage by six of his brother-officers, and the armed party which preceded the carriage was also furnished by his Regiment.

COMMENTS.

The little 7-pr. guns which we have seen in such general use in the Kafir Wars found little favour with the Royal Artillery in Zululand. The country and the enemy were of a different class, and the gunners had a wider experience. They realized that, owing to the low muzzle velocity of the guns, their shapnel fire had little effect beyond 800 yards—all very well for bush-fighting, but not good enough with Zulus in the open. Then the small bursting charges of the common shell rendered their destructive power insignificant—it was said that they would not even set a Zulu hut on fire. Double shell could only be carried in small quantities, and their range was small. In Zululand most of the guns were on "Kaffraria" carriages instead of the low, narrow-track, carriages described in the last chapter. These were practically the same as the service 9-pr. field artillery carriages, though slightly lightened, and the general opinion of the artillery officers was that, in a country so easy for wheeled traffic, they might just as well have had 9-prs., whose fire would have been far more effective.

Organization
and
Armament.

CHAPTER
XXIII.

Organiza-
tion
and
Armament.

As regards ammunition supply, the 7-prs. had no ammuniton wagons, and the one 9-pr. battery had to leave theirs behind during the march to Ulundi owing to the difficulty of getting them across the drifts. In this respect, therefore, as well as in the value of heavier metal, the experience of the Zulu War coincided with that of the Indian Frontier, the Afghan War, and the Egyptian Expedition of 1882. The usual arrangement was to have one scotch cart and one mule wagon per section, with a further reserve in an ox wagon with the baggage.

Ammunition columns were provided by the simple, but unsound, expedient of converting a field battery.

All batteries carried rockets—a trough per section (usually in the scotch cart) with a dozen rockets, and fifty more in the ox-wagon. They do not appear to have proved as alarming as was anticipated—the Zulus indeed are reported to have shown the utmost contempt for them.

The "Rocket Battery" under Major Russell carried 9-pr. rockets with three troughs on mules. Except for Major Russell and one bombardier, its *personnel* were all drawn from the 24th Regiment. The Navy had enormous 24-pr. rockets, fired from tubes instead of troughs, but these are said to have caused as much anxiety to friend as foe.

The "Gatling Battery" was manned by 10/7 from Mauritius. The guns were mounted on miniature field carriages with limbers, and were drawn by "Cape tats" (ponies). The ammunition was carried in mule wagons. At Ulundi the Gatlings jammed in the same way that they were to do, a few months later, under Roberts at Charasia. The Royal Artillery were certainly not fortunate in their first experiences with machine guns.

The English horses of N/6 kept their condition wonderfully throughout. From landing at Durban in March until return there on the conclusion of hostilities, the battery lost only eight, of which two were killed in action. N/5 had colonial horses, which were useful and handy for the light 7-pr. equipment, and had the further advantage that they could be driven to water in a mob.

It is interesting to note that in Zululand an order was issued that Nos. 1 should ride five yards in front of their teams to look out for ant-bear holes. It was to be many years before this became the universal rule.

MEDAL & CLASP.

A medal, similar to that for the Kafir Wars, was given, with a clasp for "1879".

LIST OF UNITS.

(With Designation in 1914.)

N/6, R.A.	... ...	19, R.F.A.
M/6, do.	... ...	76, do.
N/5, do.	... ...	86, do.
O/6, do.	... ...	89, do.
10/7, do.	... ...	93, R.G.A.
11/7, do.	... ...	94, do.
8/7, do.	... ...	95, do.

CHAPTER XXIV.

The First Boer War.

Introduction — Laing's   Nek — Ingogo — Majuba — Pretoria — Potchefstroom.

Comments.

Map 13.

CHAPTER
XXIV.

Introduc-
tion.
1877.

In the 70's affairs in the Transvaal had got into a parlous state. Recovery seemed to be hopeless, and the British Government came to the conclusion that the best solution would be annexation. In April 1877 this was formally proclaimed at Pretoria, with the goodwill of a large proportion of the inhabitants. But the loss of their independence was bitterly resented by a section of the Boers, and the destruction of Cechwayo's army in 1879 removed their fear of attack by the Zulus. Unsympathetic administration fanned the flame, there were mass meetings, and in December 1880 the "South African Republic" was proclaimed.

1880.
December.

The British troops in the Transvaal at that time consisted only of N/5[1] (less one section in Natal) and a couple of battalions. N/5 and half of one battalion were at Pretoria, half another at Lydenburg, the remainder split up into small detachments scattered about the country. At the first mutterings of the storm in November Major C. Thornhill of N/5 took a section of that battery, a few mounted infantrymen, and a company of infantry from Pretoria to Potchefstroom.

November.

The nearest troops were at Newcastle in Natal, but they only consisted of a couple of companies and the third

---

[1] N/5 had taken over the 9-pr. equipment and English horses of N/6 after the Zulu War. In 1914 it had become 86, R.F.A.

section of N/5. Captain Greer, R.A., who was in command, at once set about increasing the artillery. Two
more 9-prs. were drawn out of store at Durban, horses were
bought locally, and non-commissioned officers and men
were sent from 10/7 at Cape Town. One of the wagon
teams of N/5 was transferred, and (with its drivers to
instruct the garrison artillerymen) the section was soon
able to turn out with its two guns horsed, ox-teams in the
wagons. This was not all. Two of the 7-prs. on narrow-
track carriages, about which so much has been said in the
account of the Kafir Wars, were also found in store.
Drawn by a couple of mules in tandem, and manned by
volunteers from the 60th, they provided a third section.

Further reinforcements were hurried up, and Sir
George Colley, who was now Governor and Commander-
in-Chief in South-East Africa, decided to march with these
to the relief of the garrisons shut up in the Transvaal.
His total strength was about twelve hundred, with, for
artillery, under Captain C. Greer :—

        N/5—Section — Captain A. M. Vibart
        10/7— do. — Lieut. C. S. B. Parsons
        Rifles— do. — Lieut. W. A. Young, R.S.F.

The Boers had assembled in strength near the Transvaal border, where the main road to Pretoria crosses the
Drakensberg at Laing's Nek, and here Colley attacked
them on the 28th January 1881.

N/5 on the left opened fire on the Nek, but the new
Watkin range-finder[1] gave the range as 2,350 and the
shell all disappeared over the ridge. Eventually they got
down to it at 1,900, and after half an hour the advance
was ordered. The sections manned by 10/7 and the 60th
supported the attack, and their fire compelled the burghers
to quit their position on the forward crest, but only to fall

---

[1] This instrument had only just been introduced and was imper-
fectly understood. The experience of Afghanistan and Egypt (1882)
may be compared with that related above.

back to the rear edge of the plateau. . Here their field of
fire was limited but they were secure from the dreaded
shrapnel. Meanwhile the mounted troops on the right,
told off to take a hill from which the Boers would be able
to enfilade the main attack, had failed to do so; and the
infantry under heavy fire from front and flank, were also
beaten back in spite of the efforts of the staff[1] and regimental officers. The guns covered their retreat, at ranges
from 2,200 to 1,200, but were unable to bring any fire to
bear on the Boers enfilading the infantry owing to the
failure of the mounted troops.

Colley withdrew to his camp at Mount Prospect
without interference, but a few days later the Boers
appeared between there and Newcastle, and on the 8th
February he took a small force back along the road to
where it crossed the Ingogo, only two or three miles from
camp. The 7-pr. section was left on the hither side of
the river, the section of N/5 crossing with the rest of the
force to the high ground a couple of miles beyond. There
the enemy were met, and the guns galloped into action
and opened with shrapnel at 1,200. But the mounted
troops were driven in, and the Boers were soon creeping
up the slopes of the plateau on which the guns were in
action in the open. It was not long before Greer was
killed by a bullet through the head while superintending
the boring of fuzes at the trail of one of the guns.
Soon they were firing case, and, when that was exhausted,
shrapnel reversed. Nearly all the artillerymen were
struck down by bullet or buck-shot, but the riflemen lent
a hand, and the Boers were never able to rush the guns.

When the General gave the order to retire there were
not sufficient gunners left to limber up, but the drivers
and the riflemen manned the wheels, and the guns were
withdrawn to a position further back, from which they

---

[1] Among the staff officers killed in the attempt to rally the infantry
was Major J. R. Poole. He had just received a brevet majority for
his services as Brigade-Major, R.A. in the Zulu War.

were able to bring their fire to bear to the right rear upon the Boers coming down from Laing's Nek into the Ingogo valley. There the fight went on until darkness put an end to it, and when Parsons was hit Serjeant-Major Toole took command and brought the guns out of action. To do so, all the riding horses had to be taken to make up the gun teams, and the wagons had to be left, after all the ammunition which could not be carried had been rendered useless by burning the cartridges and burying the fuzes.

The casualties in the section were 16 out of 27, with 14 horses, and next day the General paid a glowing tribute to the conduct of the officers and men of the artillery and the 60th for the way they had served the guns.

Reinforcements were on the way under Sir Evelyn Wood, and these included C/1 from home and F/3 from India, but they never got further than Newcastle. For on the 26th—27th February Colley had made his throw and lost. In the tragedy of Majuba the Royal Artillery had no share, although N/5 accompanied the force to the foot of the hill. Then it was seen that to get guns to the top would be quite impossible, and they were sent back to camp. They were, however, able to check the attempts of the Boers to cut off the retreat of the survivors on the morning of the 27th.

There is no more to be said of the war in the field. Before the end of March peace had been made, but the fate of the garrisons shut up in the Transvaal has yet to be told.

After the departure of Major Thornhill's party to Potchefstroom there were left to garrison Pretoria about a thousand men (Regulars and Volunteers) with four guns —the remaining section of N/5 and two 7-prs. manned by Volunteers. The town was wisely abandoned and a military camp formed outside, whence there were sallies on several occasions in which one or two guns of

N/5 always took part. But no great success can be claimed for these efforts (save on one occasion when the Boers in position at Pienaars River were forced to surrender) the task of the guns being usually to cover the withdrawal to camp. The camp was never attacked, and as soon as peace was declared the town was re-occupied. The last troops to leave the Transvaal were the section of N/5 from Pretoria.

The little force from Pretoria under Major Thornhill[1] reached Potchefstroom on the 18th November, having covered rather more than a hundred miles in rather less than a hundred hours. On the 20th another company arrived, bringing the strength of the garrison up to two hundred men, all told. The straggling town was quite indefensible, but the court-house and prison were put into a state of defence, and a square redoubt was constructed in the flat open plain. The guns were in gun-pits thrown out at the corner nearest the town so as to allow of their supporting the defence of the court-house and prison, as well as flanking the two faces of the redoubt.

On the 15th December the Boers arrived in force, and as soon as the buildings in the town had fallen into their hands they turned their attention to the redoubt. The cemetery, with a loopholed wall only four hundred yards from the fort, was occupied, and from this they kept up a dropping fire day and night. The length of the enemy's line made it impossible for the guns to keep this fire down with the amount of ammunition available, but whenever the Boers massed they were dispersed. They were determined to get the guns, however, and on the 1st of January 1881, their numbers having been made up to upwards of two thousand, a general attack was launched. On this occasion they brought a 3-pr. smooth-bore into action, supported by heavy musketry fire along the whole

---

[1] Thornhill was recalled before the investment to take up duty as C.R.A. South Africa, leaving Lieut. H. M. L. Rundle in command of the artillery.

line, but under the fire of the 9-prs. the effort came to nothing. The gun was a nuisance to the garrison for it was amply supplied with ammunition, and as it was loaded under cover and only run out to fire, it could never be located. One shot from it came through the embrasure of the gun-pit and killed the number sponging out, severely wounding the man who was serving the vent.

There was great difficulty about water at first for the Boers cut off the supply. Water-carts were sent out to a stream under escort of 25 mounted artillerymen, but such an enterprise could not be repeated. A competition in well-sinking was started, and won by the Royal Artillery, after which there was no further shortage of water. The Regiment made another noteworthy contribution to the defence in the provision of a Union Jack to fly over the fort. This was improvised in the gun-pits from the linings, white and red respectively, of the officers' and serjeants' cloaks and an infantry blue serge.

But food was running short, and on the 21st March, when all that remained were four sacks of mealies, the garrison surrendered, marching out with all the honours of war—flags flying, bugles sounding—treated with every respect by the Boer Commander, General Cronje. The battery was allowed to keep the flag they had made, and it is now in the Institution.

COMMENTS.

The war ended in an inglorious peace. For twenty years it was to be a bitter memory for the British Army. But for the Royal Artillery it had provided some fine examples of resource and courage, upon which the Regiment can look back with pride, remembering always what they owed to the infantry who helped them so gallantly in their need. In his despatch after Laing's Nek General Colley was generous in his acknowledgment:—"Much credit is due to Captain Greer and Lieut. Parsons for the efficiency of my artillery force, seeing that there were originally only two guns properly manned and equipped in this country, the other four having been equipped and horsed locally, and manned partly by artillerymen from a garrison brigade and partly by volunteers from the 60th Rifles.'' And no finer tribute has ever been paid to the soldierly conduct of all ranks than appeared in orders on the day after the Ingogo :—

"The Major-General Commanding desires to express his high appreciation of the conduct of the officers and men of the Royal Artillery and 60th Rifles in the action fought yesterday against vastly superior numbers. The R.A. well maintained the splendid reputation of that corps by the way they served their guns under a murderous fire and brought them out of action, notwithstanding their heavy losses in men and horses . . . the Major-General has to deplore the loss of Captain Greer, R.A., who was killed at his guns, setting a noble example, worthily followed by the men under him. . . . The Major-General feels sure that the force under him will join with him in specially recognising the distinguished conduct of Lieut. Parsons, R.A., who directed the fire of the artillery in a most exposed position until two-thirds of his men were killed or wounded, and he was consequently ordered to retire. . . . I also desire to mention specially the conduct of Sergeant-Major T. Toole, R.A., who well seconded Captain Greer and Lieut. Parsons in working the guns until the end of the action, and on whom the arrangements for the withdrawal devolved, when the latter officer was wounded".

PART VI.

THE SOUTH AFRICAN WAR.

CHAPTER XXV.

THE SOUTH AFRICAN WAR—INTRODUCTORY.

The Theatre of War—The Political Situation— The Military Preparations.

MAP 13.

ANY detailed description of the vast area which was to become the theatre of war would require far more space than can be afforded here. But to make the following account comprehensible to a generation far removed from the memories of the South African War, some general idea of the country must be given.

Broadly speaking then, the territories of the Transvaal and the Orange Free State consist of an immense table-land, separated on the east and south from the coastal belts by a rugged mountain barrier, and on the west merging into desert tracts sloping slowly to the sea. Broadly speaking again, this table-land, or "High-Veld", is, in the Free State, a rolling, treeless, plain crossed by low ranges of hills, with here and there a farm-stead with its dam and clump of trees. In the Transvaal it is generally more broken and wooded, with considerable mountain features, and, in the west, rich farm lands and pretty little towns. North of the Delagoa Bay railway and the Magaliesberg range it becomes the "Bush-Veld", an inhospitable waste extending northwards to the valley of the Limpopo, the boundary between the Transvaal and Rhodesia.

The rocky barrier enclosing this table-land stretches out north-easterly and south-westerly from the great mountain mass of Basutoland—the Switzerland of South Africa. North-easterly, under the general name of the

CHAPTER
XXV.

The Theatre
of War.

Drakensberg, it follows the coast line of the Indian Ocean, a hundred miles or so inland, until it approaches the valley of the Limpopo. South-westerly it extends right across Cape Colony, sinking eventually into the sand-dunes of the Atlantic sea-board. Above Natal the Drakensberg towers like a mighty rampart; in Cape Colony the mountains descend in a series of gigantic terraces, between which lie the arid plains known as the "Karroos". Inland on the contrary the crests of the mountains are but the rim of the table-land, and in consequence the spurs thrown out to the west are far less pronounced than those to the east and south. But scattered everywhere are the "kopjes", rising suddenly and steeply out of the plain, sometimes in a series forming a sort of broken ridge, sometimes in groups of three or four, sometimes singly. And in the Transvaal there are transverse ridges, or "rands", of some importance—the Witwatersrand of Johannesburg "the Rand" *par excellence,* and the Magaliesberg just north of Pretoria. Further north still the country becomes a jumble of mountains before sinking to the Limpopo.

The rivers flow for the most part in deep beds, their existence indicated only by the fringe of tree-tops showing above the banks. Fordable in dry weather, they become raging torrents after rain, with the drifts impassable, and bridges few and far between.

The roads were merely the tracks worn by traffic across the veld, quite serviceable in dry weather, but apt to become sloughs in wet.

Five railways ran into the theatre of war from the coast, and in view of the prominent part they played throughout the war, they are enumerated here :—

i. *From Cape Town to Buluwayo, via* de Aar, Orange River Station, Kimberley, and Mafeking—crossing the Orange River at Orange River Station, and keeping just outside the western boundaries of the Free State and Transvaal.

ii. *From Port Elizabeth to Pretoria viâ Bloemfontein, with a continuation from Pretoria to Pietersburg.*

iii. *From East London, to join the above at Springfontein.*

iv. *From Durban to Johannesburg, with a branch from Ladysmith to Harrismith.*

v. *From Delagoa Bay to Pretoria.*

Connecting the three railways running through Cape Colony was a line running from de Aar to Stormberg *via* Naauwpoort. There was no direct communication by rail between Cape Colony and Natal.

All the above were single lines of only 3′ 6″ gauge, with very steep gradients, so that their carrying capacity was small.

The "High-Veld", screened by the Drakensberg from the warm wet wind of the Indian Ocean, but lying open to the breezes blowing over the western deserts, possesses a dry, clear atmosphere, with bright sunshine. Nights are cool however hot the days, and even with its violent rain storms and bitter winds, it is a good climate for campaigning[1].

The "Conventions"[2] concluded between the British Government and that of the Transvaal after the war of 1881 were so-called "settlements". But they laid the foundation of future quarrels in the feelings they engendered—of bitter resentment in the British army; of humiliation throughout the British-born South Africans; of arrogance among the Dutch. And to these was soon added the disturbing influence of gold. Its existence in the Transvaal had been known for some time, but it was

CHAPTER
XXV.

The Theatre
of War.

The
Political
Situation.
1881-4.

1886.

---

[1] The hot weather—generally including the rainy season—from October to March: the dry, cold weather from April to September.

[2] Of Pretoria in 1881, of London in 1884.

CHAPTER
XXV.

The
Political
Situation.

1895-6.

1899.

August.

September.

October.

The Military
Prepara-
tions.

only in 1886 that the value of the deposits on the Witwatersrand was realised. Adventurers thronged into the country, and soon the population of the mining centres became greater than the whole burgher community. This population was treated by the Boer Government with such contumely and injustice that, after much peaceful agitation, they began to think that they would never obtain redress for their grievances unless they won it for themselves. So came about the abortive Jameson Raid of December 1895. It brought in its train evil consequences —the estrangement of the Free State, hitherto always friendly—the famous telegram from the Kaiser to President Kruger—the hardening of the latter's heart towards the *Uitlanders*. So intolerable became their situation that in 1899 they appealed to Queen Victoria for protection. The British Government made every effort to obtain an agreement, and the cold weather slipped away in vain attempts at mediation, while the attitude of President Kruger grew ever more unbending. He was waiting for the grass to spring up so that his commandos might be free to move. There were not sufficient British troops in South Africa to guard the vital points on the frontiers of Cape Colony and Natal, but a couple of battalions were sent out, and on the 8th September a Cabinet Council sanctioned the despatch of ten thousand men from the Mediterranean and India. A month later, as evidence accumulated of the scale on which military preparations were being pressed on in the Transvaal, the mobilization of the Army Corps and Cavalry Division was ordered. President Kruger replied with an ultimatum, handed to the British Agent in Pretoria on the 9th October, in which he demanded the withdrawal of the British troops from the frontiers of the Colonies, and the recall of those at sea. He was informed that the demands were "such as Her Majesty's Government deem it impossible to discuss".

The first Boer war had shown the Republics that some military preparations were advisable. The provision of

artillery, in especial, was called for, since it was for want
of guns that the Transvaalers had been unable to reduce
the little British garrisons scattered about their country
in 1881. A certain number of Krupp field guns were
obtained by both, but it was not until after the Jameson
Raid that any great vigour' was shown in these preparations. No time was then lost, in the Transvaal at any
rate. Rifles and ammunition were imported in enormous
quantities, guns—fortress and field—of the latest design
were purchased from Creusot and Krupp, and formidable
forts were built to protect Pretoria and overawe Johannesburg.

Every burgher was liable to turn out, with horse and
rifle, when called upon, for service under his Field-Cornet
in the District Commando; and the growth of towns had
not yet materially impaired the fighting spirit acquired in
native wars and the chase. Accustomed to the use of
fire-arms, to traverse long distances on horse-back, to the
management of mule and ox transport, the Boers were
formidable antagonists in their own country. Even their
lack of discipline, and disinclination to take the offensive,
were compensated for by their skill in the use of cover,
and their power of withdrawing from position to position
without loss of *moral*. And, in addition to the burgher
commandos, which would amount to not less than sixty
thousand men, they could put into the field some three
thousand thoroughly drilled and disciplined regular troops
in the Artillery Corps of the two States, and the Transvaal
Police generally known as the "Zarps."

The Orange River and the Drakensberg marked the
boundaries of Cape Colony and Natal, but these natural
features did not coincide with the racial divisions of the
population. In both British colonies a numerous and
active Dutch element was established, and the Boers were
convinced that an invasion of Natal and the capture of
Kimberley would be the signal for a general rising in Cape
Colony, and that this would lead to intervention by one or
more of the European powers. Their plan of campaign

was, therefore, based on an offensive on three fronts. To the East, the main forces of both Republics were to sweep down upon northern Natal, thrust forward so conveniently between them. To the west, Kimberley, lying apparently defenceless just across the border, was to be seized. To the south, commandos were to hold the Orange River bridges, and provide rallying points for the disaffected population.

As we have seen the Jameson Raid gave the Boers a good excuse for hurrying on their miltary preparations, and it may well be asked why the garrison of British South Africa was not increased proportionately? The reply is that, when a brigade of field artillery was sent out in 1897, the British Government was at once charged by the Opposition with initiating a policy of provocation. Whether wisely or not, it was decided, therefore, that the risk must be run. And so, all through the fateful winter of 1899, the garrison consisted of only about seven thousand regular troops, with the following artillery :—

> one field artillery brigade ⎫
> one mountain battery ⎬ in Natal.
> two coast defence companies in Cape Town.

There were also among the local corps in both colonies many batteries which will be referred to more particularly as they take the field.

In August the first reinforcements arrived in the shape of a couple of battalions, and in September there sailed from India a cavalry, a field artillery, and an infantry brigade.[1] These were followed a little later in the month

---

[1] It was only on the 8th of September that the Cabinet called upon India for the despatch of this force, yet the three batteries sailed from Bombay on the 17th, 18th and 19th of that month. The prompt and efficient organization of the Indian army and sea transport saved, as Lord Milner bore witness, not only Natal but the whole South African position. But in view of the large native population in South Africa, it was decided that it must be a white man's war, and that the Indian Army should not take part. Much help was given, however, in other ways, as for instance by the gift of 300 artillery horses by the Maharaja Scindia.

by an infantry brigade from the Mediterranean, with
another field artillery brigade from home. The strength
of the garrison was thus brought up to rather over twenty
thousand men, with three field artillery brigades, one
mountain battery, and two coast defence companies.

On the 7th October the mobilization of the I Army
Corps and Cavalry Division was ordered, and the troops
began to sail on the 20th. But by this time war had been
declared, and the first battles were being fought.

CHAPTER
XXV.

The Military
Prepara-
tions.

NOTE.

The separation between the mounted and dismounted
branches of the Royal Artillery took place just before the
war, and so in it all units belonged to either the R.H.A.,
the R.F.A., or the R.G.A. Further changes were the
brigading of batteries in the R.H. and R.F.A., and the
numbering of companies R.G.A. in one sequence throughout, with the abolition of territorial titles. These latter
measures were only brought into force while the war was
in progress, but to avoid changes of designation during the
course of operations, all units appear in the narrative under
the title they bore at its conclusion—and retained until
after the Great War. The designations borne at the
beginning of the war are, however, added in brackets in the
list of units given in Appendix J.

A word of explanation may also be advisable regarding
the designation of staff officers. With the exception of
those on the personal staff, all such officers in whatever
branch of the staff their actual duties might lie, bore at
this period the titles now reserved for the Adjutant-
General's Department—D.A.G., A.A.G., or D.A.A.G.

All comments on the organization, armament, and employment of the artillery in the war have been deferred
to Chapter XXXVII, where, also, will be found some
particulars of the Boer artillery. Lists of units, commanders and staff officers, etc,, will be found in the
Appendix.

CHAPTER XXVI.

The Defence of Natal.

The Preliminary Moves—Talana Hill—Elandslaagte—Rietfontein—Lombard's Kop—Nicholson's Nek—The Position in Natal—The Defence of Ladysmith—The Bombardment—The Sorties—Caesar's Camp & Wagon Hill—The Relief—Section A—Expenditure of Ammunition.

Maps 13, 14, & 15.

Chapter XXVI.

The Preliminary Moves.

The geographical position of Natal *vis-à-vis* to the two Boer Republics made its defence a problem of extreme difficulty. The shape of the frontier, with its apex thrust forward into a dangerous salient, laid it open to attack from both sides. And this disadvantage was accentuated by the fact that the Drakensberg range formed a screen behind which the hostile forces could gather unseen, and pour down through the passes on the flank and rear of the defenders. There were political objections, however, to abandoning the north of the Colony to the enemy, and such a policy was equally distasteful to the military instincts of Sir W. Penn Symons, the general officer commanding in Natal. He fell in readily, therefore, with the suggestion of the Governor to send a detachment to cover the coalfield at Dundee.

Ladysmith, situated at the bifurcation of the railways and roads leading to the Transvaal and Orange Free State had been selected for the cantonment of the British troops sent out in 1897, and Colonel E. O. Hay, the commander of the field artillery brigade, was senior officer there until called to the War Office in 1899. As such he had had much to do with such preparations as were allowed—the reconnaissance of the country, the accumulation of stores, and so forth—but on his departure it was deemed advisable to make the command a whole time appointment, with the rank of Colonel-on-the-Staff. For this another gunner—

1897.

1899.

Colonel W. G. Knox—was selected, and on his arrival he lost no time in pressing for the completion of such measures as he considered necessary. His urgent appeals met with little success, "But", as he wrote, "I had the troops to play with and a grand country to manœuvre over, and within ten days of my arrival I had the garrison out on a five-day bivouac manœuvre scheme, with a mobile force to represent the nature of a force which all hoped to meet shortly". Conducted with his usual thoroughness, these manœuvres did much to bring home to the regiments, which had been long on colonial service, the practical lessons they needed. It was indeed a blow to the Colonel-on-the-Staff when a cipher telegram on the 24th September directed him to despatch to Dundee two of his battalions and the same number of batteries.

This unexpected denuding of the garrison left the defence of the town and the vast accumulation of stores to a very inadequate force, and Knox at once set the few remaining troops to erect such defences as would enable him to meet a surprise attack and hold the enemy at bay until reinforcements could arrive.

The beginning of October brought relaxation to the tension with the first arrivals from India. The batteries of the IInd brigade, R.F.A. were in Ladysmith well within a month of receiving orders at their Indian stations, and on the 7th October Sir George White landed at Durban, and assumed command of the forces in Natal. He knew nothing beforehand of the arrangement which had been made by General Penn Symons, and on reviewing the situation he determined to concentrate his forces at Ladysmith where he would be in a position to strike with his full strength against any invaders as they debouched from the mountain passes. But when the Governor represented the disastrous effect a withdrawal from Dundee at this moment might have on the mixed population of the Colony, he felt bound to accept the *fait accompli*, and sent General Penn-Symons off to take command there, proceeding himself to establish his headquarters at Ladysmith.

CHAPTER
XXVI.

1899.

The
Preliminary
Moves.

September.

October.

CHAPTER
XXVI.

1899.

The
Preliminary
Moves.

11th Oct.
On the 11th October war was declared, and on the 12th the depôts, women, and children of the units in garrison were evacuated to Pieter Maritzburg, followed by all the sick. On the 20th Colonel C. M. H. Downing arrived to take up the command of the artillery, bringing with him from home a "Special Ammunition Column" —the brigade from India had brought its own.

The troops in Natal consisted of a cavalry brigade, a mounted brigade (local), two infantry brigades, some divisional troops and the following artillery :—

C.R.A.   —Colonel C. M. H. Downing.
Staff Officer   —Captain E. S. E. W. Russell.
Orderly Officer   —Captain R. A. Bright.
I Brigade, R.F.A.   —Lieut.-Colonel E. H. Pickwoad.
Adjutant   —Captain J. A. Tyler.
13th Battery   —Major J. W. G. Dawkins.
67th   do.   —Major J. F. Manifold.
69th   do.   —Major F. D. V. Wing.
Amtn. Column.   —Major E. S. May.

II Brigade, R.F.A.   —Lieut.-Colonel J. A. Coxhead.
Adjutant   —Captain A. L. Walker.
21st Battery   —Major W. E. Blewitt.
42nd   do.   —Major C. E. Goulburn.
53rd   do.   —Major A. J. Abdy.
Amtn. Column.   —Captain R. G. Ouseley.

No. 10 Mountain Battery—Major G. E. Bryant.

The Natal Field Battery—Major D. W. G. Taylor.

The Natal Naval Volunteers—Commander G. E.
                                        Tatum.

Of the above one cavalry regiment, one infantry brigade, and the Ist brigade, R.F.A., were at Dundee, and we must now see how they were faring.

General Joubert, the Boer Commander-in-Chief, had planned to cut off the detachment at Dundee by an enveloping attack on front and flanks. But the movements

were badly timed, and the Transvaalers from Vryheid
reached their goal before the other forces were in a position
to co-operate. On the 20th their approach was signalised
by shells from Talana Hill falling into the British camp
at Dundee, fortunately sinking into the sodden ground
without doing much damage. The troops were just
breaking off, after standing to arms at dawn, and the 67th
battery, whose horses had filed off to water, replied from
the gun-park. But the range was 5,000 yards, and they
only fired a few rounds. The 69th, whose horses were
still harnessed up, galloped out to the flank and came
into action at a range of about 3,700 yards. The 13th
joined them a few minutes later; and in little more than
half-an-hour from the first shot, the State Artillery had
given up the unequal contest. Leaving the 67th and a
battalion to hold the camp, the general led out his infantry
to the attack.

The British camp lay just to the west of the little
mining town, which was commanded on the other side by
a ridge known as Talana and Lennox Hills. Between the
town and this ridge ran a spruit or donga, beyond which
half-a-mile of grass land led up to a group of farm
buildings surrounded by plantations. Above these again
came the farm enclosures, divided by stone walls running
up the hillside to a natural terrace along which lay the
boundary wall. From this terrace to the summit the slope
rose steep and rough for a hundred and fifty yards or so.

By seven o'clock the infantry were on the line of the
spruit, but directly they emerged from its shelter they
came under fire, not only from Talana Hill but from
Lennox Hill on their right. They reached the woods,
without a pause, however, and penetrated to the further
side, but here their advance was checked. It was about
eight o'clock, and the 13th and 69th batteries, moving
through the suburbs, now took up a second position on
some rising ground between the town and the spruit, from
which the 13th engaged Lennox Hill at 2,500 yards, the
69th Talana at 2,300. Their fire was accurate, but they

had only 154 rounds per gun so had to go slow,[1] and after half-an-hour of this the patience of the general was exhausted. Riding up to the woods with his staff, he dismounted, and was walking along the firing line when he was shot in the stomach. The wound was mortal, but he showed no sign until he had remounted and ridden through the troops. His presence had restored the momentum of the attack, and the transverse wall which ran along the terrace was won, but it was not until another couple of hours had elapsed that sufficient men had reached it to warrant a further effort. At eleven o'clock Colonel Pickwoad received a signal message that an assault was about to be delivered. He ceased firing, but nothing happened, and he re-opened. At 11.30 the order was repeated, and once more he stopped the fire of his batteries to no purpose, before leading them forward to a third positon on the other side of the spruit, from which they were able to bring the crest under fire at only 1,450 yards. Under this the musketry gradually died down, and, for the third time, the artillery got the signal to cease fire. This time the infantry left the shelter of the wall, but though some got forward, the assault as a whole failed. Aware only of the general result, and not seeing how close to the Boer line the foremost infantry had got, the gunners resumed their fire against the crest, and some shell unfortunately fell among the attackers lying close to it. Both sides recoiled, and as the Boers dropped back across the flat top of the hill the shrapnel followed them. It was too much for the burghers, and before they reached the shelter of the reverse slope they were in full flight. Colonel Pickwoad lost no time in crowning the position with his guns, but to the amazement of all they remained silent. Doubt had crept into his mind, for, just before the final assault, a request for a truce, purporting to come from the Boer commandant, had passed through his hands

---

[1] As will be remembered it was only on this very day that the ammunition column sent out from home for this brigade reached Ladysmith.

on its way to the general, and he did not know whether or not it had been granted. At this moment a bystander exclaimed that the Boer hospital was before them, and thinking that he saw red-cross flags, he refrained from opening fire. So the burghers got away without further loss. By inadvertence the artillery had done damage to their comrades of the infantry, and had missed such a chance of turning a retreat into a rout as seldom falls to their lot. But they had taken their full share in winning the victory, and a dozen years later General A. J. Murray, who had been one of the staff on that occasion, bore witness, at a General Staff Conference, that the infantry "had the closest possible assistance from the artillery".

Joubert's plan of encirclement had miscarried, and the British had gained a tactical victory, but General Yule (who had succeeded to the command) had no illusions as to the strategical situation. Next day he fell back to a more defensible position, to be greeted again by Boer shell, and this time of a weight never expected in the field. At midnight he moved out of range, and next day received orders to return to Ladysmith. The little column slipped away while the Boers were busy looting the deserted camp and the defenceless town. It was a trying march, in torrential rain and tropical heat, but it was skilfully conducted, and on the 26th October they were safe in Ladysmith.

While the invaders from the Transvaal attempted to surround the detached force thrust forward so dangerously to Dundee, the Free Staters came down on the railway behind, so as to cut it off from Ladysmith. Before dawn on the 21st October General French set off with a few squadrons and the Natal field battery to clear the line, and found the burghers busy looting a supply train which they had captured at Elandslaagte station. The battery soon shelled them out of the station buildings, but their little 2·5″ guns were no match for the 12½-pdrs. that the Boers then brought into action on the hill behind.

CHAPTER
XXVI.

1899.

Elands-
laagte.

21st Oct.
French broke off the fight and telegraphed to Ladysmith
for instructions and reinforcements. Colonel Coxhead
was sent out with some cavalry and the 42nd battery.
But by this time it had been possible to form a better
estimate of the enemy's strength, and further troops
were called for. Three battalions were despatched by
train, and the 21st battery trotted the twelve miles
without a halt. On their arrival French launched his
force to the attack, one battalion holding the front, the
others against the flank, the artillery between the two.

After the 42nd had quelled the fire of some Boers who
were annoying the cavalry on the left, the two batteries
advanced into action. It was a hazardous movement up
a long and steep grassy slope in the open, under the fire
of the Boer guns. Though extraordinarily accurate in
range, this fire was almost innocuous owing to their
inability to manage their time fuzes. The object of the
British batteries was to get within their time-shrapnel
range, but they were forced into action at 4,400 by rifle-
fire in an advanced post. The 21st turned on to this
and soon drove away the riflemen, after which it did not
take long to silence the Boer guns, though only percussion
shrapnel was available.[1]

The infantry had been pushing forward, meanwhile,
and, as soon as they got to their first fire position, the
guns advanced to a second position at 3,200, under a
heavy fire from the Boer guns, which reopened as soon
as they were seen to be limbering up. Once in action
the British time-shrapnel soon silenced them again, and
the gunners were then able to devote all their attention to
searching the defences, while the infantry climbed the
hill. It was a long pull in the face of torrents of rain,
but the batteries pushed forward alternately to 2,400
and 2,100, the closest that the ground allowed. From
there they were able to maintain their fire until the
final assault, and as darkness fell the position was carried,

---

[1] The Boer fuzes ranged to 5,000 yards, the British only to 4,000.

though many of the burghers fought to the last, clinging
stubbornly to the further edge of the plateau. But they
couldn't get their guns away, and the cavalry sweeping
round the hill rode twice through the fugitives. It had
been a "sealed pattern" attack, in which all arms had
played their parts in accordance with the accepted doctrine,
and it had been absolutely successful. There could, however, be no idea of holding the ground so gallantly won,
and the troops were back in Ladysmith next day, with
the two captured guns, which were to prove their value
before many days were over.

Determined to ward off any attempt of the Boers to
interfere with the retreat of the Dundee column, Sir
George White sallied out again on the 24th, taking the
42nd and 53rd batteries, this time and No. 10 mountain
battery in addition. The enemy were found in position
on the Intintanyoni hills near Rietfontein, and it was
essential to hold them there until General Yule could
bring his troops into Ladysmith. So, as soon as the Boer
guns opened fire, Sir George deployed and advanced
straight to the attack. The batteries had to drop into
column of route to pass a level crossing, and suffered some
casualties from two guns which had the range to a nicety,
but they were soon silenced by the 53rd which crossed first
and unlimbered a few hundred yards further on.

As the infantry advanced the three batteries came into
action at 2,000 yards with the mountain battery in the
centre. Here they remained for a couple of hours, nipping
in the bud every attempt of the enemy to get their guns
into action again, while the two infantries indulged in a
prolonged fire fight. The attack had, from the first, been
designed only to hold the enemy to their ground, so as
soon as it was certain that they could no longer interfere
with Yule's movements, the British troops were quietly
withdrawn, having accomplished the object of their
intervention. It had been a gunner's day.

Cavalry reconnaissances on the 27th and 29th October

CHAPTER
XXVI.

1899.

Elands-
laagte.

21st Oct.

Rietfontein.

24th Oct

showed Sir George White that the commandos were closing
in upon Ladysmith, and he determined to strike a blow
at the Transvaalers before the assembly was completed.
On the afternoon of the 29th Downing rode out with his
brigade commanders and adjutants to look for gun
positions, but was chased home by Boer patrols. Late
that night the whole of the troops were in motion—a
couple of battalions and the mountain battery to seize
Nicholson's Nek, six miles to the north, and cover the
left; the cavalry to Lombard's Kop to guard the right;
while the main body struck at Long Hill.

30th Oct.

At dawn the two field artillery brigades, under Colonel
Downing, opened fire from Limit Hill, I Brigade on Long
Hill, II Brigade on Pepworth. From the latter came the
boom of a heavy gun, and the brigade had its first experience of 96-lb. shell,[1] soon supplemented by those of long
range field guns. Sending Coxhead straight to the front
until within shrapnel range, Downing directed Pickwoad to
change front to the left and move forward until his brigade
could also bear upon Pepworth. This concentrated
shrapnel fire was more than the State Artillery could compete with. The gunners fell fast, and the survivors were
unable to serve their guns under the storm. All seemed
going well, but about 7 a.m. fresh hostile guns began to
appear on the surrounding hills, and the crackle of
musketry grew ever louder. A vivid picture of the battle-
field at this moment, as seen from the enemy's position,
has been given by one then fighting against us :—

"From the sweep of the hills we looked down as from an
amphitheatre at every movement of the troops on the plain
below—infantry advancing in successive waves, guns being
galloped up, and all the bustle and activity of a battle
shaping before our eyes. The soldiers paying little heed
to the shells that dropped amongst them, advanced with-

---

[1] These were the Creusot 155 ™ siege guns bought for the armament of the forts at Pretoria, soon to become familiar to all under
their nick-name of "Long-Toms". The field guns were 75 ™ also
from Creusot.

out a halt, although many now fell dead and wounded,
while in the rear, battery after battery unlimbered. We
saw the horse-teams ridden back, and then, to cover the
progress of their troops, heavy fire was opened. . . . .
We were well protected, but the guns on Pepworth Hill
were being severely punished. The English batteries
concentrated upon silencing our pieces there, and we could
see that the people were making heavy weather on the
hill, for its summit was covered in smoke and flame, and
the roar of the bursting shells shook the ground even
where we lay. By the time the foremost infantryman
came within 1,200 yards many fallen dotted the veld,
and their advance wavered before the hail of bullets.
They did not run away, but we saw them taking cover
behind ant-heaps and such other shelter as the ground
afforded. Their progress was definitely stayed".

It was not surprising, for the enemy's line was being
extended round our right. Their guns on Pepworth
were by this time so far silenced that Downing deemed
one battery from each brigade would be sufficient to
keep them quiet, and he could use the remainder to
relieve the pressure on the other portions of the line.
Sir George White who was on the spot agreed, and the
batteries were dispersed.

On the extreme right, Wing, handling his battery
boldly, checked the Boer extension by his enfilade fire,
but drew upon his battery in turn such heavy fire that
its position in front of the cavalry could not be maintained
without risk of destruction. He accordingly withdrew
to the shelter of the nek between Lombard's Kop and
Bulwana, where he remained in action for the rest of the
day.

The 21st and 53rd went to the support of the infantry
in the centre, but had not been long in action before there
came a call from French for further artillery support on
the right. The 21st was sent off, and Blewitt found a
a place in a gap in the hills between the cavalry and
infantry. The 13th had come up meanwhile somewhat

CHAPTER
XXVI.

1899.

Lombard's
Kop.

30th Oct.

CHAPTER
XXVI.

1899.

Lombard's
Kop.

30th Oct.

to the left of the 53rd, and taken on the guns which were enfilading the infantry from that side.

For the next couple of hours the battle resolved itself into an artillery duel between Boer guns scattered over a curved front of nearly seven miles, and the six dispersed British batteries :—

69 firing south-east from Bulwana Nek.

21 ,, east from the northern side of Lombard's Kop.

53 ,, east from behind the right of the infantry.

13 ,, north-east from behind the left of the infantry

67 ⎱ still in the first position, facing north, and suffering considerably from the guns on Pepworth which

42 ⎰ had sprung to life again as batteries were withdrawn from the position facing them.

From time to time the British succeeded in driving the Boer gunners away from their guns, but only to come back again as soon as the fire was diverted to another direction, while the Boer riflemen kept the cavalry and infantry pinned to the ground. The artillery escort of half a battalion, lying out in the open between the original gun position and the 13th battery, was the only reserve left, all the rest having been gradually absorbed into the firing line, and about eleven o'clock Sir George White made up his mind to withdraw his troops into Ladysmith. To cover the movement of the infantry the 13th battery galloped up to close in rear of their line, drawing upon itself, as the infantry fell back, the enemy's fire from guns on both flanks, and even from the big Creusot in rear. It looked as if the battery must be sacrificed, but Major Abdy marked its perilous plight and brought the 53rd into action alongside under heavy rifle fire. The 13th then retired by order of Sir A. Hunter, the Chief of the Staff, who was on the spot, and as soon as the 13th· was in action again the 53rd followed. But a shell had blown to pieces the wheels of one of its limbers so a gun had to be left.

There it stood, conspicuous in the open—but not for long. Under a furious fire, Captain W. Thwaites came up with a wagon limber and brought it in, to the relief of all.

Under cover of the fire of the two batteries, retiring alternately, the withdrawal of the infantry was safely continued, and a little later the cavalry were brought back, covered to the last in a similar manner by the 21st and 69th.

CHAPTER
XXVI.
1899.
Lombard's
Kop.
30th Oct.

And now we must hark back to the little column which had been despatched on the evening of the 29th to cover the left flank of the army by seizing Nicholson's Nek. Marching through the night they were scrambling up its steep side when some accident stampeded the mules of the infantry transport. In the darkness before the dawn the mules came rushing down through the ranks, and the panic spread to those of the battery, so that when the troops pulled themselves together they found that they were bereft not only of their guns but of the reserves of ammunition for their rifles. And their heliograph as well as their guns had been carried off by the mules, so that they had no means of communicating their plight to the army in full view below them. We need not pursue the subject. No. 10 mountain battery lost its commander, all three subalterns, and 84 men; but fortunately its captain (T. R. C. Hudson) had only reached Ladysmith —straight from the Staff College—after the column had left About 4 a.m. he was called up by a non-commissioned officer who had got away from Nicholson's Nek with the stampeding mules, and set to work at once reorganising what was left of the battery. From the debris he was able to equip a section, and before 6 a.m. it was in action with the little force left under Knox for the defence of the town.

Nicholson's
Nek.
30th Oct.

Between the crossing of the frontier by the commandos from the Transvaal and Orange Free State on the 12th October, and the end of that month, the Natal Field Force had fought five actions in defence of the Colony. The north had now been over-run by the invaders, and Sir

The
Position
in Natal.

CHAPTER
XXVI.

1899.

The
Position
in Natal.

November.

George White was confronted with the problem of how to save the south.

It would be out of place here to follow the arguments that have been adduced for and against Sir George White's conclusion : suffice it to say that he decided to hold out where he was, as the best, if not the only, means of "covering the vitals of Natal".

Ladysmith lies along the left bank of the Klip River, which there runs in an easterly direction though with many windings. Immediately below the town it turns south and makes a great bend, enclosing in its loop an expanse of flat grassland, a couple of miles across, used for race-course and show-ground. This plain and the town of Ladysmith are enclosed on the north by a rough horse-shoe of kopjes of no great elevation—perhaps a hundred to two hundred feet—while right across the southern side of the plain stretches a long table-topped ridge some hundred feet higher. On the west this ridge is connected with the kopjes by some insignificant features; on the east there is a broad gap between the most southerly of the horse-shoe kopjes and the end of the ridge, but the plain is protected by the Klip River flowing along this side.

Such was the perimeter selected for occupation. Beyond it again rose an outer circle of much more imposing heights, for the most part within 8,000 yards of the town, looking into the rear of any defences which might be con-structed, and commanding the whole of the grazing ground on the flats fringing the river. With the troops at his disposal Sir George White could neither occupy these heights himself nor deny them to the enemy. As·it was the perimeter to be defended extended to fourteen miles, and his force amounted to little more than thirteen thousand men.

We can now take stock of his artillery. The C.R.A. was Colonel C. M. H. Downing, who had only arrived from England on the 20th October. Under him were the Ist

CHAPTER
XXVI.

1899.

The
Defence of
Ladysmith.

November.

and IInd Brigades, R.F.A.; the two guns saved from the wreck of No. 10 Mountain Battery; a "Maxim-Nordenfeldt Battery" consisting of the two 12½-prs. captured at Elandslaagte, manned by horse artillerymen who had come from India with remounts under Lieut. K. Kincaid-Smith; and two old 6·3″ R.M.L. howitzers unearthed from store in Cape Colony, manned by gunners of the mountain battery under Captain W. H. Christie.[1]

In addition to the above there was the Naval Brigade under Captain the Hon. Hedworth Lambton, that had arrived at Ladysmith in the nick of time. It was indeed fortunate that on hearing from General Yule of the arrival in his camp, on the day after Talana, of 96-lb. shells, Sir George White had appealed to the Admiral for assistance. "In view of heavy guns being brought by General Joubert from the north", he suggested "the sending of a detachment of bluejackets with long-range guns firing heavy projectiles". In ninety-six hours from his message reaching Simons Town mountings had been designed and constructed there, brought round to Durban in the "Powerful", landed and railed up to Ladysmith, arriving in time to take part in the battle. There was thus added to the artillery of the garrison two 4·7″, three 12-prs. of 12-cwt., and one 12-pr. of 8-cwt., as well as two 3-pr. Hotchkiss manned by the Naval Volunteers.

---

[1] The "State of Divisional Troops" shows :—

Detail.	Officers	R. & File	Guns	
10 M.B.	2	80	2	
Maxim-Nordenfeldt ...	1	18	2	
Howitzer Detachm't,	—	—	2	Manned by 10 M.B.

Kincaid-Smith, as well as one of the Maxim-Nordenfeldt's (now in the Rotunda) had been in the Jameson Raid.

CHAPTER
XXVI.

1899.

The
Defence of
Ladysmith.

The perimeter which Sir George White had decided to
hold was divided into sections, and guns allotted to each,
as follows :—

A. Section—Colonel Knox. The eastern half of the
horse-shoe round the town, from Helpmakaar Hill to Cove
Redoubt, with the section of No. 10 mountain battery,
the Maxim-Nordenfeldt battery, and the howitzer detachment.

B. Section—Major-General Howard. The western half
of the horse-shoe from Leicester Post to Range Post, with
the 69th battery.

C. Section—Colonel I. Hamilton. The western and
southern sides of the plain, Wagon Hill and Cæsar's Camp,
with the 42nd battery, and the two 3-pr. Hotchkiss.

D. Section—Colonel Royston. The eastern side of the
plain, along the Klip River from Helpmakaar Hill to
Cæsar's Camp.

The Naval Brigade held the kopjes immediately north
of the town with two 4·7″s on Junction Hill and Cove
Redoubt, and the four 12-prs. on Gordon Hill.

The *Mobile Reserve* or *Flying Column* comprised the
two field artillery brigades (less one battery each) which
"for purposes of manœuvre and defence will be under the
orders of Colonel Downing, who will see that they are
ready for immediate action at daybreak daily, and
occasionally mass them at some central place as if for
immediate action".

13th Nov.

On the 13th November the 13th and 67th batteries
supported the mounted troops in an attempt to turn the
Boers off Rifleman's Ridge, but it was no more successful
than the enemy's efforts of the previous week. On the
28th Cæsar's Camp was again the centre of interest. In
order to enfilade the ridge a "Long Tom" was mounted
on Middle Hill, and in their anxiety to get the maximum
effect from its fire, a position was selected within range of
the 6·3″ howitzers. These were at once brought over by

28th Nov.

night to the nek between Wagon Hill and Wagon Point.
A two-day's duel was enough for the big gun, after which
it disappeared in the night, to reappear on Telegraph
Hill a fortnight later, but this time at a more respectful
distance. The howitzers followed it round to the nearest
point, but the range was too much for them, and they
had not succeeded in reducing it to silence before it was
taken away to oppose the relieving force at Vaal Krantz.

We have seen how the "Long Tom" on Pepworth Hill
put a few shell into the town during the course of the
battle of the 30th October. Soon the Boer guns could be
seen by the garrison crowning all the heights that surrounded them. In all they brought into action during
the siege four 155$^{m/m}$ Creusot guns, four 120$^{m/m}$ Krupp
howitzers, six 75$^{m/m}$ Creusot field guns, and a dozen less
up-to-date pieces. The range of those specially mentioned
—10,000, 6,000 and 5,000 yards respectively—enabled
them to keep out of reach of anything in Ladysmith except
the naval guns. As regards the remainder, their
inferiority in numbers prevented their being brought
within effective range of the 15-prs.

The bombardment, from which the Boers expected
great things, and for which they performed "Repository
Exercises" which would have astounded the founder of that
School, proved in fact singularly ineffective. The number
of guns employed would have been quite inadequate even
if intelligently directed. Being aimlessly used at the
whim of individuals they lost what little effect might have
been secured. They were, moreover, all outside shrapnel
range, and their common shell were for the most part
either blind or smothered in the soft ground.[1] So the
inhabitants—civil and military—soon learnt how danger·
was to be avoided. The camp of the IInd brigade,
R.F.A., for instance, was exposed to daily shelling, it con-

---

[1] The "duds" were much sought after as souvenirs, and gunners
were in request to remove the fuzes—an occupation much discouraged
by their officers.

W

CHAPTER
XXVI.

1899.

The Bom-
bardment.

tained six hundred men, the same number of horses, and many mules and oxen, yet in four months the only casualties in camp were four men wounded, one horse and six mules killed. This is not, of course, to say that the Boer artillery did no damage : there were cases of single shell causing many casualties. Perhaps, however, their finest effort was the stampeding, and then shepherding by shell fire into their own lines, nearly three hundred oxen —a gunnery feat which caused the immediate reduction of rations in the garrison.

The Sorties.

The two sorties which followed are of great regimental interest, for they led to an investigation into the responsibility for the destruction of captured guns, and the appearance of a paragraph in the Regulations definitely assigning this duty to the Royal Artillery.

24th Nov.

The big gun which had been "shifted" by the Navy from Pepworth reappeared on Gun Hill on the 24th November, and immediately began to make itself obnoxious to Section A, whose defences it enfiladed. Colonel Knox suggested a sortie, Sir George White agreed—provided that it was commanded by his Chief of the Staff, Sir Archibald

8th—9th
Nov.

Hunter. On the night of the 8th-9th December the latter took out five hundred men[1] and rushed the position with complete success. It was found to contain a $120^{m/m}$ howitzer as well as the $155^{m/m}$ gun, and gun-cotton charges were exploded in the breech and muzzle of both, blowing off the muzzle[2] of the gun and completely destroying the howitzer.

10th—11th
Nov.

A couple of nights later a similar enterprise against a howitzer on Surprise Hill also effected its purpose, though owing to some delay in firing the charge, the retreat was interrupted, and several casualties incurred.

The inspiring effect of these two successful sorties was soon over-shadowed by the failure of the first attempt at

[1] Including a dozen mountain gunners as a "spiking party".
[2] This is now in the R.A. Institution. The gun was sent to Pretoria where it was shortened and provided with a new breech before being sent to join the force investing Kimberley.

relief. The whole condition of the defence was altered by
Colenso. All energy had now to be devoted to strengthening the works in view of heavier bombardment, and
keeping up the strength and spirit of the troops to meet
more vigorous attacks. And yet the scale of rations had
to be reduced all round. Hard food for the horses was
almost exhausted and there was little sustenance in the
withered grass. The garrison was ceasing to be a "Field
Force".

CHAPTER
XXVI.

1899.

The Sorties.

    The bombardment had failed to bring about the sur-
render of Ladysmith so confidently anticipated, and the
Boers knew that the transports were crowding into Table
Bay. They had beaten back Buller's relieving force for
the time being, but French was keeping Schoeman on the
qui vive about Colesberg; Kimberley and Mafeking still
held out; and Roberts was coming. Something must be
done at Ladysmith, and the plan of an attack on the long
ridge which stretched right across the southern side of the
valley, and dominated the whole of the defences, was
resuscitated.

Cæsar's
Camp and
Wagon Hill.

1900.

    This ridge was nearly three miles long, with a flat top
a quarter to half a mile broad. The eastern portion, for
about two-thirds of its total length, was known as Cæsar's
Camp, the western portion as Wagon Hill. At the
extreme western extremity was a little knoll called Wagon
Point, separated from Wagon Hill by a nek.
    The defences consisted of four closed redoubts, all sited
along the rear edge of the flat top, with a line of picquets
along the forward crest. Four of the guns of the 42nd
battery were on the left of the principal redoubt near the
centre of Cæsar's Camp, the other two towards the eastern
end. The gun-pits had been continually strengthened and
improved until they became a very important feature of
the defence. On the right of the principal redoubt was a
naval 12-pr., which had been added to Goulburn's command, and the two 3-pr. Hotchkiss were in the picquet
line.

CHAPTER
XXVI.

1900.

Cæsar's
Camp and
Wagon Hill.

5th—6th Jan.

On the night of the 5th-6th January the Naval
Brigade were bringing up a 4·7″, and its emplacement was
being prepared on Wagon Point, when the picquet in the
nek between that and Wagon Hill became aware of Boers
climbing up the bed of the donga: it was about 3 a.m.
The picquet opened fire, the Hotchkiss joined in with case,
but there was a good deal of confusion, and the defenders
were driven back—taking the 3-pr. with them—to a
small fort near the west end of Wagon Hill. Here they
held the Boers at bay while the little party of blue-
jackets, sappers, and others round the gun emplacements
on Wagon Point repelled all attempts to work round that
feature.[1]

On Cæsar's Camp the attack came just as the reliefs
were in progress so that the picquet line was doubly
manned, and the attackers soon gave up any idea of
rushing it, lying down instead among the rocks and bushes
on the slope. A flanking party, however, working round
the eastern extremity of the ridge undetected, and
emerging suddenly on the summit, surprised the post on
that flank.

Thus when daylight came the Boers had secured a
foothold on the ridge at both ends. Reinforcements were
hurried up, but could make little progress until they
received artillery support. This was forthcoming as soon
as the day broke, for Colonel Coxhead despatched his two
reserve batteries, the 21st to the right, the 53rd to the
left. Blewitt coming into action near Range Post, at a
range of 3,400 yards from Mounted Infantry Hill,
effectually reduced the fire of the Boers who were supporting the attack on Wagon Hill, and by keeping under fire
the ground below its western extremity, prevented any
reinforcements reaching those who had gained a footing
on its summit.

On the left Major Abdy found a position for his battery
in the valley of the Klip where the guns were screened

---

[1] These were the unfinished emplacement for the 4·7″, and the gun
pits used by the 6·3″ howitzers in November.

from Bulwana by mimosa scrub. From it, at a range of only just over 2,000 yards, he was able to sweep the eastern slope of Cæsar's Camp. The infantry pressed forward close behind the shrapnel as he lengthened range and fuze, and the Boers fell back in confusion.

CHAPTER
XXVI.

1900.

Cæsar's
Camp and
Wagon Hill.

6th Jan.

Both batteries drew a hot fire upon themselves from the guns which had been shelling the ridge, but all accounts agree in extolling the regularity and accuracy of their fire. One incident created a great impression. Serjeant James Boseley of the 53rd had his left arm and leg carried away by a shell, and it was thought that he was dead until his voice was heard calling to his detachment to "roll me out of the way and get on with your work".[1]

The 42nd on Cæsar's Camp had been exposed to a convergent shell fire from the "Long Toms" on Bulwana and Telegraph Hill and from many smaller pieces. In this bombardment of the ridge, which continued all day with varying intensity, the gun-pits were frequently hit, but the solid labour put into them came through the test triumphantly.

As the hours passed the musketry gradually died down though with outbursts of activity now and then, but it was not until the light was fading that the famous charge of the Devons cleared the top of Wagon Hill, and a general advance by the defenders of Cæsar's Camp drove the Boers who had been clinging to its slopes in flight across the valley beyond. It was the chance for which the 42nd had been waiting, and they took full toll of the fugitives.

There was no fear of any renewal of the attack, but the garrison had to endure nearly two months more of ever increasing weakness and disappointment, as rations were reduced[2] and Buller's guns died away. There were, of

---

[1] This incident was selected by Sir George White as an example of individual bravery in his address on the siege. Marvellous to say Serjeant Boseley recovered.

[2] The gunners had the additional distress of seeing their horses turned out to pick up what food they could on the withered grass, saved from death by starvation only by butchery. Horseflesh began to be issued a month before the relief.

CHAPTER
XXVI.

1900.

Cæsar's
Camp and
Wagon Hill.
course, many readjustments of the garrison, and the casualties among infantry officers led to a call upon the artillery to fill their places. Gradually the idea of taking the field with a "Flying Column" evaporated, and more and more of the guns were dug in, but some batteries remained to the end in reserve—hooked in every morning by 4 a.m.

The final allotment of the guns is perhaps worthy of record :—

Section A.

Cemetery Hill	— 2 Maxim-Nordenfeldts.
Tunnel Hill	— 2 guns 13th battery.
Railway Cutting	— 4 do. do. do.
Liverpool Castle to	1 4·7".
Junction Hill —	1 12-pr. of 8-cwt.

Section B.

Gordon Post	· — 1 12-pr. of 12-cwt.
Observation Hill	— { 2 guns 69th battery. 1 6·3" howitzer.
Cove Redoubt	— 1 4·7".
Gordon Hill	— 1 12-pr. of 12-cwt.
Rifleman's Post	— 4 guns 69th battery.
King's Post	— 4 guns 67th battery.

Section C.

Wagon Hill	— { 1 3-pr. Hotchkiss. 1 6·3" howitzer.
Cæsar's Camp—centre	{ 1 12-pr. of 12-cwt. 4 guns 42nd battery.
Cæsar's Camp—left	— { 2 guns 42nd battery. 1 3-pr. Hotchkiss.

General Reserve.

Klip River Camp	— 21st & 53rd batteries.
Ladysmith Camp	— 67th battery (less 4 guns temporarily in Section B.)
Railway Cutting	— 2 guns No. 10 mountain battery.

On the 27th February came the news of Cronje's surrender to Lord Roberts, and of the progress of the Relieving Force on the Tugela, and rations were restored. Early next morning observers on Wagon Hill saw a continuous stream of Boers trekking northwards, and at sunset Dundonald's mounted troops rode into the town. Thus ended the 118 days' Defence of Ladysmith. And it ended worthily. For though the 1st of March brought no sign of Buller's army, Knox led out all of the garrison still able to march to harass the burghers. But Botha had got together a rear-guard on the hills of Modder Spruit station, and to press an attack without the co-operation of the relieving force was more than the little column were equal to, though the batteries[1] managed to trot into action.

On the 3rd of March Buller and his Army made their formal entry between the ranks of the defenders—so weakened by sickness and privation that they could hardly lift their helmets to cheer.

CHAPTER
XXVI.
The Relief.

28th Feb.

1st March.

3rd March.

SECTION A.

Before closing this short account of the share taken by the Royal Artillery in the Defence of Ladysmith, a word must be said of the personal services of one of the best-known officers of the Regiment. The arrival of Colonel Knox in Ladysmith, and the steps he took to prepare for the war that was so soon to come, have been already mentioned. When Sir George White established the Headquarters of the Natal Field Force in the town, Colonel Knox still remained "O.C. Ladysmith", and as such took

---

[1] 53rd and 67th. The 53rd was made up by 13 and 69, the 67th had 84 horses not yet eaten, and the ammunition column turned out 18 pairs for spare. For some time the horses had only been getting 2 lb. of grain.

command of all units left for the defence whenever the
Field Force moved out. But it is as the Commander of
"Section A" that he will be best remembered. This
section consisted of the chain of kopjes forming the north-
eastern portion of the defences. It included the Supply
Depôt and the Railway Station, Junction Hill, Gloucester
Post, Tunnel Hill, Cemetery Hill, Helpmakar Hill, and
Devonshire Post. It was the most vulnerable part of the
defences—low-lying, exposed to reverse and enfilade fire
by heavy guns, and approached most nearly by cover
for attackers. It looked indefensible, and yet it was never
seriously attacked. The chief credit for this immunity
was undoubtedly due to its commander.

With two brother subalterns[1] of the Regiment, Knox
had gone to Turkey during the Russo-Turkish War of 1877,
and had seen in the Shipka Pass what could be done by
entrenchments. The lesson was not forgotten, and when
he was entrusted with the defence of Section A he
applied it with the full force of his personality. No precaution was neglected—observers, alarm signals, clear-
ances, obstacles, communications—all were provided for.
Exposed to the converging fire of 96-lb. shell, *protection* was all important. Bales of bhoosa for parapets,
canvas screens for defilade, must serve until time and
material allowed of more solid construction—then
every man not under arms must be at work building.
When parapets were eight foot thick it was time to bring
them up to twelve.

Eventually the whole Section became a continuous
fortification, but this was only effected by the exercise of
unceasing viligance and relentless determination. "The
Colonel-on-the-Staff has noticed" heralded many a sharp
reproof for slackness or stupidity, but good work was as
freely recognized. Posts were officially named after the

---

[1] Major-General Sir F. Eustace and Brig.-General Lord Playfair.
They had a regimental friend at court in Sir C. Dickson, who had
been specially sent to Constantinople to act as principal military
attaché.

regiments that constructed them, and in his farewell order he placed on record his high appreciation. "Begun under heavy shell fire, and continued and perfected by men enfeebled by short rations, these defences must for long remain a monument to the willing hands and stout hearts that made and manned them. The Colonel-on-the-Staff feels confident that he voices every man's mind when he expresses his regret that the enemy did not give the works the honour of a serious attack".

EXPENDITURE OF AMMUNITION.

UNIT	Talana Hill 20th Oct.	Elands-laagte 21st Oct.	Reit-fontein 24th Oct.	Lombard's Kop's[1] 30th Oct.	Defence of Ladysmith	TOTAL
Naval Brigade						
4·7″	—	—	—	46	514	560
12-pr.	—	—	—	98	784	882
3-pr.[2]	—	—	—	—	80	80
1st Brigade, R.F.A. ...						
13th Battery ...	453	—	—	467	205	1,125
67th do. ...	158	—	—	220	773	1,151
69th do. ...	626	—	—	183	1045	1,854
2nd Brigade, R.F.A.						
21st Battery ...	—	225	—	546	750	1,521
42nd do. ...	—	198	326	692	617	1,833
53rd do. ...	—	—	354	314	315	983
Section 10th Mountain Battery, R.G.A. ...	—	—	125	57	101	283
Maxim Nordenfeldts...	—	—	—	32	48	80
Howitzer Detachment	—	—	—	—	776	776
Natal Field Artillery ...	—	74	—	—	—	74
Totals ...	1,237	497	805	2,655	6,008	11,202

[1] Including 31st October and 1st November.
[2] Natal Naval Volunteers.

CHAPTER XXVII.

THE DEFENCE OF CAPE COLONY.

The Arrival of the Army Corps—Stormberg—Colesberg—Belmont —Graspan — Modder River — Magersfontein — The Defence of Kimberley.

MAPS 13, 16, & 17.

CHAPTER
XXVII.

1899.

The arrival
of the Army
Corps.

31st Oct.

ON the 31st October Sir Redvers Buller, who had come on ahead of his troops, reached Cape Town. On the previous day Sir George White had been hustled back into Ladysmith with the loss of a detached portion of his force : Natal lay open to the invaders and Boer commandos were gathering on the Orange River with the apparent intention of invading the central districts of Cape Colony; Sir Redvers decided that his plan of an ordered advance through the Free State must be abandoned, and the troops sent on, as they arrived, to the most threatened points, irrespective of the Order of Battle. Thus there went :—

To Natal—the divisional artillery of the 1st and 2nd divisions.

To East London—the divisional artillery of the 3rd division.

This left in the neighbourhood of Cape Town only the corps artillery and the horse artillery of the cavalry division.

Sir Redvers Buller decided to proceed himself to the danger spot in Natal, leaving an "appreciation of the situation" for the guidance of Sir F. Forestier-Walker. In this he explained that in view of the supreme importance of keeping Cape Colony from rebellion he had formed a strong column under Lord Methuen ready to take the field in the direction of Kimberley ; that he was forming a force of mounted men and horse artillery under French to attack Colesberg and that as soon as this could be occupied Gatacre's force should be advanced to Stormberg.

We can now consider the operations of these three forces separately, commencing, for convenience of treatment, with that on the right.

Early in November, 1899, the Boers crossed the Orange River at Bethulie, and the little garrison of Stormberg was withdrawn "according to plan". The enemy's advance was slow and hesitating, however, and General Gatacre, who had come out in command of the 3rd division, and still had with him his divisional artillery[1]— less the 79th battery—was collecting troops at Queenstown. By December he felt strong enough to take the offensive against the Boers who had contented themselves with taking up a position covering Stormberg.

A tedious railway journey to Molteno on the afternoon of the 9th was followed by a distressing night march, that (owing to a mistake on the part of the guide) amounted to not less than a dozen miles, some of it very rough going. The wheels of the 77th battery had been bound with raw hide to deaden the sound; the 74th—only included at the last moment—had no opportunity of taking this precaution. But in practice it was found to make the draught very heavy without materially reducing the noise.

The first streak of dawn found the force trudging along in column of route at the foot of a rocky ridge, very uncertain of its own whereabouts, or that of the enemy. Suddenly a rifle shot rang out, and then the ridge burst into flame. The goal had been reached—to the surprise of both sides.

Marginal notes:
CHAPTER XXVII.
1899.
The arrival of the Army Corps.
Stormberg.
9th Dec.
10th Dec.

---

[1] *VI Brigade, R.F.A.*

Commander	—	Lieut.-Colonel H. B. Jeffreys.
Adjutant	—	Capt. S. W. W. Blacker.
74th Battery	—	Major R. G. McQ. McLeod.
77th ,,	—	Major E. M. Perceval.
79th ,,	—	Major E. H. Armitage.
Ammunition Column	—	Major R. F. McCrea.

The infantry were leading, the guns behind them, the mounted infantry bringing up the rear. There was no time to issue orders for attack. The infantry formed to their right to face the ridge, which ran parallel to the track they were following: the batteries cleared off the road to the left, and then, wheeling by sub-divisions to the right again, were led forward by Colonel Jeffreys in column of sections so as to get to the flank of the infantry. Their need for immediate support was too pressing to allow of the delay which would have resulted from falling back to the only good artillery position, which was at some distance behind.

The 74th took the first chance that offered, and swung "Right Wheel into Line" close on the left of the infantry. But a donga, five or six feet deep, lay across their line of advance, and one gun failed to clear it—to be at once smothered under a hail of bullets. Major Perceval took the 77th further on, to what appeared to be an isolated kopje, in the hope of enfilading the enemy's line. It turned out, however, to be only a spur from the ridge, and as he rode on to the crest he was shot at close range by the Boer picquet. Lieut. Clark, following with the leading section, man-handled his guns into action, and cleared the ground with a few rounds of case.[1] Five guns were ultimately got up, though the slope was steep and rough, and shrapnel fire was opened against the Boer position. But the sun was just rising over the hill behind it, full in the gunners' eyes, and against the blaze of light the whole foreground was blacked out.

The prompt action of the artillery had covered the dangerous moments while the infantry were shaking out, but the positions hurriedly taken up by the batteries left much to be desired. Moreover, the infantry were finding their advance barred, as much by the difficulties of the ground as by the fire of its defenders. Sheer scarps, only climbable in places, ran across the face of the ascent,

---

[1] The 74th had also fired a few rounds of case at a Boer picquet on their spur.

and proved a severe task to men just off a month's voyage, and exhausted by a trying night march. After the guns had been in action for half-an-hour or so, General Gatacre decided to break off the action, and Colonel Jeffreys accordingly withdrew them to the more central position mentioned above, from which they would be able to cover the withdrawal of the infantry. The move was carried out by the batteries alternately, and they came into action in line, the 77th on the right, 74th on the left. Here the brigade faced the centre of the enemy's position at a range of about a couple of thousand yards, and under their shrapnel the musketry of the riflemen lining the ridge soon slackened. A gun opened, firing segment very slowly with black powder charges, but did little harm.

Of the two battalions in the attack, the left was climbing steadily, while the other had just received the order to retire, when the 74th came into action. The battery opened fire at once, taking range and fuze from the 77th, but unfortunately the part of the ridge opposite the battery was that where the leading ranks of the left battalion were nearing the crest, and their figures among the rocks and scrub were taken for the Boers—as had happened at Talana Hill. The mistake was realised in the 77th, and an attempt made to correct it before a gun was fired, but just too late, and, by an unhappy chance, the shell burst over the battalion commander, mortally wounding him and causing several other casualties. The result was a general recoil. The infantry on the left rallied behind the guns, the rest dribbled back, but many, completely exhausted, lay down under the shelter of the rocks on the hillside, or crowded into the donga at its foot.

It was evident that the infantry were not in a condition to renew the attack. Fatigue and want of sleep had taken the heart out of many of them, and General Gatacre ordered a withdrawal to Molteno, covered by the fire of the artillery brigade. This held the enemy off while the column got under weigh, and the batteries had been a couple of hours in their second position before they got the order to fall back in their turn—alternately as before. The 77th had just moved off when it was confronted by a fresh commando riding up on to a ridge facing the battery. There was no time to look for a position, it was a case of "Halt, Action Front". The bullets were soon knocking up the dust round the guns, Major Perceval gave the order to serve them kneeling—probably for the first time in field artillery—and the shrapnel at twelve hundred yards was more than the burghers could stand. It had been a fair duel between guns and riflemen, for the battery[1] was without escort and right in the open, and the guns had won.

Meanwhile the 74th had dropped back in its turn, and Gatacre himself gave the order for the 77th to follow. Unfortunately one gun was caught in a quicksand when crossing a river bed and abandoned (by order) before a second team could be brought back to pull it out. Otherwise the withdrawal was effected without incident. The two batteries had fired 516 rounds, and in the words of the Official History, "By their steadiness and the excellence of their practice had held the enemy at bay".

Back at Sterkstroom reinforcements were received, including the third battery of the brigade; and the Boers, contented with their repulse of Gatacre's attack, left him unmolested to recover strength.

About the same time that the Boers who had crossed the Orange River at Bethulie advanced to Stormberg, a similar commando seized the bridge at Norval's Pont. Here again their advance was cautious and hesitating, although the little garrison had been withdrawn from Naauwpoort. It was only on the 14th November that the Boers reached Colesberg, but it was plain that they must not be allowed to penetrate further into Cape Colony if the spread of disaffection was to be avoided. Before leaving for Natal, Sir Redvers Buller sent what few troops he could

---

[1] Five guns only for one had been dropped on a ridge to cover the move of the 74th, and was firing in the opposite direction.

scrape together to re-occupy Naauwpoort, and entrusted the
task of organizing the defence on the central line to
General French,[1] who had come out in command of the
cavalry division. His force was chiefly composed of
mounted troops with the horse artillery of that division :—

Commander	—Lieut.Colonel F. J. W. Eustace.
Adjutant	—Captain A. d'A. King.
"O" Battery	—Major Sir J. H. Jervis-White-Jervis.
"R" do.	—Major B. Burton.
Ammunition Column	—Major C. E. Maberly.
do. do.	—Major S. Belfield.

Under French's bold and energetic leadership there
were daily reconnaissances, ever increasing in audacity as
his strength grew, and this activity soon produced its effect
on the enemy. As their outposts were drawn in, French
followed close, and occupied Arundel as a pivot from which
he could strike first to right and then to left. The force
had now grown so that it could be organized as a division,
with two brigades of mounted troops—one on each side of
the railway—the infantry, with the section of 68/R.G.A.
(9-prs.) holding the centre, and the horse artillery and
New Zealand mounted rifles as a reserve in the hands of
the general. There were many brisk encounters in which
the horse artillery took their full share—sometimes as a
brigade, sometimes by batteries, frequently as sections.

The little town of Colesberg lies in a hollow, surrounded
on all sides except the north by hills. On the south side
these formed a very strong position, and French had no
intention of knocking his head against it, but he ushered
in the New Year by the capture of a hill on the south-west
from which it seemed a way might be found to the open
ground on the north. The hill was taken by the infantry
under McCracken—whose name it was henceforth to bear
—and the cavalry pushed on, while for three hours the

---

[1] General French had slipped out of Ladysmith, by Buller's order.
in the last train that got through. His staff officer was Major D.
Haig.

CHAPTER
XXVII.
1900.
Colesberg.
4th Jan.
6th Jan.
horse artillery brigade out in the open had to endure the fire of a 15-pr. and a couple of pom-poms snugly ensconced among the rocks on the kopjes several hundred feet above them. There was hot fighting again on the 4th January, 1900, and on the 6th an attempt to rush at night a hill commanding the Boer communications failed somewhat disastrously. But on the same day the reinforcements included the 4th battery, R.F.A., and before he had been a week with the force Major A. E. A. Butcher had astonished friend and foe by placing one of his guns on the top of Coles Kop. This was a solitary hill, so steep as to be almost precipitous, which stood up 800 feet above the plain three miles west of Colesberg. It was a formidable climb for an unencumbered man, but, with the assistance of the Royal Engineers and fifty men of the Essex Regiment, the gun was got up on the 11th, a second followed it on the 16th, and a wireway was then constructed for hauling up ammunition, stores, etc. From Coles Kop the town, and the Boer laagers round it, were in full view, and great was the amazement of the burghers when they found shells bursting among them.

15th Jan.
A few days later they made a determined attack on our post at Slingersfontein, which threatened their communications with Norvals Pont, and were on the point of success when a brilliant bayonet charge by a party of New Zealanders led by Captain W. R. N. Madocks, R.A., cleared them out. Right up to the end of January there was constant activity, but on the 29th of that month General French was summoned to Cape Town to be told his part February. in the new phase about to open. When he returned it was only to break up his command and take away with him the horse artillery and nearly all the regular cavalry.

Clements was left with a remnant of the mounted troops, an infantry brigade, "J" battery R.H.A. (Major P. H. Enthoven) just arrived from India, the 4th battery, R.F.A., and sections of 37/R.F.A. and 68/R.G.A. The front which he took over with this attenuated force was about 25 miles in length, and the Boers soon found out

how thinly it was now held.   Fighting became continuous
all along the line, and Clements was hard put to it to find
troops to reinforce each threatened point.[1]   On the 9th
February the pressure on the right about Slingersfontein
became particularly severe, and a party of Australians
under Captain H. G. Moore, R.A., emulated the feat of
the New Zealanders under another gunner, just mentioned.   "J" battery was kept busy throughout the day,
four guns firing over 500 rounds, and the enemy's attempt
at envelopment was arrested.   But the unequal contest
could not be maintained indefinitely, and on the 12th
Clements decided to draw in his flanks.   This was
effected during the night, although it involved bringing
down the two guns of the 4th battery from their eyrie on
Cole's Kop.[2]   The delicate operation of lowering the first
was successfully accomplished, but time ran out, and the
second had to be thrown over the cliff.

On the following night the withdrawal was resumed to
Arundel, where Clements took post to cover the railway
junction at Naauwpoort.   Here reinforcements began to
arrive, including the 2nd and 39th batteries, R.F.A.
(Major T. Slee and Captain H. J. Brock), and half the
57th company, R.G.A. (Captain R. E. Home) with two
5″ guns.   It was now possible to resume active operations,
and in these Major Butcher soon gained the confidence of
all arms for his leading of detached forces.   But we must
now leave the central line by Colesberg, and turn our
attention to the western.

Of Methuen's original division (the 1st) only one
brigade (the Guards) was left, but the division was reconstituted, with this as a nucleus.   The IIIrd brigade
R.F.A. (which had come out with the last reinforcements
before the mobilization of the Army Corps) became his

---

[1] On one occasion the two howitzers of the 37th (Lieut. H. R. W. Smith) were brought safely across from left to right within range of the Boer guns, disguised under tarpaulins and drawn by oxen.

[2] They had fired a couple of thousand rounds from the top.

divisional artillery, with in addition a naval brigade armed with four 12-prs. of 12-cwt.

IIIRD BRIGADE, R.F.A.

Commander	—Lieut.-Colonel F. H. Hall.
Adjutant	—Captain F. B. Johnstone.
18th Battery	—Major A. B. Scott.
62nd do.	—Major E. J. Granet.
75th do.	—Major W. F. L. Lindsay.
Ammunition Column	—Major N. E. Young.

Belmont.

On the 21st November Methuen marched out of Orange River Station, leaving the 62nd battery and an infantry brigade to guard his communications. He had not far to go to find the enemy : they were waiting for him on a line of kopjes flanking his line of advance near Belmont station. Under cover of a reconnaissance the troops were moved up to striking distance on the afternoon of the 22nd Nov. 22nd, and during this operation an unfortunate incident occurred that was to cripple the mobility of the two field batteries throughout the fighting which followed. An advanced party was busy preparing the bivouac when a Boer gun opened fire upon them, and the artillery was called for. Colonel Hall, anxious to relieve the annoyance, trotted the whole way. It was not very far, but the horses were still weak from their long sea voyage and the change of seasons, and the day was very hot. Five died from exhaustion, and the remainder were long in recovering their full powers.

23rd Nov. The line of kopjes occupied by the Boers ran parallel to the railway a couple of miles to the east, with a lower ridge, as an outwork, between. This advanced line was attacked by the two brigades side by side, the 18th with the right, the 75th with the left, the naval brigade in the centre. Starting before dawn, the troops were met by a violent burst of musketry with the first glimmer of light. There was no hesitation, however, and the first rush carried the centre of the position. On the left the resistance was sturdier, and the 75th battery working round the flank came under heavy musketry fire.

Meanwhile the battalions on the extreme right had overlapped the covering ridge, and pressed on to the attack of the main position, supported by the 18th battery at short range—1,375 yards. By the time they had gained a footing on the kopjes the rest of the attackers were pouring across the valley between the ridges, and the burghers were galloping off across the plain.

The field batteries could not see what was happening in the centre, and the naval guns were too heavy to get up on to the kopjes to turn the retreat into a rout. The 18th tried to get round the end of the ridge, but the horses had had no water all day, and could not bring the guns along in time.

Only eleven miles further on the enemy barred the way again—this time on a semi-circle of small kopjes astride the railway. Once again the troops were moved up in the afternoon, and once again the artillery horses were the sufferers, for the water supply at the bivouac was so scanty that none could be spared for them.

Under the impression that the enemy were in small force and could be shelled out of their position, the guns were pushed into action early—those of the naval brigade[1] which had been brought up by train, alongside the line at 5,000 yards, the 75th on their right, the 18th further out on that flank. But the Boers had received considerable reinforcements during the night, including three Krupp field guns and a couple of pom-poms under Major Albrecht, commandant of the Free State Artillery, and had no intention of abandoning their position without a fight. At 7 a.m. Lord Methuen issued orders for a formal attack. On the left the naval guns and the 75th battery advanced parallel to the line, and engaged the Boer guns from successive positions at 4,000 and 2,300 yards. On the

CHAPTER
XXVII.

1899.

Belmont.

23rd Nov.

Graspan.

25th Nov.

---

[1] Only two could be brought into action as the greater part of the naval brigade was formed into a battalion of infantry.

right the 18th battery shelled the centre and left of the enemy's position at ranges of 2,975, 2,425, and 1,425, detaching one section still further to the right. The 75th inclined towards the 18th so as to increase the artillery support of what was becoming the main attack, and the infantry dashed for the crest. But the burghers had not waited for them, and when Hall got his guns up on to the ridge they were out of range. As at Belmont, the 18th was sent forward, and made a great effort to stop the escape of the Boer laager, but only got up in time to fire a dozen rounds at 4,800 yards as the last wagons disappeared.

As at Belmont the enemy's position had been gallantly carried by a direct frontal attack, but the infantry had suffered severely, and the lack of cavalry and horse artillery had again denied them much of the fruits of victory.

After a badly needed rest on Sunday, the advance was resumed on the 27th November. The enemy were soon found in position again, but this time not on the kopjes—which made such an admirable target for the British guns—but on a river line. On the far side the village of Modder River clustered round the railway station, with that of Rosmead a mile or so to the west, and these, it was thought, formed the Boer position. No one suspected that the main force of the defence was lying hidden in carefully concealed entrenchments all along the river, which itself was only distinguishable by the line of bushes fringing its banks. It could be approached only across a plain sloping gently downwards—a bare expanse, dotted with ant-bear heaps, absolutely destitute of cover.

Advancing side by side to the attack, as on previous occasions, the two brigades were met by a violent outburst of musketry as soon as they got within a thousand yards of the river. But on this occasion the fire, directed from the level of the plain, and sweeping horizontally over it, proved far deadlier than before. Further advance became

impossible, and there the infantry lay prone—engaged in
a desultory fire-fight with an unseen enemy. The turn
of the artillery had come.

The naval brigade, brought up as at Graspan by train,
engaged the enemy's artillery from a knoll on the left of
the line at a range of about 5,000 yards. The two field
batteries dashed to the front, and came into action on the
other side of the railway, first at 2,000 and 2,200, then
at 1,200 and 1,400 yards. In order to relieve the infantry
by keeping down the fire from which they were suffering,
the batteries shelled the buildings and enclosures in
which the riflemen were thought to be, and engaged
the guns whenever they could be located. The latter
were constantly shifting position, and the British guns,
though so superior in number, were out in the open and
under musketry fire; so the contest was by no means
one-sided. And there the two batteries stood until four
o'clock, when, owing to casualties[1] and want of ammunition, the 75th was ordered to fall back out of rifle range.

As it was doing so it met the third battery of the
brigade, the 62nd, which staggered on to the field, a
welcome reinforcement, having covered 62 miles in 28
hours.[2]

Meanwhile attempts had been made to work round the
Boer flanks, and on the left a few men got across the river,
assisted by a section of the 18th[3] sent by Colonel Hall to
help the infantry. Gradually more men were collected
and sent over, and Rosmead, which had been evacuated by
the Boers under the fire of the section of the 18th, was
occupied. Further advance along the north bank was,
however, found impossible in face especially of three guns
under Major Albrecht which were fought with great deter-

---

[1] The battery had lost four officers, 19 other ranks, and 22 horses.
[2] Starting from Orange River Station at 10 a.m. on the 27th, they
had reached their halting place on the 28th, when they got news of the
battle and pressed on. Six horses fell dead in the traces, forty more
never recovered.
[3] Under Captain G. T. Forestier-Walker.

CHAPTER
XXVII.

1899.

Modder
River.

28th Nov.

mination even after they had been deserted by their escort. But during the night an infantry brigade was slipped into Rosmead, and this threat to their line of retreat was more than the burghers could endure. When morning broke the position was empty.

The despatch bore generous tribute to the work of the 18th and 75th. "If I can mention one arm particularly it is the two batteries of artillery". They had each fired over a thousand rounds. The expenditure was :—

> 18th Battery —1,029.
> 62nd do. — 247.
> 75th do. —1,008.

Lord Methuen was now only twenty miles from Kimberley, and learnt that the town could hold out for another six weeks at least. With his anxiety on this account allayed, he felt that he could give his men and horses a rest. They had been marching and fighting almost continuously for a week, over a poorly watered country in midsummer heat, and needed it sorely.

By the 10th December he was ready to move on again. A temporary bridge had been constructed to take the place of that blown up by the Boers, field works had been thrown up to cover it, and important re-inforcements had been received. These included the infantry brigade left behind on the lines of communication, a 4·7″ gun for the naval brigade, and two batteries from the corps artillery :—

"G" battery, R.H.A.—Major R. Bannatine-Allason.

65th (howitzer) battery, R.F.A.—Major W. Tylden.

Only half-a-dozen miles from the British camp at Modder River a line of rising ground stretched diagonally across the front. The first three miles from the right consisted of a low scrub-covered ridge : then came Magersfontein Hill, a grim, rock-bound kopje, which dominated the plain between it and the river ; and from it away to the north-west the ridge continued for miles but at a lesser

elevation. From Modder River to Magersfontein Hill the
plain rose gradually with two well-marked knolls—Headquarters Hill and Horse Artillery Hill, as they came to
be called. The ground was generally open, but with
patches of scrub.

For reasons which need not delay us, Lord Methuen
decided upon an assault at dawn on the key of the position
—Magersfontein Hill—preceded by an artillery bombardment on the previous afternoon. Between three and four
o'clock, therefore, on the 10th December the whole of the
guns (except those naval 12-prs. which were mounted on
the works guarding the bridge) moved out into action,
duly escorted. The 4·7″, just to the west of the railway,
shelled the kopjes at 7,000 yards. The howitzer battery,
on the other side of the line near Headquarter Hill, and
the three batteries of the IIIrd brigade, somewhat more
forward on the right, shelled what was thought to be the
Boer position. No one knew that a hundred yards from
the foot of Magersfontein Hill a trench had been dug,
and the earth levelled off so that it was practically
invisible. At 6.30 p.m. the fire ceased—there had been
no reply.

During a night of pitchy darkness and torrential rain
the Highland Brigade stumbled forward, pointing for
Magersfontein Hill, and were in the act of deploying at
about 4 a.m. when they were struck by heavy rifle fire.
The brigadier fell among the first, and there was some
disorder, as was inevitable, but there was a general surge
towards the enemy, and some gallant efforts to storm the
position. From one cause or another these all failed, and
the brigade gradually dissolved into an irregular line of
men lying out on the plain, some four or five hundred
yards from the foot of the hill.

"The accurate and well-sustained shooting of the
artillery now saved the brigade from destruction".[1] As
soon as it was light enough to lay their guns the IIIrd

---

[1] Official Account.

CHAPTER
XXVII.

1899.

Magers-
fontein.

10th Dec.

11th Dec.

CHAPTER
XXVII.

1899.

Magers-
fontein.

11th Dec.

brigade opened fire on Magersfontein Hill, dropping to
1,700 yards as soon as they discovered that the trenches
were at its foot. The effect was soon evident in a
diminution of the musketry fire, and Colonel Hall then
sent the 18th battery in to close quarters, supporting it
shortly afterwards with the 62nd. There these two batteries remained in action for the rest of the day at a
range of 1,300 to 1,400 from the *schanzen* half-way up
the hill, 1,100 to 1,200 from the trench at its foot.

It is time now to turn our attention to events further
eastward, in which direction General Babington had taken
the cavalry and mounted infantry, supported by "G"
battery, in the hope of turning the enemy's left. The
sudden roar of musketry was the first indication he had of
the fate of the infantry attack, and he at once despatched
"G" in that direction, with orders not to come into action
until he was stopped by bullets or could get a clear view
of what was going on. As the battery trotted forward
Major Bannatine-Allason was able to glean a good idea of
the situation from the scattered men who kept getting up
out of the scrub "like partridges out of roots", and
accordingly directed his course towards the knoll since
known as "horse artillery hill". To reach this the wire
fence marking the frontier of the Free State had to be
crossed, and while it was being cut the battery came
under rifle fire, which became very hot as the guns were
unlimbered on the crest. Fortunately the recoil soon took
them back a little down the reverse slope, and they were
not run up again.[1] From here "G" shelled Magersfontein
Hill at 2,200 yards until joined by the 75th, when Major
Bannatine-Allason turned his fire on to the "low ridge" on
the right that overlooked his position at only 1,400 yards.
By this time the cavalry and mounted infantry had been
reinforced by the Guards, and these troops held up the
Boers on the ridge for the rest of the day, assisted by the
howitzers of the 65th battery which, thanks to balloon

---

[1] Thanks to this cover the battery was able to remain in action for
24 hours with trifling loss.

observation, were able to search the ground behind the ridge.

In the centre affairs had assumed a serious aspect. A little before mid-day the Highlanders had begun to dribble back, but were re-formed behind the two field batteries under cover of the rapid fire of the guns. When, however, some shell fell among them they broke again, and for the rest of the day the sole defence consisted of the line of guns and the battalion of the Guards that formed their escort. All night they held the ground.

The expenditure of ammunition was :—

"G" Battery —1179 rounds.

III Brigade {
18th Battery — 940 do.
62nd do. —1000 do.
75th do. — 721 do.
65th do. —
}

After his repulse at Magersfontein Lord Methuen remained strictly on the defensive. On the 10th January Lord Roberts wrote to say that he might have to withdraw some of his troops, but would comply with his request for four 4·7″ siege guns "which should make his lines practically impregnable". Sure enough one of the siege train companies (92/R.G.A.—Major E. G. Nicholls) duly arrived, and there were daily duels with the Boer guns which were dotted here and there along their miles of entrenchments—constantly changing their position from one emplacement to another. The field howitzers occasionally joined in, with the help of the balloon to discover concealed targets, and in February a 6″ gun was brought up on the railway. It is time to turn our attention to affairs in Kimberley.

It was obvious that one of the first acts of the Boers would be an attempt to capture Kimberley, and in September Colonel Kekewich was sent to take such precautionary measures as were possible. On the 20th of that month the local forces were reinforced by half his battalion

Chapter
XXVII.

1899.

The Defence
of
Kimberley.

September.

and 31/R.G.A., which, as we have seen, had been converted from a "coast defence company" into a "provisional field battery"—though armed with mountain guns. The local artillery consisted of the artillery troop of the Diamond Fields Horse, and, thanks to Major W. H. F. Taylor, R.A., commanding the artillery of the Cape Colonial Forces, this battery had recently exchanged its four old 7-prs. for six 2·5″ guns. The guns of the 31st were installed in pairs in the defences, with also a pair of 7-prs. brought in by one of the police detachments. The Diamond Fields battery was kept in reserve and used with the mobile force. But for the twelve 2·5″ there were only 2,600 rounds.

The defences consisted of a circle of redoubts which had been hastily thrown up when war was seen to be inevitable, in accordance with the original defence scheme. The length of the perimeter was even more out of proportion to the strength of the garrison than that at Ladysmith, but the burghers never made even a pretence of assaulting.

From an artillery point of view there is little to be said about the defence, except that Major Chamier (of the 31st), with the local rank of lieut.-colonel, was frequently in command of troops in the reconnaissances and demonstrations which were a feature of the defence. The enemy's bombardment was fitful and even less effective than at Ladysmith. They had only nine guns, and those field guns, for most of the time, and to deal with these the de Beers workshops designed and made a 4·1″ B.L. with a 28-lb. shell and a range of 8,000 yards. Named the "Long Cecil" in compliment to Mr. Rhodes, it opened fire on the 19th January 1900,[1] and fired altogether 255 rounds. But the enemy were not to be outdone, and they brought up the "Long Tom" which had been disabled in

----

[1] The designer of Long Cecil, Mr. Labram, Chief Engineer of the de Beers company was unfortunately killed by a shell only a few days before the relief. There is a model of the gun in the museum of the Royal United Service Institution.

the Ladysmith sortie, and repaired at Pretoria. Fortunately it did not get into action until relief was almost in sight, for the great 96-lb. shell proved somewhat trying to the nerves of the inhabitants accustomed only to those of 9-prs.

On the 13th February the welcome sight of burghers trekking homewards gladdened the eyes of the garrison; on the 15th French's cavalry came riding in, and Kekewich sallied forth to try and cut off the retreat of the Boer guns.

CHAPTER
XXVII.

1900.

The Defence
of
Kimberley.

15th Feb.

CHAPTER XXVIII.

The Invasion of the Orange Free State.

The arrival of Lord Roberts—The Plan of Campaign—The Army of Invasion—The Relief of Kimberley—Waterval Drift—The Pursuit of Cronje—Paardeberg—Poplar Grove—Driefontein—The Occupation of Bloemfontein—Cape Colony.

MAPS 13 & 19.

CHAPTER
XXVIII.

The Arrival
of Lord
Roberts.

In the course of the "Black Week"—10th to 17th December, 1899—the British arms had suffered three severe defeats at Magersfontein, Stormberg, and Colenso. It was decided to send the veteran Field-Marshal, Lord Roberts, Colonel-Commandant R.A., to take the supreme command. The artillery staff of the Army Corps brought out by Buller, were still at Cape Town where they joined him.

1900.

10th Jan.

The new Commander-in-Chief arrived in Cape Town on the 10th January, 1900, and found the situation had improved during his time at sea. In Natal the 5th division had arrived and Buller was preparing for another attempt to break through to Ladysmith, where the garrison had just shown its undaunted spirit by repulsing a serious attack. In Cape Colony, Gatacre was firmly established at Sterkstroom, and the Boers seemed contented to leave him unmolested. At Colesberg French's ceaseless activity had foiled any attempt to advance into the central districts. In the west Methuen and Cronje were facing each other in their lines, inactive except for artillery duels. Kimberley and Mafeking were still holding out, and Plumer was collecting a force in Rhodesia for the relief of the latter.

The position as regards British reinforcements was that a brigade of three batteries of horse artillery was on its way from home, and two batteries from India; the 6th division was on the point of embarkation; and the mo-

Headquarter Staff, R.A., South Africa, April 1900.

Capt. C. D. Kirby, Lieut.-Col. H. C. Slater, Major J. E. W. Headlam and
Major-General G. H. Marshall.

bilization of the 7th division had been ordered. On landing Lord Roberts received a telegram stating that, in addition to the above the despatch of the following further artillery reinforcements had been arranged :—

Four brigades (one howitzer) of field artillery.

Four companies of Militia artillery.

The Elswick Volunteer field battery.

The City Imperial Volunteers, including a field battery.

Colonial contingents, including a field artillery brigade from Canada, and a field battery from Australia.

The 4th cavalry brigade and 8th division to follow if required.

The brigades and batteries of horse and field artillery would provide for the colonial forces, and the regular divisions would, of course, bring their own field artillery.[1] But heavy artillery was necessary to meet the long-range guns brought into the field so unexpectedly by the enemy. The Navy had come to the rescue when this want was first discovered, now it was to be the turn of the Royal Garrison Artillery. During December, January, and February there were sent out to supplement the naval guns, and to take them over when the sailors were recalled to their ships, a siege train and a dozen companies.

Before he had left England Lord Roberts had decided on an advance by the western line, which had been taken by Lord Methuen. This was the only one of the three railways from Cape Colony on which the bridge over the Orange River was in our hands, and by this way alone could the relief of Kimberley be effected in time. But the Commander-in-Chief had no intention of being tied to the railway. His plan was to cut loose from it, and strike across country into the Free State when the

---

[1] When it came to the embarkation of the 8th division it was decided that its divisional artillery must be kept in England for home defence as being the only brigade fit for service.

CHAPTER
XXVIII.

1900.

The Plan of
Campaign.

The Army
of Invasion.

February.

appropriate moment arrived. To attain this freedom of movement it was essential that the transport should be re-organized, and to this task Lord Kitchener applied himself as soon as he landed, assisted by Sir William Nicholson, who became Director of Transport, with Captain W. T. Furse, R.A., as staff officer.

By early in February everything was in order, and Lord Roberts left Cape Town, arriving at Modder River on the 8th. The army of invasion was gradually assembling about the scenes of Methuen's early victories at Belmont and Graspan. In round numbers it consisted of 8,000 horse and 25,000 foot, with just on a hundred guns. The artillery was as follows :—

G.O.C.—Major-General G. H. Marshall.

A.D.C.—Captain A. D. Kirby.

A.A.G.—Lieut.-Colonel H. C. Sclater.

D.A.A.G.—Captain J. E. W. Headlam.

Cavalry Division.

R.H.A., 1st Cavalry Brigade.

Commander   —Lieut.-Colonel A. N. Rochfort.
Adjutant   —Captain J. C. Wray.
Q Battery   —Major E. J. Phipps-Hornby.
T   do.   — do.   F. B. Lecky.
U   do.   — do.   P. B. Taylor.
Ammunition Column — do.   E. C. Holland.

R.H.A., 2nd Cavalry Brigade.

Commander   —Lieut.-Colonel W. L. Davidson.
Adjutant   —Captain H. F. Askwith.
G Battery   —Major R. Bannatine-Allason.
P   do.   — do.   Sir G. V. Thomas.
Ammunition Column —Captain J. P. Du Cane.

R.H.A., 3rd Cavalry Brigade.

Commander —Lieut.-Colonel F. J. W. Eustace.[1]

Adjutant —Captain A. d'A. King.

O Battery —Major Sir J. H. Jervis-White-Jervis.

R do. — do. B. Burton.

Ammunition Column — do. C. E. Maberly.

CHAPTER
XXVIII.

1900.

The Army
of Invasion.

February.

6TH DIVISION.

X Brigade, R.F.A.[2]

Commander —Lieut.-Colonel J. M. McDonnell.

Adjutant —Captain W. H. Onslow.

76th Battery —Major R. A. G. Harrison.

81st do. — do. H. A. Chapman.

Ammunition Column — do. A. Bell-Irving.

Two 12-prs. of 12 cwt.—Naval Brigade.

7TH DIVISION.

III Brigade, R.F.A.

Commander —Lieut.-Colonel F. H. Hall.

Adjutant —Captain F. B. Johnstone.

18th Battery —Major A. B. Scott.

62nd do. — do. E. J. Granet.

75th do. — do. W. F. L. Lindsay.

Ammunition Column — do. N. E. Young.

9TH DIVISION.

65th (How.) Battery —Major W. Tylden.

82nd do. — do. A. S. Pratt.

Ammunition Column —Lieut. A. A. Montgomery.

Two 4·7″—Naval Brigade.

CORPS TROOPS.

91st Company, R.G.A.—Major J. R. H. Allen.

---

[1] Until appointed A.A.G., R.H.A., on the staff of the Cavalry Division.

[2] Less 82nd Battery temporarily allotted to the 9th Division, whose artillery had not yet arrived.

CHAPTER
XXVIII.

1900.

The Army
of Invasion.

February.

This left the following artillery with Methuen :—

20th Battery, R.F.A. —Major C. H. Blount.

37th   do.[1]   do. — do. R. A. K. Montgomery.

38th   do.   do. — do. H. E. Oldfield.

92nd Company, R.G.A.— do. E. G. Nicholls.

With two 4·7″ and two 12-prs. (12 cwt.) of the Naval
Brigade.

As starting point for his invasion Lord Roberts
selected Ramdam, a farm with a good water supply just
across the Free State frontier east of Graspan. Alarmist
messages from Kimberley, whose inhabitants found the
bombardment by the newly arrived "Long Tom" more
than they were prepared to put up with, rather hurried on

the preparations, and in the small hours of the 12th
February the cavalry division rode out of Ramdan by
moonlight. French had orders to "relieve Kimberley at
all costs". Striking due east he seized the drifts of the

Riet river without difficulty, and next day turned north for
those on the Modder. The drifts were gained, and the
rising ground beyond occupied, after a skirmish, but the
heat was intense, and the march had been long and waterless. The effect on the horses was almost disastrous—the
cavalry had over four hundred *hors de combat*, and the

horse artillery were no better off. The 14th had to be
devoted to rest and such measures as were possible. For
the artillery this meant taking the ammunition column
horses for the batteries, and going on with only one
column—that of the 2nd cavalry brigade—made up with
mules.

At nine o'clock on the morning of the 15th February
the cavalry division started on its forty-mile dash for
Kimberley, moving off in column of brigade masses, the
3rd leading, with the 1st and 2nd following in that order.
the guns in battery columns on the left of their brigades.

---

[1] Less one section with Clements.

Some three miles from the river the Boers had lined with
riflemen two ridges which converged to a nek through
which the way led, where they had also a couple of guns.
It was a difficult situation, for, if time was given, the
enemy would be reinforced and the position would be
entrenched.   French had to face very much the same
problem as that which had confronted Roberts at Charasia
twenty years before, and like him he chose the bold course.
The nek was about threequarters of a mile wide, and the
ground sloped up to it without wire fences or other
obstacles.   He ordered the leading brigade to gallop
straight for it in open order, the 2nd brigade, with its
horse artillery, to follow in support, the other two horse
artillery brigades to cover the charge by shelling the
ridges on either side.   This bold stroke was fully justified
by the result.   In spite of the heavy cross fire the position
was won with a loss of only fifteen in the whole division
exclusive of the horse artillery.   Rochfort's brigade on the
left had come under a very effective shrapnel fire and
suffered somewhat severely, two teams being knocked
over as they were going into action, and losing altogether
4 officers, 21 other ranks, and 36 horses—the latter including those of the brigade commander and adjutant.

The country was now open to Kimberley, and as the
cavalry approached, the investing force cleared out of the
way.   At six o'clock French rode in to the town.

The few hundred men that Kekewich had sent after
the Boers when they were seen to be moving off to the
north had been checked at nightfall, and next morning
French took the 1st and 3rd cavalry brigades to their
assistance.   The enemy were strongly posted, and the
wide turning movements tried the horses highly, but
somehow or other, Eustace's guns were got into action
and the final position was carried.   Nothing more could
be done, and for the moment we must leave the cavalry
plodding their weary way back to Kimberley in order to
glance at the progress of events further back.

The first incident to be recorded is a disaster which

CHAPTER
XXVIII.

1900.

Waterval
Drift.

16th Feb.

brought the name of "de Wet" into prominence. The three divisions had followed the cavalry at such intervals as were necessitated by the waterless nature of the country, and on the 15th February there remained at the drifts of the Riet river only the great convoy of ox-wagons carrying the reserve supplies for the army—the oxen resting and grazing after the labour of getting the wagons through the drift. On this prize de Wet swooped down. The little escort defended the laager stoutly, and troops (including the 18th battery) were sent back from the 7th division. But the Boers also were reinforced, and they had taken up a position from which they must be driven before the wagons could be moved. Lord Roberts decided to abandon the convoy rather than upset his plans by sending back more troops to save it. And so nearly two hundred wagons with an enormous mass of supplies fell into the hands of the enemy—and the army was soon very hungry. The writer, just arrived from India, and riding with his groom and a third charger to overtake Headquarters, ran right into the Boers, and got away only by hard riding. It was a fortunate encounter for he was able to turn back the whole party of foreign military attachés who were following, and would otherwise have been captured *en bloc*.

When the movement of the cavalry division from Ramdam was reported to General Cronje he refused to budge from his laboriously fortified Magersfontein lines, insisting that it was but a feint, like that on the other flank in the previous week. He was firmly convinced that the British army could not move away from the railway, and that the main attempt to relieve Kimberley must take the form of a frontal attack on his position. But on the 15th the cavalry were in Kimberley, infantry were on the Modder, and other masses were reported to be moving in the same direction. If he did not move now he would be cut off from the Free State. He had to give in, and, at 10 o'clock on the night of the 15th, his motley gathering of burghers, women, and children, began to move, making for the gap between the cavalry and the foremost infantry. Next morning the advanced troops of the 6th division, at Klip drift, saw the tail of the immense train of wagons trailing east across their front.

CHAPTER
XXVIII.

1900.

The Pursuit
of Cronje.

16th Feb.

Every effort was made to interrupt their passage, but the Boer rear-guard put up a good fight, falling back from ridge to ridge, and the horses of the mounted infantry and field batteries were exhausted. Several in the 76th battery dropped dead in their traces, and on one occasion the escort was driven in and the battery nearly captured. But the Boers were kept on the north of the Modder, which they must cross to get to Bloemfontein, and the other divisions were coming up.

It is time to return to the cavalry. We left French on the evening of the 16th back at Kimberley. There at last he got information as to the situation in general, and what was expected of him. He saw at once the importance of Lord Kitchener's suggestion that he should make straight for Koodoo drift where the main Kimberley— Bloemfontein road crossed the Modder. The brigades which had been in action that day were quite unfit to move, but the 2nd brigade was available, and that night he started off with it. It was a tedious trek of nearly thirty miles with weary horses, but one can imagine the feelings of the commander when he rode on to the ridge behind which the march had been concealed, and saw the whole of Cronje's commando lined along the north bank of the river, its head just approaching the drift. Once across, they would have a clear run to Bloemfontein, and the horse artillery was at once ordered into action. The range was only just over two thousand yards, and the first shell burst over the leading wagons as they plunged down the bank to the drift. The surprise was complete, and something like panic ensued. It was fortunate for the Boers that they had with them a disciplined unit in

CHAPTER
XXVIII.

1900.

The Pursuit
of Cronje.

17th Feb.
the Free State Artillery. Major Albrecht[1] brought his
four guns into action, and, although it only took Davidson's brigade ten minutes to silence them, the diversion
enabled the burghers to pull themselves together, and
even to threaten an offensive. There was a good deal of
skirmishing, but the wagons never moved, and French
clung on, determined that there should be no escape—
though his men and horses were starving.

Paardeberg. While the cavalry from the north were heading off
the Boers at Koodoo Drift, the 6th and 9th divisions were
pressing up the valley behind them, and, by exertions no
whit inferior to those of the cavalry, were now within
striking distance. No doubt Cronje could have cut his
way out with his fighting men, but rather than desert their
women and children—and lose their wagons—the burghers
determined to fight it out where they were. Their repulse
of the British at the battle known as "Modder River" had
shown the defensive strength of a position in the bed of
a river, and the place where the convoy had been
brought to a standstill was admirably adapted for such
a purpose. The stream ran at the bottom of a bed about
fifty yards wide and thirty foot deep, its sloping sides
covered with willow and mimosa. From the top of the
banks the plain sloping gently upwards could be swept
with fire for a thousand yards or more. Numerous dry
watercourses with which the banks were scored presented
successive lines of defence against attacks along the valley.
Caves dug in the steep banks of the river bed offered a
secure refuge from gun or rifle fire.

18th Feb. Lord Kitchener who had been placed in command—
more or less—of all the troops on the spot, had two courses
open to him—attack or investment. An attack, even if
unsuccessful, would immobilise the defenders, but it was
doubtful whether the forces then available were sufficient
to draw an investing line so tight as to preclude the
possibility of the Boers breaking out at night. He

[1] He had served in the German artillery in the Franco-German
war.

decided on attack. Owing however, to his having no regular staff, and to the timely interventions of the Boers from outside, this developed into a succession of disjointed efforts. It is not necessary to follow these in any detail, but it may be stated that the general intention was for the 6th division to hold the enemy in front while he was attacked from both flanks along the line of the river. The attack from the east was supported by the 76th and 81st batteries, that from the west by the 65th and 82nd, while the remainder of the artillery bombarded the laager from the rising ground on the south bank.

Scarcely had the 81st got into action when they came under shell fire from the rear—the first of the Boer reinforcements to arrive from Natal had seized the ridge behind them. At once the guns were turned into the new direction, and the intruders silenced. A much more serious interruption was to follow. The ubiquitous de Wet, returning from the loot of the convoy at Waterval drift, rushed the small detachment on what came to be called "Kitchener's Kopje", and opened fire on the artillery posted between the kopje and the laager. The 81st battery was just limbering up after disposing of the first interloper when this second surprise came, and its three sections were soon firing in three directions—one against Kitchener's kopje, one against a part of deWet's force which had occupied a hill further east, and one against a party of burghers who were creeping up to the battery under cover of a herd of cattle. The 76th also turned on to Kitchener's kopje to protect the rear of the 6th division, but all attempts to dislodge the Boers proved unavailing. De Wet's bold stroke had thrown the arrangements for a combined attack against Cronje out of gear, and it failed with heavy loss.[1] But Cronje had been pinned to the ground.

---

[1] Captain Lennox of the 81st battery was among the "missing" and in spite of every effort his fate could not be discovered. It was supposed that in the dark he had ridden into the Boer lines, by mistake for a field hospital, when in quest of assistance for a gunner too severely wounded to be removed.

CHAPTER XXVIII.

1900.

Paardeberg.

19th Feb.

Early in the morning of the 19th February, Lord Roberts arrived, bringing with him the 7th division and the remainder of the naval brigade which he had called up from Modder River. He decided to proceed by siege methods, bombarding the laager by day, sapping up to it from both ends by night, meanwhile investing it closely. The whole of the artillery was placed at the disposal of the G.O.C., R.A., and was distributed as follows :—

> *North Bank*—Naval Brigade—three 4·7″ and one 12-pr., Xth Brigade, R.F.A. and 65th (howitzer) battery.

> *South Bank*—Naval Brigade—four 4·7″ and three 12-prs., IIIrd Brigade, R.F.A.

On the north bank the position on Gun Hill was at about 3,000 yards from the laager, but that on the rising ground south of the river was only 1,200, and the Boer snipers were annoying—movement drawing fire at once. The effect of the bombardment by forty guns and howitzers appeared great, but, as the official historian suggests, Boer wagons standing out in full view above the ford possibly drew more than their share of the shells, while the burghers, sheltered under the banks suffered less than was imagined. The destruction of the ammunition and food in the wagons was, however, important, and the howitzers of the 65th Battery were told off for this duty until the arrival of the naval brigade, whose guns were so admirably suited for the task. Their high velocity and flat trajectory proved somewhat embarassing, however, owing to the tendency of their shell to ricochet into the investing lines. Indignant deputations, bearing fragments of 4·7″ and 12-pr. shell, were frequent visitors to artillery headquarters.

20th Feb.

The next event was the manœuvring of de Wet off Kitchener's Kopje by the 2nd and 3rd cavalry brigades, with their horse artillery, assisted by the IIIrd brigade, R.F.A. He did not go far, however, and reinforcements were gathering to him, so that it became necessary to provide an outer circle of defence as well as the inner ring round the laager. The infantry in the latter pushed their approaches forward each night, and the artillery fired occasional salvos by batteries detailed for the purpose, or on some nights all batteries fired at stated times into the laager. The artillery dispositions were also varied to get at the supposed hiding places of the Boers, as intelligence came in. But owing to the delay in the arrival of ammunition convoys the artillery bombardment had to be greatly curtailed. Heavy storms had made the tracks over the veld almost impassable, and had also made the crossing of the Modder a matter of some difficulty, causing considerable inconvenience and some anxiety.

One incident which occurred at this period is perhaps worthy of mention in regimental history. Batteries were told in turn to stand ready at night to move if required, and in one case, though teams were hooked in, the poles were let down. All went well until a cow, looking for her calf, wandered lowing through the lines. The alarming noise in the dark was too much for the horses, and the whole battery stampeded, causing considerable alarm. But all were recovered next day, little damaged.

On the 23rd the enemy made another attempt to recapture Kitchener's kopje, but were repulsed, the IIIrd brigade, R.F.A., being chiefly engaged. On that day the balloon detachment arrived, co-operating with the artillery in their bombardement on the 24th; and although communication between battery and balloon was in its infancy much useful information was received,

On the 26th the artillery received notable reinforcement, Colonel Perrott bringing one of the batteries of the Siege Train, (91/R.G.A., Major J. R. H. Allen), with four 6″ howitzers, and Major P. J. R. Crampton the first three of the pom-poms he had been sent out to organize. Lord Roberts came over himself to the bivouac of the artillery staff to concert a scheme for crushing the resistance of the Boers. There was to be a bombardment next

CHAPTER
XXVIII.

1900.

Paardeberg.

22nd Feb.

23rd Feb.

24th Feb.

26th Feb.

CHAPTER
XXVIII.

1900.

Paardeberg.

26th Feb.

27th Feb.

morning by the whole of the artillery, and in order to let the enemy know what was in store for them the 6″ howitzers were to fire 20 rounds that afternoon. The bombardment never came off, for the 120 lb. lyddite shells, bursting in their cramped quarters in the river bed, broke at last the spirit of the burghers. They would not agree even to Cronje's plea that they should hold out over the anniversary of Majuba, but insisted on surrender at six o'clock next morning—and set to work to get their white flags ready. They did not wait so long, for the attack of Smith-Dorrien's brigade before dawn on the 27th brought the white flags out fluttering everywhere.

The G.O.C., R.A., was bidden to meet General and Mrs. Cronje at breakfast with Lord Roberts, while his staff arranged for an artillery team to bring their carriage out of the laager and take it to the railway at Modder River. The adjutant of the brigade who went into the laager in quest of the carriage, found the burghers ready to greet him with a civil good morning, and to lend a hand in splicing the broken pole. But the writer has a vivid recollection of the very different reception he met with from the commander to whom he had to communicate the order for the horses. As also of the hours spent in the laager collecting the artillery *materiél*. The whole river bed was a mass of decomposing bodies—of horses chiefly —killed by our artillery fire. There were large white flags everywhere.

With General Cronje were captured some four thousand burghers, four field guns, and a pom-pom. Three of the guns were modern 75 mm. Krupps, the fourth an older model.

The ten days' ward at Paardeberg had been a severe strain on men and horses. There were no tents, and the ground afforded no shelter. On the 22nd the rains broke with nightly thunderstorms, and though the sun shone warmly in the mornings it was midday before clothes were dry. The tracks over the veld became morasses, so that the convoys were delayed, and, as will be remembered, the army supply park had been cut off by de Wet. Men and horses were half-starved. It is true that there was meat—"trek-ox"—but nothing more beyond ration biscuits. The horses were reduced to three or four pounds of oats, and the big draught horses could not pick up a living on the veld like Boer ponies. The worst work of the rain was, however, the flooding of the river, which brought down large numbers of dead horses and oxen, and sometimes men as well as all the refuse of the laager. From earliest dawn until dusk on the 23rd they drifted past in endless procession, and men had to be posted with poles to push off the swollen carcases when they stranded. Water-carts were sent some miles for drinking-water, but all used for cooking and washing had to come from the river. In the bivouacs on the banks of the Modder round Paardeberg were laid the seeds of the terrible attack of enteric which played such havoc with the army at Bloemfontein.

Anxious as Lord Roberts was to push on to Bloemfontein, it was impossible to do so until supplies for the march had been accumulated. All that could be done for the moment was to clear the troops off the contaminated ground, and there was a general sigh of relief when the orders for the move were received. Not only did they bring the army on to cleaner ground, and restore to their proper formations units which had become detached during the fighting, they were also couched in regular form and issued as "Army Orders." This was a welcome change from the happy-go-lucky way in which disjointed notes had been issued only to those directly concerned, so that no one knew what was going on. Lieut.-Colonel J. M. Grierson, R.A. had joined the Headquarters on the 27th, and at once made his presence felt.

Reconnaissances, in which Captain G. F. Milne, R.A., was prominent, showed that de Wet had taken up a position right across the Modder valley, only half-a-dozen miles to the eastward, and as soon as the army was ready

Lord Roberts assembled the leaders, and explained his plan of attack. French with the cavalry division was to start at night, and go right round the Boer left on to their line of retreat; the 6th division to attack their left flank; the brigade of Guards, with the naval brigade and the XIth brigade, R.F.A.,[1] to attack the centre as soon as the 6th division had turned the left; while the 7th division held the enemy's troops as far as the river, and the 9th division beyond it.

All were in motion long before dawn, and when, as the sun rose, the headquarter staff surveyed the scene from a hill above the naval guns, the spectacle was magnificent. The whole field was spread out with the four great masses of troops—over a hundred guns—in full view. But it was obvious that things had not gone "according to plan." The cavalry were halted far on the right, and would certainly be too late to cut off the enemy's retreat, while the 6th division was not yet deployed for attack. They were soon set in motion, however, the two infantry brigades in line, the divisional artillery between them, the 4·7's joining in from the centre. The Boer flank was turned without difficulty, and as the 6th division started to roll up the line, the Guards joined in. The 7th were delayed by some resistance between the centre and the river, but otherwise there was little fighting, and the British found ammunition, food, kits, abandoned in the Boer lines. The sight of the great force arrayed against them had been too much for the burghers, in spite of the presence on the field of the two Presidents, but owing to the delay of the cavalry they got safely away.

It was a great disappointment, and this may perhaps excuse a digression to recount an incident in which the writer was involved during the day. He was watching the XIth brigade shelling the last stand of the enemy when a mounted orderly, pushing up, presented a visiting card. Turning, he saw behind him an officer in the green

---

[1] The XIth brigade belonged to the 9th division, but had only just reached the army and not yet joined it.

uniform and silver lace of the Russian staff, with another
officer in unfamiliar khaki beside him. The Russian
hastened to explain. He, and his Dutch comrade, were
military attachés with the Boers, but their wagon had
broken down, so they had ridden over to the British.
"There it is", he exclaimed, "the Boers are plundering it,
turn your guns on to them". "But there is a red cross
flag flying over it, we can't fire on that". "Oh, that is
nothing, we just put that for safety"! The attachés were
treated with the greatest courtesy, and returned to the
Boers—via Cape Town and Delagoa Bay. In after years
Lord Roberts was fond of recalling, with a twinkle in his
eye, how unfortunate they had been in missing their con-
nections, so that it must have been months before they
were back with their hosts. Curiously enough it fell to the
writer to recall the incident to the Russian—then Chief of
the Staff—in Petrograd during the Great War.

In spite of the poor spirit shown by his commandos at
Poplar Grove, de Wet took up his stand again to obstruct
the advance on Bloemfontein. This time he selected for
his position three groups of kopjes—Abraham's Kraal,
Damvallei, and Driefontein—stretching across the line of
advance in that order from the left. Against these Lord
Roberts advanced on the 10th March in three columns.
On the left French had the 6th division and the 1st cavalry
brigade. In the centre the Field-Marshal himself commanded the 9th division, the 2nd cavalry brigade, and all
the heavy guns and howitzers. On the right Tucker had
his own 7th division, and the 3rd cavalry brigade.

The advance was made in echelon from the left, and
as soon as his cavalry came up against the Boer flank,
strongly posted on Abraham's Kraal, French sheered off
to his right, leaving a detachment (with "U" battery) to
hold the enemy in front.

The Boer centre was on the Damvallei kopjes, and
against this the other two batteries of Rochfort's brigade
came into action, while the cavalry pushed on to turn the

flank, followed by the 6th division, less the 82nd battery which was left to take the place of "U" at Abraham's Kraal. This movement across their front caused the enemy to reinforce their left on Driefontein. Leaving "Q" to hold those on Damvallei, Rochfort took "T" and "U" to support the attacks on Driefontein, where they were joined by the divisional artillery of the 6th division, less the 82nd battery.

Covered by the fire of these four batteries the infantry of the 6th division deployed for attack. The enemy fell back from their first position to one further back, the advance against which exposed the infantry to fire from two field guns behind Damvallei which made excellent practice at over 6,000 yards. The horse artillery moved forward so as to give the infantry closer support, the 81st moved out into the plain so as to engage the Damvallei guns at a range more within the power of their 15-prs., and the enemy fell back again. The horse artillery had got the range and as the Boers disappeared over the sky line the fire was continued with lengthened range and fuze.[1] This time the withdrawal was to a still stronger position, where they were supported by pom-poms on a knoll behind their left as well as by the Damvallei guns on their right. The horse artillery were called away to rejoin their cavalry brigade working round the extreme southern flank, but the infantry pushed on resolutely, though now supported only by the 76th and 81st batteries. The forward crest was gained, but here they were checked by the steady fire of the defenders of the summit, who were drawn from the ranks of that highly disciplined force the Transvaal Police, generally known as the "Zarps". Colonel Barker[2] took his batteries forward to a position from which they could support an assault; the infantry swept forward regardless of the rifles in front and the guns that raked their flanks; and the last of the daylight showed their

---

[1] The bodies of fifty burghers and double that number of ponies were counted on the reverse slope with the shrapnel all among them.

[2] Who had succeeded McDonnell wounded at Paardeberg.

bayonets crowning the crest. It was the end. Everywhere the burghers were leaving their trenches. Broadwood, who had got far round on the right with the 2nd cavalry brigade, brought his horse artillery into action and tried to pursue, but it was beyond the power of his exhausted horses.

Three days later, the Field-Marshal rode into Bloemfontein, and the Union Jack was hoisted with all due ceremony over the capital of the Orange Free State—as he had seen it hoisted over the capital of Afghanistan twenty years before.

The following table shows the artillery that marched into Bloemfontein.

Formation.	Unit.	6-in. How.	4.7 in. R.N.	12-pr. R.N.	5-in. How.	15-pr.	12-pr.	Pom-poms.
1st Cavalry Bde.	Q, T, & U, R.H.A	—	—	—	—	—	18	—
2nd do. do.	G & P, do.	—	—	—	—	—	12	—
3rd do do.	O & R, do.	—	—	—	—	—	12	—
Heavy Artillery	Naval Brigade	—	4	3	—	—	—	—
	91/R.G.A.	4	—	—	—	—	—	—
6th Division ...	Xth Brigade 76, 81, 82, R.F.A.	—	—	—	—	18	—	—
7th do. ...	IIIrd Brigade 18, 62, 75, R.F.A.	—	—	—	—	18	—	—
	Pom-pom Section	—	—	—	—	—	—	3
9th do. ...	XIth Brigade 83, 84, 85, R.F.A.	—	—	—	—	18	—	—
	65th (How.) Batty.	—	—	—	6	—	—	—
	Total ...	4	4	3	6	54	42	3

Each artillery brigade had an ammunition column, and there was also an "ox ammunition reserve" for the army.

Lord Roberts entered Bloemfontein on the 13th March; on the 15th, Generals Gatacre and Clements crossed the Orange River and invaded the Orange Free State from the south. In the last chapter we left them facing the

Boer invaders of Cape Colony, and a few words must now be said as to their doings in the interval.

The Boer commando which had defeated Gatacre at Stormberg left him unmolested to recover strength at Sterkstroom. As soon as Lord Roberts' advance drew away the Boers in front of him he advanced, and on the 10th March his leading troops (74th battery) reached Bethulie too late to save the railway bridge but in time to prevent the destruction of the road bridge. The guns were brought into action on the south bank, to deny to the enemy—who still held the north bank—further interference with the bridge. By the 15th his main body had arrived and crossed without opposition. Meanwhile a Colonial Division[1] of mounted troops, which had been formed under General Brabant, working through the unsettled districts of Cape Colony on his right reached the Orange River at Aliwal North, and seized the bridge there before it could be blown up. It was then held in the same way as that at Bethulie until the Boers withdrew.

We left General Clements resuming the offensive against Colesberg. By the 8th March he had occupied the hills above Norvals Pont, the enemy falling back before him. But they still held the north bank, the river was in flood, and the bridge blown up. It was not until the 12th that the pontoons arrived, and at dawn on the 15th the bridge was thrown without opposition. On the 16th General Pole-Carew, who had been despatched from Bloemfontein with a small column for the purpose, established connection with the two forces from Cape Colony.

---

[1] It had for artillery the Cape Mounted Rifles battery, sometimes reinforced by a section of the 79th.

CHAPTER XXIX.

THE RELIEF OF LADYSMITH.

Colenso—The Loss of the Guns—Spion Kop—Vaal Krantz—Tugela
Heights—Cingolo—Monte Cristo—Wynne's Hill—Hart's Hill—
Pieter's Hill — The Artillery Action — The Expenditure of Am-
munition.

MAPS 13, 18, 20, & 21.

DURING the latter half of November 1899 the troops
destined for the Relief of Ladysmith were gradually
assembling in Natal. In considering their subsequent
action it has always to be remembered :—

CHAPTER
XXIX.

1899.

Colenso.

November.

(i). That reconnaissance could not be pushed over the
Tugela.

(ii). That a range of heights, fringing the far bank,
concealed all movements of the enemy, while
exposing those of the British to their view.

(iii). That the only map available was one pieced
together when war became imminent from farm
surveys.

(iv). That the whole of the original Intelligence Staff
in Natal, with their scouts and guides, were shut
up in Ladysmith.

Lieut.-General Sir F. Clery was in command, and had
with him one infantry brigade and the artillery of his own
division (the 2nd), the same of the 1st division, and both
infantry brigades of the 3rd, with, in addition, a naval
brigade, a mounted brigade and some battalions for the
lines of communication—altogether perhaps twenty thousand men.   The artillery was as follows :—

C.R.A.—Colonel C. J. Long.

Staff Officer—Captain G. F. Herbert.

IV Brigade, R.F.A.

Commander	—Lieut.-Colonel H. V. Hunt.
Adjutant	—Captain H. D. White-Thomson.
7th Battery	—Major C. G. Henshaw.
14th do.	—Major A. C. Bailward.
66th do.	—Major W. Y. Foster.
Ammunition Column	—Major W. A. Smith.

V Brigade, R.F.A.[1]

Commander	—Lieut.-Colonel L. W. Parsons.
Adjutant	—Captain R. W. Boger.
64th Battery	—Major C. E. Coghill.
73rd do.	—Major C. M. Barlow.
Ammunition Column	—Major N. D. Findlay.

The naval brigade under Captain E. P. Jones had two 4·7″ on wheeled mountings and ten 12-prs. of 12-cwt.

Buller had come on himself to Natal at the end of November. He had, however, brought with him only his personal staff, and left Clery in nominal command, with instructions to force the passage at Colenso where the hills giving back a little from the river formed a rough amphitheatre, across which stretched a line of hillocks known as the "Colenso Kopjes". Conspicuous on the nearest of these stood out Fort Wylie constructed by the Naval volunteers when they held the passage. Not so noticeable were the trenches which now seamed the lower slopes of the hills, and the many emplacements for guns to command the crossings. The appproach to the river on the British side was over a bare plain sloping gradually downwards, the only cover being that afforded by some shallow dongas and by the houses round Colenso station.

The troops moved up to Chieveley; a rough survey was made with the assistance of the artillery range-takers; and the naval guns were brought into action on "Shooters

---

[1] The third battery (the 63rd) had been shipwrecked and was being re-equipped at Cape Town.

CHAPTER
XXIX.

1899.

Colenso.

13th—14th
Dec.

Hill''. All the 13th and 14th December they rained lyddite on what were supposed to be Boer positions six thousand yards away. Fort Wylie was knocked about, but there was no reply.

There were two points at which it was thought that the Tugela could be crossed—the road bridge at Colenso,[1] and Bridle drift, some five or six thousand yards above it. The plan of action was for a brigade to force the passage at each of these points, with another brigade between to support either as required. The mounted brigade was to occupy Hlangwane Hill on the right, and the remaining infantry brigade was to take up a position from which it could support it or the Colenso attack.

Parsons' artillery brigade was to support the attack on Bridle drift; Hunt's brigade (less one battery), accompanied by six naval 12-prs., to prepare the crossing at Colenso; the 7th battery to accompany the mounted brigade.

Two 4·7s'' and four 12-prs of the naval brigade were to advance to a position on the left of the railway, 4,000 yards from Fort Wylie; the remaining two 12-prs. to remain in action on Shooters Hill.

On the left, owing to the deficiencies of the map or the mistake of the Kafir guide, the infantry of the 5th brigade took a wrong direction and blundered into the long loop made by the river in its tortuous course. Here, exposed to fire from both flanks as well as from the front, they were at the mercy of the Boers, with whom they were unable to get to grips owing to their failure to find the ford.

Colonel Parsons, marching on the right rear of the infantry, brought his brigade into action on the right bank of the Doornkop Spruit to shell a kraal which faced the infantry. He also attempted to silence the guns from whose fire the infantry were suffering severely, but owing to the range had to fall back on percussion shrapnel which proved ineffective. In order to find a position within time

---

[1] Generally known as the "Iron Bridge"—the railway bridge had been blown up by the Boers.

shrapnel range, it was necessary to cross the spruit, but
as the brigade started to do so a shell upset the leading
gun, killing two horses and completely blocking the way.

Buller had observed the mistake of the infantry, and
now sent orders to the brigadier to withdraw the infantry
from the loop, and to Parsons to cross the spruit and cover
their retirement. Finding a practicable spot this time,
he got the guns across and advanced to a low ridge close
behind the infantry. Here the batteries were only 1,200
yards from the river, and under rifle fire, but the Boer
trenches were well searched, and with the assistance of the
navy the troublesome guns were silenced. Thanks to this,
and to the advance of a couple of battalions from the
4th brigade, the infantry were able to effect their
retirement unmolested. Parsons then took his brigade to
a position near the naval guns, where they remained in
action until the general withdrawal.

On the right the mounted brigade found Hlangwane
occupied by the enemy. Their advance against it was supported by the 7th battery, from a position between Hlangwane and Colenso, at 2,400 to 2,600 yards.[1] But the
defenders were quite as strong in numbers as the attackers,
and the latter made little progress. Before long they were
brought to a standstill, and so remained until the orders
for withdrawal were received. The well-directed fire of
the battery was then of considerable service in checking
the Boers creeping through the bush to harass the
retreating troops.

The chief interest lies, however, in the centre, the
attack on Colenso, and more particularly in the action
of the artillery supporting that attack.

Colonel Long, though in command of the whole of the
artillery, had been directed by Buller to supervise personally the action of that portion told off to support the
attack on the Colenso bridge. His command consisted of

---

[1] The left section, and at first the centre also, were directed against
Fort Wylie at 3,100 yards.

Hunt's field artillery brigade and six naval 12-prs. under Lieut. Ogilvy; and the whole moved off with the infantry of the 6th brigade, an hour before dawn. The field artillery horses outpaced the foot-soldiers, however, so that the batteries had to halt now and then to let the infantry close up. The naval guns, on the contrary, being drawn by ox-teams, fell behind.

Long's intention was to come into action at medium range (2,000—2,500 yards), but as he was approaching the intended position a sudden outburst of fire from the direction of the Bridle drift was heard. It was plain that the left attack was having a warm reception, and to Long this meant one thing only. So he rode on with Hunt, accompanied by his staff officer, Captain Herbert, and Captain White-Thomson the adjutant of the brigade.

After crossing a deep-cut donga, with a solitary road across it (referred to later as the "large" donga), they were about abreast of Colenso station when Long pointed out to Hunt a slight rise in the ground, just beyond a "small" donga, as the place for his guns.

At the moment the batteries were halted a couple of hundred yards back, while patrols—two from each battery —rode on to the river bank. Leaving his adjutant to mark the position and to take ranges, Hunt went back to lead his brigade into action. The batteries came up at a trot, and at once a hot fire broke out, which caused some casualties. No teams were stopped, however, and the guns were brought into action in excellent line on the exact spot selected—66th on the right, 14th on the left. The limbers were taken to the rear and the wagons brought up in the orthodox manner.[1] It was about six o'clock.

At the time the fire broke out the naval 12-prs. were crossing the "large" donga a quarter of a mile in rear. The two leading guns, which had got across, came into action, but on the arrival of the first shell the Kafir

---

[1] There was some delay with the wagons of the 14th battery, and pending their arrival limber supply was ordered and the teams sent back.

drivers bolted, and there was a jam in the drift. So the other four guns were brought into action behind the donga as soon as they could be extricated from it.

To return to the field artillery. The "objective" (as it was then termed) given by Colonel Long had been the hill behind Fort Wylie on which the enemy's works were visible. The range to this was found to be 1,825, but White-Thomson came to the conclusion that the fire was coming from Fort Wylie, the range to which was only 1,200, and he informed his colonel accordingly as soon as he came up. This order was therefore given to the batteries, and the whole of the fire was directed on Fort Wylie. Range and fuze were soon found, and a rapid "section fire" opened. The Boer musketry was, however, very heavy, and the shrapnel from several small guns[1] were bursting very accurately in front of the batteries. Casualties soon began to mount up. Long, walking up and down behind the guns, was among the first to fall, severely wounded by a shrapnel bullet. Hunt soon followed, and other officers and men were dropping fast. But there was no interruption to the rapid and accurate fire of the batteries which soon silenced the enemy's guns, and caused their rifle fire to slacken, until it became almost insignificant.[2]

After being in action for about an hour, as near as can be judged, the ammunition began to run short, although the second line of wagons had been brought up.[3] The infantry lines could be seen lying down on the plain, three-quarters to a mile in rear, instead of advancing as expected. Major Bailward, who had succeeded to the command, came to the conclusion that the best thing to do would be to keep the five or six rounds left with each gun

---

[1] Apparently the heavier guns could not be brought to bear upon the batteries, but they were firing heavily upon the wagon line and the naval guns.

[2] The fire of the naval guns and the 7th battery which were also directed on Fort Wylie no doubt contributed to this result—as also that of the infantry escort lying down extended on the flanks of the guns.

[3] Batteries had six wagons which were brought up three at a time

until the infantry attack developed. Meanwhile to avoid further casualties by withdrawing the detachments into the "small" donga, where they could find cover, yet be close at hand, ready to re-open fire as soon as required. The gunners were accordingly fallen in and marched back, and all the wounded carefully carried under cover. There was no thought of abandoning the guns, and therefore no steps were taken to render them unserviceable. A little later Captain Herbert was sent back by Colonel Long to inform General Clery of the situation, and on his return brought with him Captain Babtie, R.A.M.C., who had volunteered to attend to the wounded.

Meanwhile General Buller, who had been watching the left attack, had felt some anxiety about Long's guns as they were not where he expected to see them. About 8 o'clock therefore he rode across, meeting Captain Herbert on the way. After hearing his report he sent him with orders to Hildyard to occupy Colenso village with two battalions and rode on himself to the "large" donga. By this time the burghers had taken advantage of the cessation of fire from the two field batteries to reoccupy the positions from which they had been driven, and the space between the donga and the guns was swept with gun and rifle fire. Buller came to the conclusion that any attempt to re-open fire was out of the question, and stopped the wagons which were on their way from the ammunition column,[1] ordering at the same time the withdrawal of the naval 12-prs. There was some difficulty about this owing to the desertion of the Kafir drivers with their ox-teams, but it was effected with the help of horses from the wagon lines of the 14th and 66th batteries which had been established in the donga.

Then occurred the episode which so profoundly stirred the imagination of all. Buller called on Captain Schofield, R.A., to try what could be done to save some at least of the guns. Accompanied by two other aide-de-camps— Captain Congreve and Lieut. Roberts—he took up two

CHAPTER XXIX.

1899.

Colenso.

15th Dec.

[1] Captain Herbert had been sent back again to hurry them up, and Major Smith had sent forward 9 wagons and asked the naval brigade to cover their movements by fire on Fort Wylie.

limbers with volunteers from the 66th battery under Corporal Nurse. Two guns were saved, but Congreve was wounded, and the son of the Field-Marshal was killed. Another attempt was made by Lieuts. Schreiber and Grylls of the 66th, but Schreiber was killed and Grylls wounded, and nothing could be done. Finally Reed, the captain of the 7th, appeared from his wagon line with three teams: but the fire was then too heavy, and his teams were crippled before he could get near the guns. He himself, half his men and two-thirds of his horses were hit, and Buller forbade any further effort. He formed the opinion, also, that the infantry were too much exhausted with the extreme heat to be kept out all day—with the probability of a hard fight for the guns as soon as darkness allowed the Boers to cross the river—and decided to abandon the guns and withdraw his whole force to camp.[1]

The two battalions of Hildyard's brigade sent forward to Colenso by Buller's order after he had met Herbert, succeeded in establishing themselves in the station and surrounding buildings, and Colonel Bullock went to the "small" donga to tell Hunt that the guns were perfectly safe as his battalion was just to the left and would protect them. There the infantry hung on until the general withdrawal was ordered. Even then several officers protested vigorously, expressing their readiness to remain until nightfall so as to bring the guns in. But to no avail—the orders were explicit, and so those in the donga saw a retirement beginning on their left, and also away on their right rear. Then, for the first time, the gunners began to be apprehensive for the safety of their guns.

Between two and three o'clock the British Army left the battle-field, bare save for the ten guns standing so silent, and the little handful of gunners and infantrymen in the donga.

---

[1] The Victoria Cross was awarded to Captains Babtie, R.A.M.C., Congreve, R.B., Reed and Schofield, R.A., Lieut. Hon. F. Roberts, K.R.R., Corporal C. E. Nurse, 66th Battery, Pte. C. Ravenhill, R.S. Fusiliers. Nineteen N.C.O's and men of the 7th and 66th Batteries received the Distinguished Conduct Medal.

Then at last the enemy ventured out to seize their prey. Botha had telegraphed to Pretoria at 12.40 p.m. "We cannot go and fetch the guns as the enemy commands the bridge with their artillery". It was not until the naval guns under Captain Jones had gone that they came out. But the last scene is better told in the words of the adjutant, who was present :—

"Then came a long lull in the firing, and someone in our donga looked out and saw some half-dozen Boers in amongst the guns. These we fired at and knocked over a man and one or two ponies. A party came to bring the wounded man back, and Colonel Bullock ordered our men not to fire at the men helping the wounded, but at the same time shouted out 'Keep away from those guns or I shall fire'. Presently about 150 Boers rode all round us and in amongst the stretcher-bearers, who were just arriving on the scene. Some of the Boers were very truculent and threatened to shoot us all, others were very good and got water and cigarettes for our wounded. I saw Long off on a stretcher, and Hunt[1] put on one, and then walked off myself to the ambulance station".

Those not too severely wounded were marched off as prisoners, the guns taken away as trophies. The mounted brigade on their way back to camp saw the Boers swarming round the guns, and the 7th came into action, but before they could open fire men in khaki were seen among the burghers and British ambulances drove up—nothing could be done.

The loss of the guns at Colenso created a profound sensation although, as pointed out in the German Official Account, such surprises as those of Hart's infantry and Long's artillery must be expected during the first engagements in a war. Be that as it may, Long was made a scapegoat, and his action, therefore, demands careful consideration in regimental history.

---

[1] The Boers hearing him addressed as "Colonel" decided to take him a prisoner though he was severely wounded.

CHAPTER
XXIX.

1899.

The Loss of
the Guns.

15th Dec.

In a memorandum issued before the battle, Buller had stated that at first only the naval guns could render effective aid, as it was impossible to bring the field batteries into action without risk.   This condemnation to inactivity of the arm to which he was devoted touched Long's gallant nature to the quick.   He was well aware that the Boer guns outranged his, but his idea of equalising matters was to take the 15-prs. in to a range at which they would be able to develop their special qualities to the utmost. · It may be admitted that he overdid it.   Their fire would have been just as effective if he had brought them into action a thousand yards further back as he had first intended.   But he had been brought up on the old gunner maxim "The greater the difficulties of the infantry the closer should be the support of the artillery", and when he heard the burst of fire which greeted the brigade on his left his first instinct was to get in to their aid.   It was only a year since, as commander of the Anglo-Egyptian artillery, he had seen at the Atbara and Omdurman the annihilating effect of shrapnel at decisive ranges.

At the same time it must be admitted that, whatever blame may be attached to his action in exposing the batteries to the risk of being silenced, the loss of their guns cannot justly be attributed to him.   These were in no danger of capture as they lay out on the plain.   There were plenty of other guns to render any approach to them by daylight impossible, and if, as Buller thought, the infantry that had been in the attacks was too exhausted to fight for them after dark, this was not the opinion of the infantry officers themselves.   Moreover, two of his four infantry brigades had not fired a shot.   But the most interesting comment on Long's action is that of General Botha, who was in command of the Boers at Colenso.   Sir Percy Fitzgerald, who visited the battlefield with him in 1908, gives the following account of what he said on the subject.

"It was one man who spoilt it all; one man who saved the British Army that day.   They say he had no orders, or he disobeyed orders, and they broke him for what he did; but it is the simple truth I tell you, that when Colonel Long rushed his field artillery out into the open and began shelling the woods and slopes of Hlangwane, where he must have seen our men, he upset the whole plan. I don't know if any of our men were premature and revealed their presence by shooting, but whatever it was it was Colonel Long who saw them and realized that our force on Hlangwane was already across the river and there was grave danger of a flank attack, and he made it so hot that they had to open fire all along and so gave the whole plan away.

That day I saw the maddest, bravest thing I have ever seen : I was on the hill above the bridge there and with the field glasses could see it all. All our people were watching : it was a terrible thing to see, like looking down at a play from the gallery. When the teams and the men were shot down, just swept away by our fire, for it was at very close range, Long brought his guns up as close as he could; and when we saw another lot of men and more teams dash out to work or save the guns we held our breath; it was madness, nothing could live there. Then came another lot, and another and another. My God, it was awful. I think it was six, seven, eight times, perhaps more, that fresh men dashed out to save the guns. I was sick with horror that such bravery should be useless. God, I turned away and could not look; and yet I had to look again. It was too wonderful. Lord Roberts' only son was one who lost his life in that mad effort; he was killed at the guns. They saved a few—three I think—and then someone must have stopped them. Colonel Long was shot down in the first lot but not killed. They blamed him for the failure, blamed him for risking and losing his guns, but that man saved the British Army that day. It was his action that exposed our plan, and forced us to fight, and then the whole battle turned that way

CHAPTER
XXIX.

1899.

The Loss of
the Guns.

15th Dec.

CHAPTER
XXIX.

1899.

The Loss of
the Guns.

15th Dec.

and Buller's army never advanced across the river by the road bridge, as it was intended, to where we had planned to crush it, in that little flat across the river under the hills. It was a great disappointment for us. The advanced guard had already crossed and if Long had not exposed us on Hlangwane, the whole force would have been in our hands. That would have been the greatest disaster that has ever befallen the British Army".[1]

Some resentment was caused in the Regiment when it was discovered that a monument had been erected on the spot where the guns had stood with an inscription stating that they "were abandoned" there. The natural impression given by such words was that they had been abandoned by their gunners. The matter was taken up with the Government of Natal, and the following inscription substituted for that to which exception had been taken—"Near this spot on 15th December, 1899, the 14th and 66th Field Batteries, Royal Artillery, came into action during the attack on Colenso. Two guns of the 66th Battery were rescued under a galling fire, the remainder could not be withdrawn on the failure of the attack, and fell into the hands of the Boers".

During the month which followed the failure at Colenso the British force in Natal received very considerable reinforcements, including the 5th division under Sir Charles Warren. Its divisional artillery consisting of :—

IX BRIGADE, R.F.A.[2]

Commander	—Lieut.-Col. A. J. Montgomery.
Adjutant	—Captain C. N. B. Ballard.
19th Battery	—Major H. A. D. Curtis.
28th do.	—Major A. Stokes.
Ammunition Column	—Major J. R. Foster.

[1] *South African Memories.* By Sir J. Percy Fitzpatrick. (Cassel & Co., Ltd.).
[2] The third battery, the 20th, had been landed at Cape Town and sent up to strengthen the force under Methuen.

The artillery in Natal was also strengthened by the 63rd[1] (Major W. L. H. Paget), the 78th (Major D. C. Carter), and the 61st (Major A. H. Gordon). The 14th and 66th were also refitting at Pieter Maritzburg.

With the increase to his strength, General Buller determined to make another attempt to force the passage of the Tugela—this time at Potgeiter's drift some miles above Colenso. But there had been heavy rain, and the troops, with their immense train of ox-wagons, were floundering through the mud for nearly a week before they reached their point of concentration. And when Buller saw how awkwardly the drift was situated, at the end of a re-entrant, faced by an amphitheatre of hills now occupied by the enemy, he decided to try still further up stream so as to get round their flank if possible. The point now selected was Trickhardt's Drift, five miles above Potgeiter's, where the opposing heights drew back from the river, and left a couple of miles or so of manœuvring space on the far side. To force the passage here he gave Warren the mounted brigade, three infantry brigades, and six field batteries. To hold the enemy in front while this movement was in progress, he left Lyttelton at Potgeiter's Drift with two infantry brigades, the 61st (howitzer) and 64th batteries, and two 4·7″ and eight 12-prs. of the naval brigade.

The troops were put in motion on the 16th January, and by the 19th both parties had effected lodgments on the enemy's side of the river. There was little fighting except for a successful skirmish by the mounted troops near Acton Holmes on the extreme left.

Next day Warren moved forward with the object of securing an artillery position from which an attack on the

CHAPTER
XXIX.

1900.

Spion Kop.

January.

10th—15th
Jan.

19th Jan.

20th Jan.

---

[1] The 63rd had been wrecked in the "Ismore" on the way to Cape Town. The men of the battery, and of the squadron 10th Hussars on board, behaved splendidly, but it was impossible to work the derrick to hoist out the horses owing to the water in the engine room, and ports were too small to get any but the smallest through. So when the ship broke in two only eight artillery and twenty hussar horses were saved out of the 306 on board.

Boer trenches could be adequately supported. This was found on "Three Tree Hill", at a range of a little over two thousand yards, and was occupied by the 7th, 63rd, and 73rd batteries[1] although the ground was cramped and so steep and boulder-strewn that the guns had to be man-handled into action and the wagons left below. For the other three batteries—the 19th, 28th, and 78th—a place was found in a mealie field on the right. Far more important, however, than any defects of the ground, was the fact that the guns were close under the enemy's position, and so could see nothing beyond the forward crest. The true crest, where the main defence lay, was thus secure from direct artillery fire.

Supported by the thirty-six guns firing over their heads, the infantry pressed forward, their front covered by a smoke-screen formed by the grass on the hill-sides set alight by the shrapnel. The lower slopes were quickly passed, and then the forward crest was carried, but before them now stretched a thousand yards of bare slope, and here the artillery could give no assistance. The attempt to cross it without such support was judged too hazardous, so the battalions dug in where they were, and there they lay for the next four days, mere spectators of the tragedy about to be enacted.

Next day the 19th and 28th batteries were moved to the left flank to support an attempt to work further round that way. But nothing came of it, and Warren asked Buller for "four howitzers whose indirect fire alone seems likely to be of any service against the concealed position of the enemy". On the 22nd two sections of the 61st were sent, and one went to each of the artillery positions in the hope that they would be able to get at the guns which were raking the infantry on the crest. A German observer with the Boers was quick to note their arrival—"I see our people are coming away from the position. It is almost enfiladed by the howitzers, which simply blow away

---

[1] The divisional organization was to a large extent broken up, infantry brigades re-grouped, and the same with batteries and artillery brigades.

the entrenchments". All day guns and rifles blazed but there was no movement.

Warren realised that it was no good trying to get round the flank of an enemy possessed not only of interior lines, but of superior mobility, and decided to seize instead a prominent hill which jutted out from the range between him and Lyttleton, and appeared to command the whole Boer position. It had been named "Spion Kop", or "Look-out Hill", by the first Boers who gazed from its summit over the rich land of Natal.

The 23rd was devoted to preparation. The batteries near Three Tree Hill were to be ready to support the attack by directing their fire towards the lines of advance of reinforcements, which would be illuminated by star shell. The 19th battery was brought across from the left to a position from which they could command the nek connecting Spion Kop with the main range. Guns were laid on points in the enemy's line and all preparations made for night-firing.

As soon as it was dark the stormers started, and their arrival on the summit came as a complete surprise to the Boer picquet stationed there. There was no opposition, and the gunners acknowledged with a salvo of star shell the cheers which announced the success of the enterprise. But when the morning mist rose, those on the plateau at the top found themselves the target for both gun and rifle fire from three sides And all that the artillery could do to help them was to prevent reinforcements reaching the burghers who were climbing up the steep slopes as at Majuba. The field batteries dropped shrapnel over the approaches, the naval guns from Mount Alice swept the slopes on the right. All that terrible day both sides stubbornly clung to their ground. The incessant rifle-fire continued hour after hour, and to this was added a converging shell fire, from front and flanks, which wrought terrible havoc in the crowded ranks. It was probably the deciding factor. All messages from the summit called for the silencing of the guns. "It is impossible to permanently hold this place as long as the enemy's guns can play on this hill" wrote Thornycroft. And so the gunners persisted in their futile efforts to silence the invisible weapons, securely tucked away behind the hills. If they could have turned their guns on to the Boer riflemen lining the outer edge of the plateau they might have rendered great service, but no one knew where they were. Gordon had sent his observing party to the top, but the signallers were commandeered by the staff, and he got no information from them. He always maintained that if he had been told where the enemy were he could have dropped his shell upon them, and probably changed the fortune of the day.

Could guns have been got on to the plateau? Captain Hanwell, who had been sent up to investigate the possibility, reported that heavy guns could not be got up before dawn and that, in any case, the nature of the ground would make the construction of emplacements impossible. But Warren decided that an attempt must be made. A mountain battery[1] was on its way to the front, and under the guidance of Captain Kelly, R.A., A.D.C. to Warren, pressed on as far up the hill as was possible before dark. It was waiting until the moon rose to light the further way, when the troops from the top began coming down upon it. Two naval 12-prs. were also ordered up from Potgeiter's, and the Sappers were set to work to prepare a way for them. But they did not reach the foot of the hill until midnight by which time all was over.

Buller had, however, by no means given up the idea of forcing a way through the Boer position on this flank. The eastern arm of the amphitheatre of hills which faced Potgeiter's drift came right down to the river on the right, ending up with a flat-topped hill called Vaal Krantz. It was thought that this hill, if captured, would provide a

---

[1] No. 4 Mountain Battery (Major H. C. C. D. Simpson). It was sent out from home as soon as news was received of the disaster to No. 10 at Nicholson's Nek.

position from which guns would be able to command the road to Ladysmith. Moreover there was here a height on the near side—Zwart Kop—which would provide a dominating position for the British guns in the attack of Vaal Krantz.

CHAPTER XXIX.

1900.

Vaal Krantz.

Preparations were at once put in hand, and by the 4th February six naval 12-prs., two guns of the 64th field battery, and the mountain battery had been got into action, concealed among the bushes and trees on Zwart Kop— sailors, gunners, and sappers working together. Two 5″ guns of 57/R.G.A. (Major C. E. Callwell) which arrived opportunely, were too heavy to be got to the summit, but a place was found for them on a spur to the westward. The naval 4·7″ and 12-prs. were still on Mount Alice. And among the reinforcements arriving at this time was "A" Battery, R.H.A.[1] from India which joined the cavalry brigade, now formed of regular regiments. Colonel Nutt, sent out to succeed Colonel Long as C.R.A., was put in command of the great battery on Zwart Kop, and of a battalion left there as escort.[2]

4th Feb.

It was Buller's intention to tie the Boers to their trenches on the hills facing Potgeiter's by a threat of attack from the bridge-head kopjes, and then to capture Vaal Krantz by surprise. For the demonstration Parsons took out the two field artillery brigades and the howitzer battery, and they came into action in the plain in short echelon of batteries from the left—63, 28, 19; then 78, 73, 7; with 61 to the left rear. At 7 a.m. fire was opened, the naval guns on Mount Alice joining in, and the infantry advanced on a broad front. There was no reply until the infantry had got within a thousand yards of the trenches,

5th Feb.

---

[1] Major E. A. Burrows. It had 15-pr. guns, for the 12-pr. 6 cwt. horse artillery equipment had not reached India.

[2] His orders were comprehensive. "The O.C. R.A., will direct the action of the 14 guns on Zwart Kop and of the two 5″ below. His task will be to prevent the enemy from bringing any reinforcements from the right towards the rear; further to fire on all localities from which rifle fire is directed on our infantry when it advances, and to subdue the fire of every gun which the enemy brings into action.

when three guns and a pom-pom opened fire on the long line of batteries. The infantry lay down, and for an hour and a half the duel went on without appreciable damage on either side.

Meanwhile the left battery (63) slipped quietly away to cover the throwing of a new bridge facing Vaal Krantz, and when this was completed the remaining batteries followed in succession from the left at ten minutes interval. It was a risky manœuvre, for they were out in the open in full view of the Boer gunners, and as the batteries limbered up the fire was turned on each in turn. "Linesman", gives us a vivid picture of the scene :—

"The gallant 'feinters' withdrew themselves from the zone of fire, their retirement being covered by the continuous fire of the imperturbable batteries. . . . . It seemed impossible that those guns could get away . . . . something like a groan burst from all the waiting and watching thousands. But it was quickly changed to a roar of applause as out of that tornado, quietly and in order, trotted those incredible gunners, with not a sign of hurry . . . whilst overhead and on every side yelled and roared the projectiles from the angry Dutch guns. Magnificent and war too".

The demonstration was over, it is time to see how the real attack was faring. As soon as the 63rd battery arrived, and took up its position to cover the building of the bridge, the work started, and at the same time the guns on Zwart Kop were unmasked, bringing an overwhelming fire on to the Vaal Krantz ridge. The 63rd was joined by the 64th, and they were kept busy dealing with the burghers scattered among the mealie fields by the river who were firing on the bridge-builders. The other batteries, as they arrived, found positions in the bend of the river north of Zwart Kop, and joined in the fire upon Vaal Krantz. At two o'clock the infantry crossed the bridge, and by four had gained the summit with little loss owing to the concentrated artillery fire. All had gone well until then, but now came disappointment. So steep and so encumbered were the sides of the ridge that the mounting of guns upon it was out of the question— and yet this was the object for which it had been taken.

During the night both sides were re-adjusting their artillery, and when dawn broke on the 6th February the Boer guns were clustering round Vaal Krantz, on the summit of which the infantry lay exposed to fire from front, rear, and flanks. The first shot was fired from Doorn Kop on the extreme right by a 155-mm. Creusot— the very "Long Tom" from Telegraph Hill which had fared so badly in its duel with the 6·3″ howitzers at Ladysmith.

All this day and the next the fire went on as at Spion Kop, but here the defenders had been able to throw up some protection under which they lay close. No object was any longer to be gained, however, by clinging to the ridge, and after nightfall on the 7th the delicate operation of withdrawal was successfully performed.

The gunners had been kept very busy during the fighting on the Upper Tugela, and had expended a prodigious amount of ammunition, without having very much to show for it :—

R.N.—2,393 rounds.

R.A.—7,296 rounds.

It had been very hard work, day after day, dawn to dusk, and the field batteries had been subjected to a good deal of sniping, as well as being shelled by guns hopelessly beyond their range, which they never even saw. There was nothing to be done except to grin and bear it. Fortunately most of the shell buried themselves harmlessly and the men consoled themselves by digging them up as trophies.

The "heavies" were better off, and we may perhaps follow the fortunes of the 57th, the first company R.G.A., to appear in this new rôle. On the 5th they fired only 79 rounds, chiefly devoted to an attempt to find the Boer guns which were enfilading the long line of field batteries. On the 6th and 7th they fired 136 and 181 rounds respectively, all at long range, trying to keep down the fire of the guns which were playing on the infantry on Vaal Krantz—especially the big Creusot. During the night of the 5th they received an urgent summons to take on this gun which had just opened fire. Passing along the defile between Zwart Kop and the river in the dark a wheel slipped over the edge, an awkward business with so heavy a gun as a 5″. But after an hour's delay, the two guns got into action on a spur to the north, whence they reduced the Long Tom to quiescence, though it was situated fifteen hundred feet above them and at a range of six thousand yards.[1]

Sir Redvers Bullers had failed to break through the centre of the Boer lines at Colenso, and he had failed again in two attempts to turn their right flank, but he lost no time before trying his luck again. This time he decided that his first effort should be directed against their left flank, which they had now advanced across the Tugela.

Below Colenso the Tugela changes its course to the north for some miles, before turning eastward again, and, in the big bend thus formed, lies a rough horse-shoe of hills enclosing an elevated plateau. The toe of this horse-shoe is to the north at Clump Hill, just where the river turns east, the heels at Hlangwane and Green Hill lying opposite each other, with the Gomba Spruit running right across at their feet. From it to Clump Hill the length of the horse-shoe is approximately five miles : the width of the heel between Green Hill and Hlangwane, perhaps two miles. On the eastern side of the horse-shoe, northward from Green Hill, stretches a long ridge, known as Monte Cristo, reaching almost to Clump Hill. Outside the horseshoe, rises the isolated, rocky hill of Cingolo.

---

[1] The gun was not damaged, but the earthwork which sheltered it was much knocked about, and on the 7th a lucky shell blew up some of its ammunition. Ten years later the Boer officer who commanded it, was attached to the Royal Artillery at Potchefstroom and recalled his annoyance at the incident.

While the army was gathering together at Chieveley for the fresh effort, Buller made further alterations in the organization of his troops and especially in that of the artillery. Colonel Parsons was given the general control, with Majors Paget and W. A. Smith in command of newly organised brigades.[1]

Naval Brigade[2]— one 6″ and one 4·7″ on railway trucks.

two 4·7″
ten 12-prs. } on wheeled mountings.

2nd Division — 7, 63, 64 R.F.A. under Major Paget.

5th Division — 28, 73, 78 R.F.A. under Major W. A. Smith.

Corps Troops — 19, and 61 (How.) R.F.A.

No. 4 (Mtn.) and 57 (Heavy) R.G.A.

In addition to the above, the Chestnut Troop and two naval 12-prs. were with the cavalry brigade guarding against any movement of the enemy from the upper Tugela; the Natal field battery and two naval 12-prs. (of 8 cwt.) with the troops covering the right.

Hussar Hill lay half-way between the two forces, and this was occupied by the 5th division on the 14th February, with the 2nd on its right. The guns (including four naval 12-prs.) were brought up, the field batteries to the forward slope, the heavy and the mountain battery on the hill, the howitzers behind. Opposite stretched the enemy's trenches from Hlangwane to Green Hill and back to Monte Cristo, and on these a slow and steady fire was maintained for the afternoon of the 14th and the two following days. From its place behind a spur of Hussar Hill the 61st commanded the whole position at ranges of from three to five thousand yards, and in order to enable him to carry out without

---

[1] It must be remembered that brigades were not yet recognised officially as permanent units.

[2] Three more 4·7″ arrived before the operations commenced. One was brought into action on the 19th, the others not until the 25th.

CHAPTER
XXIX.
1900.

delay changes of target beyond the limits of the deflection leaf, Major Gordon devised wooden "gun-arcs". These became the first of the "indirect laying stores" introduced into the service after the war.

Cingolo.

While the enemy's attention was thus kept on their position from Hlangwane to Green Hill the infantry of the 2nd division were gradually extending their right so as to encircle Cingolo, and the mounted brigade working round

17th Feb.
it. At 6 a.m. on the 17th fifty guns opened fire, the infantry crossed the Gomba, and before dark, the whole summit was occupied, the Boers falling back to Monte Cristo. The northern portion of this long ridge was out of range of the field guns on Hussar Hill, so the 64th had been pushed forward on the previous evening in anticipation of its being required. It was now ordered up on to Cingolo, but the infantry had to work hard all night to make a way, so steep and rough were the sides. They were to be well repaid for their labour.

Monte
Cristo.

18th Feb.

The occupation of Cingolo paved the way for the capture of Monte Cristo, and no time was lost in pressing on. At dawn on the 18th February, 57/R.G.A.—with its four guns together at last[1]—prepared the way by shelling at 8,000 to 8,500 yards the trenches which were out of range of the field guns. The infantry swung round supported by the four guns of the 64th battery that they had helped to drag into position on Cingolo, soon reinforced by those of the 7th battery. By 10.30 they had effected a lodgment on the southern end, by noon the whole ridge was won, and the attack was swinging round again against Green Hill, supported this time by the howitzers of the 61st battery searching the lower features. By two o'clock Green Hill was over-run, and Hlangwane became untenable.

---

[1] The left half-battery had been sent round from Cape Colony by Lord Roberts at the special request of Sir Redvers Buller.

The afternoon was spent by the artillery in shelling the parties of burghers making for the shelter of the Tugela valley as the infantry drove them across the front of the guns. The field batteries had their chance and took it. Before night Parsons had brought the whole of the field artillery forward to the plateau west of Monte Cristo. Here also on the next morning came the heavies, joined in the afternoon by the two wheeled 4·7″ from Chieveley, which had been dragged by sailors and soldiers when the oxen gave out. Meanwhile two of the 12-prs. from Hussar Hill had moved further still, to Clump Hill looking north across the river, and here also one of the 4·7″ was got during the night after a reconnaissance by Captain Jones, R.N. and Colonel Parsons.

On the 20th there was a good deal of artillery action, chiefly in order to keep down the fire of the enemy's guns directed against the infantry consolidating the position, clearing roads, getting up supplies, and getting into position for the next phase. And here the naval guns in the advanced position proved of great value in dealing with the guns on Pieter's Hill.

Throughout the week's operations the batteries remained continuously in action, the gunners sleeping round their guns, the drivers by their horses, sometimes hooked in all night. Their chief difficulty was water, which often necessitated taking the horses a mile or two over rocky ground at night—and there was no moon to lighten the path.

By the morning of the 20th February, Buller was in full possession of all the ground on the right bank. His next step must be the crossing of the river and the attack of the barrier hills beyond. Some further details regarding the course of the river and the nature of its valley are therefore called for.

A mile below Colenso the Tugela, which up to this flows in an easterly direction, takes a sharp bend to the

CHAPTER XXIX.

1900.

Monte Cristo.

19th Feb.

20th Feb.

The Crossing of the Tugela

north-west for a couple of miles, and then turns to the north-east for four or more, before resuming its easterly course.

On the Boer side, from opposite Colenso to the turn of the river, the main range gives back from the stream, forming a rough amphitheatre, dominated by the gloomy height known as Grobelaar. Across this amphitheatre stretch the Colenso Kopjes enclosing between them and the river a space some two and a half miles long, with an average width of not more than a mile. Beyond the turn of the river the hills close in upon it, their boulder-strewn slopes partly covered with scrub. These hills are best known by the names they earned at this time—Wynne's Hill, Hart's Hill, Kitchener's Hill, and finally Pieter's Hill. Between the end of the Colenso kopjes and Wynne's Hill the deep wooded glen of the Onderbrook spruit, and between Wynne's Hill and Hart's Hill that of the Langenacht spruit, run down to the Tugela.

On the British bank, the heel of the horse-shoe at Hlangwane has a long spur running northwards to Naval Hill which abuts on the Tugela just below the turn in the course of the stream. Beyond that comes Fuzzy Hill, and finally, where the river turns east again, Clump Hill, the toe of the horse-shoe.

The first point to be considered was where the bridge should be thrown. Below its turn the bed of the river gradually deepens until it runs through a deep and rocky gorge, with sides so steep in the lower portion that they can only be scaled with difficulty. Above the turn, on the other hand, the river can be approached without difficulty, and, moreover, the bed is to a great extent defiladed. There, accordingly, Buller decided, was the place for the bridge.

Under the fire of the guns which had been moved into position on the previous evening, the bridge was completed on the morning of the 21st, and an infantry brigade, with Smith's brigade of field artillery (28, 73, 78)

crossed on to the space of low-lying land on the left bank
before mentioned.[1]   The Boers resisted manfully, but
during the afternoon a bridge-head was established, and
all night long troops were pouring across.   By the
morning of the 22nd the greater part of the infantry and
a large number of batteries were cooped up in that narrow
strip between the Colenso Kopjes and the river from Fort
Wylie to the Onderbrook Spruit.   The batteries lined the
kopjes in order from the left of 73, 28, two 12-prs., 7, 78,
with 61 concealed in the rough ground near the bridge.
Firing over their heads from the right bank were
the 5″ guns, the remainder of the 12-prs. (six), two 4·7″s.
(wheeled), 63 and 64.   The task of the latter was to keep
down the fire of the guns which were annoying the troops
in the low ground, but it was a difficult one, for the Boer
guns were widely dispersed on the hill tops, and hard to
locate.

Meanwhile the infantry had secured the ground as far
as the Onderbrook Spruit, and in the afternoon Wynne's
brigade crossed the spruit and attacked the hill which
bears his name.   The supporting artillery on the left bank
concentrated their fire on the enemy's trenches, the
naval guns coming down close to the river on the right
bank and joining in.   But as usual the Boer defence lay
along the further side of the long plateau which formed

---

[1] The 66th battery was also moved up to Colenso, its first appearance in the field after its disaster there.   It had been re-formed under
Major C. C. Owen, and Lieut. Grylls (on his recovery from the wound
received on that occasion) with the survivors among the other ranks
and any who had previously served in the battery and were to be
found in Natal.   Horses were selected from those contributed by the
Maharaja Scindia, and the guns also came from India.   At the beginning of the war there was a strict rule against British service officers in
India going to South Africa.   But after Colenso, when India was called
upon to send the equipment for three batteries, Major-General Tyler,
the Inspector-General of Artillery, took the opportunity to give some
artillery officers a chance.   He insisted that a captain must go in charge
of each battery sent, and selected Captains Headlam, Gordon, and
Sandilands.

CHAPTER
XXIX.

1900.

Wynne's
Hill.

23rd Feb.
the top, and the infantry could get no further than the forward crest.

At night the heavy guns were ordered across to support the further attack which was to take place next day. Unfortunately, however, the 4·7″ came first to the bridge, and their weight (being all concentrated on the gun wheels) was too much for the pontoons. The guns had to be taken out of their cradles before the bridge would bear them, with the result that they, and the 5″, did not get over until the morning of the 23rd. For the attack on that day Colonel Parsons distributed the artillery on the left bank in two lines. In the first, just south of the Onderbrook Spruit, were 28, 61, 64, 78, in that order from the left. In the second were 73, two 4·7″s, four 12-prs. and 57/R.G.A. On the right bank were left only four 12-prs., but they were joined during the day by 7, 19, and 63 for whom no room could be found in the limited area across the water, and the 6″ and 4·7″ on the railway at Chieveley assisted in keeping down the guns on Grobelaar.

To relieve the situation on Wynne's Hill, Hart's brigade was to attack the hill beyond, afterwards called Hart's Hill. In order to get into position to do so the infantry had to file down the bed of the river for some distance and to cross the Langenacht Spruit, losing heavily while doing so. It was late in the afternoon, therefore, before they deployed for attack. Once again the forward crest was reached, but dusk was now approaching, the gunners could not see to distinguish friend from foe, and at 5.30 they ceased firing. There were desperate charges, but all to no avail. The fall of night found the summit still in the enemy's hands, and the situation of the remnants of Hart's brigade precarious. All night long they held on doggedly to the slopes, but at dawn fell back to the line of the railway.

It had been a hard and disappointing day for the gunners. Owing to the delay at the bridge which kept the heavies out of action, the Boer guns had it all their own way in the morning. The 5″ had then been hurried into action in an unsuitable position for such guns, where at each discharge they ran back so far that it took five minutes, and much labour, to run them up again. Eventually they were run on to the forward slope, and here, though fully exposed to view, they were "false-crested", and so escaped the natural consequences. The howitzers were not so fortunate. Full advantage had been taken of their special powers to occupy a completely covered position. But, unfortunately, this lay between the field batteries of the first line and the heavies of the second, so that their secluded valley was soon catching the "overs" intended for the one, and the "shorts" intended for the other.

The batteries did their best to ease the plight of the infantry, but from their positions on the low ground by the river they could not see the tops of the hills where the Boers were holding the backward crest. They were unable, therefore, to render support where most required, and when heavy rain came on, blotting out all view, the howitzers alone were able to continue their fire.[1] Ammunition supply was also an anxiety with only the one bridge behind them, and the 5″ guns had only fifty rounds left when night fell. The field artillery wagon lines by the bridge, and the ammunition columns below Hlangwane worked well, however, and there was never any failure of supply. The strain on the corps troops column under Capt. Findlay, which had five natures to supply, was particularly severe, and his services were warmly acknowledged in Buller's despatches.

The 24th February had hardly dawned when the fight was renewed. It was indeed fortunate that the Boers would not take the offensive, for the survivors of Wynne's

---

[1] In those days they were the only batteries that could continue their fire without seeing the target.

Later on the howitzer battery was called upon to shell the upper part of the Onderbrook ravine from which an attack was anticipated. Ranging and distribution along the ravine were completed before it was quite dark, and slow firing was kept up until after 8 o'clock.

and Hart's brigade clinging doggedly to the slopes up which they had pressed so gallantly, were hardly in a condition to meet an attack. But a change was coming.

Pieter's Hill.

It will be remembered that Buller had dismissed the idea of throwing a bridge across the Tugela in the straight stretch below Fuzzy Hill, where it would have been completely screened from the view of the enemy, because the height and steepness of the right bank rendered the river bed inaccessible. On this day, however, his intelligence staff found a way down, and he decided to abandon the operations in progress, withdraw the bulk of the troops engaged in them, move the bridge to the newly discovered site, and send the troops back by it to turn the enemy's left, which rested on Pieter's Hill.

25th Feb.

Throughout the night of the 24th guns and wagons rumbled over the bridge, and by the morning of the 25th all the artillery except the 73rd battery—left on the Colenso kopjes—were once more on the right bank. During the day there was a truce for the removal of the wounded, but this did not preclude the movement of troops, and before nightfall the batteries had retrieved their baggage, got into position, and settled down to enjoy a little repose.

26th Feb.

The 26th was spent by commanders and staff officers in reconnaissances, by the batteries in registering their targets. The plan for next day's attack was for Barton's brigade[1] to descend from the plateau, cross by the new bridge, and work down stream under cover of the left bank until it reached the foot of Pieter's Hill, which it was then to attack. Kitchener (who had succeeded Wynne—wounded on the hill which bears his name) was to follow Barton, and, as soon as the latter's attack began to make itself felt, to assail the hill then known as "Railway", and since as "Kitchener's". When this attack had developed,

---

[1] Brigades had become so intermingled during the recent fighting that the term now only denotes the temporary amalgamation of three or four battalions under a brigadier.

Norcott's brigade from its sangars on the terraces of Hart's hill was to carry the summit.

To support these attacks Colonel Parsons disposed the artillery as follows :—

The northern spur of Hlangwane had three tiers of guns—on the summit of Naval Hill the two 4·7″ on wheeled carriages : a little lower four 12-prs : below them the 7th and 64th batteries.

Between the above and the edge of the bank over-hanging the river were, in order from the left, 28 and 78 R.F.A., No. 4 mountain battery less a section, and "A", R.H.A.

Fuzzy Hill held 63 R.F.A. and 57/R.G.A., the field battery in the middle with two 5″ on each flank. Beyond were 19 and 61 R.F.A. and the two 4·7″ on platform mountings.

Clump Hill held four 12-prs. and two 2·5″ of No. 4 Mountain Battery.

Back at Chieveley there were still the guns on the railway—a 6″ and a 4·7″—and two 12-prs. while the 73rd held the extreme left on the Colenso kopjes—the only guns on the far bank.

During the night of the 26th—27th the old bridge was dismantled, and by nine o'clock the next morning it was ready in its new site. Long before that Parsons had assembled brigade and battery commanders to explain the orders, and allot the targets. Scarcely had they got back to their guns when a great roar of cheering swept down the line as mounted orderlies brought copies of Lord Robert's telegram announcing the surrender of Cronje at Paardeberg.

It is not necessary for us to follow the fortunes of the attacks on the three hills individually, but before con-sidering the action of the artillery a few words must be said as to the general progress of the infantry. With varying vicissitudes, due to irregularities in the obstacles presented by the ground, the strength of the defences, and

the action of the defenders, the attackers struggled "upwards still and onward" until the false crests were
reached. Between these and the true crests, where the
main defences were, there intervened a space that, though
short, was swept with a fire under which many a gallant
charge melted away. In the end, however, the time came
on each hill in turn, when from all sides the encircling
infantry surged forward in a rush which would not be
denied.

To turn now to the gunners' part. From their great
battery of 70 pieces, high above the Tugela, they watched
the whole drama unfolding before their eyes, and at a
range where their fire would be most effective—two to
three thousand yards. It was to be no game of long bowls
this time. During the morning, while the infantry were
hidden in the deep ravine, the bombardment was slow and
chiefly directed against the enemy's right and centre, so
as to conceal as long as possible the intention to turn his
left; the heavies taking on the Boer guns whenever they
opened fire. About noon the first of the infantry began
to emerge on the slopes, and thenceforward all guns were
devoted to smoothing their path, the fire concentrated on
each hill in turn as the attack developed against it.

There was no pursuit, but that was not the gunners'
fault. Batteries were limbered up and ready, but Buller
met them at the bridge and ordered them back.

On the 1st March the army advanced, and a battery
commander wrote home an account of the event of the
day :—

> "I saw Buller and all his staff. He turned aside to
> speak to me, and said with a laugh that I might go on
> to reconnoitre, which is what I had told him I was
> doing. I guessed what was going to happen, and
> went after them (looking for howitzer positions!) until
> we rode right into Ladysmith. None of them knew
> that Buller was coming till we were in the town.
> Then suddenly round the corner dashed Sir G. White
> and all his staff. They simply rushed at us. Such

a shaking of hands and talking. It was most
interesting and I am glad I was there, though of
course I had no right to be".

Two days later came the official entry of the Relieving
Army, as told in Chapter XXVI. But Ladysmith had
been relieved on Majuba day when Pieter's Hill went up
in smoke and flame, and two short extracts from the
accounts of eyewitnesses bring the final scene vividly
before the reader.

"It is the early morning of Majuba day. A group of
senior officers are gathered round their colonel on a spur
above the deep depression in which the Tugela runs,
conning over a paper on which are written down the
General's orders for the coming battle. There are thirty
thousand fighting men scattered over the hills and slopes
within two miles of where they stand, but all still and
silent but for the dull roar of the falls below and the
occasional neigh of some lonely horse. . . . The Colonel
rolls up his paper stuff. 'Well', says he, 'you all know
what to do. You had better be getting back to your
batteries'. . . And we ride to our posts, confident
already of victory, for we like the plan, certain that the
guns at any rate will speak with no uncertain sound. . . .
All through the morning hours, while we ply the enemy
opposite with the very best we have, our thoughts are ever
with the Fusiliers filing unseen down the valley below
toward the point where they are to breast the declivities
out of the ravine and show themselves. . . . Now the
infantry are surging up out of the trough of the Tugela
right in front of us in thousands. . . . On the right
they are almost in touch with the Fusiliers, to the left
they are already lapping round the underfeatures of the
fatal kopje known as "Hart's Hill". Every gun is
speaking, naval gun and mountain gun, five-inch and
fifteen pounder. And now the leading line of khaki is
tramping steadily forward into the lyddite bursts. The
gunners at their howitzers nestling behind an eminence

CHAPTER
XXIX

1900.

Pieter's Hill.

only knowing that they are being called upon to fire very rapidly, the infantry as they sniff the fumes only know that somehow the shells always fall just a little way in front of them''.

And here is how it looked from the other side[1] :—

"When we came to the rear of Pieter's Heights we saw the ridge going up in smoke and flame. The English batteries were so concentrating on the crest that it was almost invisible under the clouds of flying earth and fumes. . . . Suddenly the gun-fire ceased, and for a space the fierce rattle of Mausers followed by British infantry swarming over the sky-line their bayonets flashing in the sun. Then a rout of burghers broke back streaming down the hill in disorderly flight with the British cheering loudly, shooting, and running down the slope. . . . The British had blasted a gap, and wherever we looked Boer horsemen, wagons, and guns were streaming to the rear in headlong retreat''.

The
Artillery
Action.

The fortnight's fighting from the 14th to the 28th February has a special interest for gunners, for it is the first example of the handling of a force, comprising every branch of field army artillery—horse, field, field howitzer, mountain, heavy—in a series of actions culminating in a notable victory.

The success of the artillery action was chiefly due to the close and constant watch kept on the infantry, so as to see when and where they needed support, and then the rapid concentration of fire upon the point where for the time being it was most needed. It may be conceded that the conformation of the ground afforded unusual facilities for the observance of these principles, but due credit must be given to the skill with which Colonel Parsons marshalled[2] his heterogeneous collection of "pieces", and

---

[1] *Commando* by Denis Reitz.

[2] e.g. the posting of the guns on Clump Hill from which the 12-prs. were able to enfilade the Boer trenches on Pieters Hill and the mountain guns to deny access to it for reinforcements.

Major-General Sir Lawrence W. Parsons, K.C.B.

directed their fire in spite of the wide dispersion of the batteries. This owed much to the way in which he had previously named every point in the enemy's position, so that his orders as to the application were always intelligible. Perhaps, however, the real secret lay in the confidence which he inspired, as well in the Commander-in-Chief and his staff, as in the batteries he commanded, and the battalions they supported. He never tired of impressing upon brigade and battery commanders the importance of keeping their eyes fixed continuously on the infantry, since if once lost to view it would be impossible to pick up the lines of skirmishers among the rough rocky hills. He urged them to maintain their fire till the last moment, and then, when it became impossible to shell the works without running grave risk of hitting our own men, to increase elevation and fuze by at least five hundred yards to harrass the retreating Boers. The infantry had thoroughly realised the importance of this artillery support, and they had learnt to let the gunners know when they were about to charge by waving their fixed bayonets high above their heads. They were known to grumble if the guns ceased too soon out of regard for their safety.

		DAYS IN FEBRUARY														
BRANCH	UNIT	14	15	16	17	18	19	20	21	22	23	24	25	26	27	TOTAL
R.N.	One 6″	—	—	—	25	30	15	17	—	12	48	26	—	6	28	207
	2 4·7″ wheeled	77	48	30	43	78	19	47	40	131	244	180	—	111	146	1,194
	4 4·7″ rail and platform	—	—	—	—	22	—	—	—	—	19	15	—	6	125	187
	10 12 pr.	31	50	43	65	110	82	65	237	318	375	226	—	337	1,133	3,072
	Total R.N.	108	98	73	133	240	116	129	277	461	686	447	—	460	1,432	4,660
R.H.A. R.F.A.	A Battery	—	—	—	—	—	—	—	—	—	326	—	—	—	239	565
	7th do.	—	11	30	31	307	107	10	82	—	464	137	—	88	272	1,539
	63rd do.	—	29	21	157	266	160	234	83	—	536	174	—	—	203	1,863
	64th do.	110	24	52	104	98	220	206	220	—	368	161	—	104	413	2,080
	28th do. ⎫ 73rd do. ⎬ 74th do. ⎭	146	—	—	249	625	376	62	964	359	—	—	—	98	450	3,329
R.G.A.	19th do.	—	—	—	—	78	—	—	343	548	424	—	—	95	284	1,772
	66th do.	—	—	—	—	—	—	—	—	—	58	—	—	—	—	58
	61st do. (How.)	110	—	—	151	140	68	43	67	261	374	268	—	66	145	1,693
	4th Mountain	—	—	—	—	—	—	—	—	—	907	—	—	5	170	1,082
	57th Heavy	—	20	79	131	236	100	7	62	130	286	220	—	76	179	1,526
	Total R.A.	366	84	182	823	1,750	1,031	562	1,821	1,298	3,743	960	—	532	2,355	15,507
	Grand Total	474	182	255	956	1,990	1,147	691	2,098	1,759	4,429	1,407	—	992	3,787	20,167

CHAPTER XXX.

The Halt at Bloemfontein.

The Position in the Orange Free State—Karee Siding—Sanna's Post—
Wepener—Cape Colony—Preparations for Advance.

Map 13.

VICTORIOUS as had been their progress the position of the British forces in the Orange Free State in March, 1900, was not without its embarassments.   Those of a military character must now be indicated : with the civil measures for the pacification of the country we are not concerned.

Lord Roberts with his army lay in and around the capital, of which Major-General G. Pretyman, R.A., was appointed Military Governor.   Gatacre and Clements had crossed the Orange River and established their forces in the south.   The burghers who had opposed the advance on Bloemfontein were so utterly demoralised that de Wet had dismissed them to their homes to recover their spirits. The commandos which had been on the Orange River were in full retreat to the north-east.

If Lord Roberts had been able to follow up his victories by immediate pursuit all resistance might have ended. But this was quite impossible.   Men and animals were exhausted by forced marches under great heat in a waterless country.   Not the least trying of their experiences had been the halt at Paardeberg.   Here were laid the seeds of the terrible outbreak of enteric which followed the arrival at Bloemfontein.   The army must have rest.

Even if the troops had been in a condition to proceed, railway communication with the sea-ports was essential for a further advance.   Supply by road from the Kimberley line had been made to serve for the dash at Bloemfontein : it was obviously inadequate for the invasion of the

CHAPTER
XXX.

1900.

The Position
in the
Orange Free
State.

March.

CHAPTER
XXX.

1900.

The Position
in the
Orange Free
State.
Transvaal. The railways from Port Elizabeth and East London were now in our hands, and only required the repair of the bridges over the Orange River, but the transfer from the Kimberley line of the masses of stores which had accumulated there was a tremendous undertaking. How this affected the artillery will be told later: it is first necessary to relate the military operations which took place while the army was waiting for its supplies.

The first of these operations was the affair known as "Karee Siding". Shortly after reaching Bloemfontein the 7th division had been pushed out to Glen, ten miles north of the capital, where the railway crossed the Modder river. The bridge had been blown up, and, in order to prevent interference while it was being repaired, it was decided to occupy Karee Siding, eight miles further on, where the Boers had taken up their position on a line of kopjes astride the railway.

For this operation the 7th division was joined by the cavalry and mounted infantry under French, who, as the senior, took command of the combined force. His plan was for the cavalry and mounted infantry to encircle the enemy's flanks, and for the 7th division then to attack in front and drive them into the arms of the mounted troops. But the latter were slow in getting to their places, and the infantry met with an obstinate resistance. There was a protracted fire-fight, in which the artillery took their full share, until the burghers were pressed back to their last position. A rocky ridge facing this was seized, but there the infantry were held up, and called for help from the artillery. A section of the 18th battery was sent, but the ridge was too steep for horses. Nothing daunted, the gunners unlimbered, and assisted by the infantry, hauled up their guns until they could poke the muzzles over the crest. The enemy's rifle fire was heavy, but the range was soon found at 1,225 yards, passed on to the infantry—who had been firing at 2,000—and the enemy cleared off. The mounted troops had, however, failed to get round, so the burghers streamed away across the plain towards Brandfort, unmolested excepted by the long-range fire of the artillery.

Earlier in the month, Broadwood, with the 2nd cavalry brigade, ("Q" and "U" batteries under Rochfort) and some attached troops, had been sent to Thabanchu, forty miles east of Bloemfontein, to clear up the situation in that part of the Free State. By this time, however, the commandos which had been opposing Gatacre and Clements on the Orange River were falling back in that direction, and he decided to withdraw to the Waterworks, half way to Bloemfontein.

At this point the Modder runs from south to north, the Waterworks being on the left bank, close to the drift where the road from Thabanchu crosses. From there the plain stretches westward for rather more than twenty miles to Bloemfontein, broken only by a solitary hill— Boesman's Kop—between which and the Waterworks runs the Koorn Spruit. There had been recent rain, and the river and spruit could only be crossed at the drifts.

Broadwood's withdrawal to the Waterworks was carried out without incident during the 30th March, and the troops were just stirring next morning when gunfire was heard from across the Modder—the direction from which they had arrived. A few shell fell among the convoy, which had been swollen by refugees from the Thabanchu district, and the Kafir drivers at once inspanned their teams and raced away in panic for the drift over the Koorn Spruit where the wagons spread out like a huge fan. The horse artillery had just hooked in, and the general told Colonel Rochfort to send one battery over to the other side of the drift to cover the movement of the convoy. "U" battery accordingly trotted on, breaking into column in order to get through, while "Q" battery followed in line at a walk.

The drift was a deep one, with very steep sides, so that the wagons, as they dipped down, disappeared from

CHAPTER
XXX.

1900.

Sanna's
Post.

31st Mar.

the sight of those following. There de Wet lay in ambush with four hundred men, who quietly surrounded and seized each wagon as it reached the bottom—except the first few which were allowed to pass unmolested. These were in view rising the further bank as the adjutant of the horse artillery rode into the block to clear a way for the guns and there was thus nothing to arouse suspicion in his mind, particularly as the Boers were indistinguishable in appearance from the refugees with the convoy. So much so that Captain Wray, looking for a mounted officer of the transport, paid no attention to a man on foot who spoke to him until the latter, laying his hand upon his knee, made him realise that he was a field-cornet arresting him.[1] The fate of "U" battery was very similar. A Boer officer appeared, and demanded their surrender, and at the same time burghers lining the bank levelled their rifles. The guns were hemmed in between the wagons of the convoy and could not turn, the men were scattered and unarmed. The battery was helpless and was captured *en bloc*, with the exception only of the battery commander and the serjeant-major. But the guns were left unattended, and the sudden outburst of fire stampeded the teams. Five were shot down by the Boers but one got away.

"Q" battery was then some two or three hundred yards behind, and was coming quietly along in line, when a man who had escaped gave the alarm. Major Phipps-Hornby wheeled his subdivisions to the left so as to get the convoy between the battery and the Boers, and, as soon as he considered they were covered, "subdivisions left wheel—gallop". The Boers, seeing that the plot was discovered, opened a furious fire which brought down a couple of teams (one gun and one wagon), but

---

[1] Burnham, the well-known scout, who was on the further side of the spruit had tried to warn Wray and had been captured in consequence. At first he was put with the officers, but, seeing how closely they were guarded, he asked Wray to get him moved, and this was effected by complaining that he was not the sort of man with whom British officers cared to associate. That night Burnham escaped.

the remaining five guns came into action near the build-
by the gun of "U" which had escaped.   They were out
in the open at little over a thousand yards from the
riflemen, who had the cover of the river bank, but here
they fought their guns hour after hour—and there were
no shields in those days.

Broadwood was in a difficult position.   Behind him the
enemy were in force with guns on the other side of the
Modder.   Before him they were holding the drift of the
Koorn Spruit, blocking the road to Bloemfontein.   One
of his batteries had been captured, the other was in action
under heavy rifle fire.   His mounted infantry were holding
the drift over the Modder to protect his rear, his cavalry
were at the Waterworks.   He decided to send the latter
to find a crossing over the Koorn Spruit further south by
which they might work round and take de Wet in rear.
The crossing was effected, but little impression was made
on the Boers in the spruit.   A line of retreat had, however, been opened, and the order was given to man-
handle the guns to the shelter of the station buildings
where they could be limbered up under cover.   By this
time the only unwounded men left with the guns were
the battery commander,[1] one serjeant, one corporal and
eight gunners, and the task was beyond their strength
unaided.   Volunteers were called for, and, with their help,
the five guns nearest shelter were brought in.   The
enemy were now alive to what was going on, and concentrated their fire upon the last.   Several attempts to
bring it in with pairs of horses were made, but every horse
was shot, and eventually it had to be abandoned.   Then
the battery moved off at a walk to the crossing which had
been found by the cavalry, and as soon as it had filed
safely past, the mounted infantry of the rearguard were
withdrawn from the Modder drift, and covered the retreat

---

[1] Colonel Rochfort, the brigade commander, had been shot while
pulling a limber up to the gun in order to have the ammunition closer
to hand.

CHAPTER
XXX.

1900.

Sanna's
Post.

31st Mar.

of the guns with great steadiness. They had paid a rare tribute to the gallantry of the gunners by springing to their feet and cheering them as they passed.

The following order was issued to the army :—

"The Field-Marshal Commanding-in-Chief, after careful consideration of the report upon the gallant and daring conduct of "Q" Battery, R.H.A., at the action of Sanna's Post, deems that where all 'were equally brave and distinguished', no special selection can be made for the grant of the Victoria Cross. Lord Roberts has therefore been pleased to direct that one officer shall be selected by the officers, one N.C. officer by the N.C. officers, and two gunners or drivers by the gunners and drivers, for the decoration under Article 13, Royal Warrant".

"The following are the selections :—

Major Edmund John Phipps-Hornby.
No. 47511 Serjeant Charles Parker.
No. 70062 Gunner Isaac Lodge.
No. 33286 Driver Henry Glascock".

The casualties in the horse artillery brigade were :—

		Killed	Wounded	Missing	Total
Officers	...	1	4	5	10
Other Ranks	...	4	29	128	161
Total	...	5	33	133	171

In pursuance of the policy of showing British troops as widely as possible, other detachments had been sent to various places. But directly the Boers appeared again in the field under leaders like de Wet, such a system had to be abandoned, and orders were issued for the withdrawal of the detachments to the line. We need not linger over the fate of those which were unprovided with artillery, but mention must be made of the defence of Wepener, for, although the guns of the garrison were not manned by the Royal Artillery, the artillery troop of the Cape

Mounted Rifles had been closely allied with the Regiment
on previous occasions, and was commanded by an officer[1]
who had gone through his artillery training at the School
of Gunnery.

At this time Wepener was held by something under
two thousand men of the Colonial division under Colonel
Dalgety. His artillery consisted of the artillery troop,
C.M.R., with two 15-prs., two naval 12-prs. (8-cwt.),
and two 2·5″ on field carriages, to which must be added
a 14-pr. Hotchkiss, the property of one of the colonial
regiments. Fortunately a strong position was available,
and time to fortify it, before de Wet brought up his force,
now grown to eight thousand men or more, with a dozen
guns.

The siege opened on the 7th April with heavy artillery
fire, supported by musketry. The bombardment was
repeated on the 10th, and was followed by a determined
night attack, but this was repulsed, and there was no
further attempt to rush the defences. For more than a
fortnight, however, they were continually under fire, and
the working parties, repairing by night the damage done
during the day, were harassed by snipers. Against an
artillery more numerous and better armed than themselves the gunners had a hard struggle, but the guns
were skilfully moved from one prepared position to
another, and were kept in action to the last, although
they had to be sparing of ammunition, as is shown by the
following statement :—

	15-pr.	12-pr.	2·5″	Hotch-kiss.
In hand at commencement of siege	850	250	250	350
Expended	750	200	100	342
Remaining when relieved	100	50	150	8

The defence of Wepener was the bright spot in the
somewhat gloomy period which succeeded the triumphs
of the invasion.

---

[1] Major H. T. Lukin, C.M. Rifles.

Lord Roberts' invasion of the Free State had caused the withdrawal of the Boer commandos which were facing Gatacre and Clements in Cape Colony. But far to the west, in the direction of Prieska, the inhabitants were ripe for rebellion, and the small columns which had been keeping order there had been absorbed into the advance on Bloemfontein. Seizing the opportunity of their absence a small Boer commando, moving through Griqualand, crossed the Orange River, and commenced distributing rifles,

So serious was the situation considered that Lord Roberts despatched his Chief of the Staff from Paardeberg, to deal with it. Lord Kitchener organised three small columns on the line about De Aar, to move through the Carnavon and Prieska districts, two of them under artillery officers— Colonels Sir Charles Parsons and J. Adye. The artillery employed consisted of the 44th battery, R.F.A., and the New South Wales Field Battery, augmented later by the 68th battery, R.F.A., and "D" battery, Royal Canadian Artillery. With the fall of Bloemfontein the heart was taken out of the rebellion, Prieska was occupied on the

21st March, and Adye was left in charge south of the Orange River, while Sir Charles Warren was appointed Governor of Griqualand. The one party of rebels and Boers in his territory was rounded up by Adye—the 44th battery still with him—and Warren with the Canadian battery and a section of the Cape Artillery cleared the country to the north.

While the measures described above for the pacification of the country were in progress, every nerve was being strained to fit the army for the invasion of the Transvaal. We have seen how heavy had been the calls upon the artillery. Men and horses, harness and stores, ammunition, all were necessary if the batteries were to regain their efficiency. The Ammunition Park was moved round from Modder River to Naauwpoort, bringing with it the whole of the baggage left behind by the artillery which had accompanied Lord Roberts, and many ammunition wagons abandoned during that advance. But there was no possibility of getting it any further by rail: the line to Bloemfontein was blocked with the accumulation of the army's requirements. Naauwpoort was, however, the remount depot, and although the number of draught horses was quite inadequate for the requirements of the artillery, there were also in the depot several hundred mules, many of considerable size and power. An artillery staff officer was sent down from headquarters, and arranged that the ammunition park should draw the horses and mules and march to Bloemfontein.

Starting on the 1st April the great convoy arrived at Bloemfontein on the 14th, with sixty vehicles of sorts, and 450 mules besides all the available horses. A sort of fair was at once held, all units sending in parties with their demands, and the artillery staff distributing the contents of the convoy between the different claimants according to the urgency of their needs—horses, mules, saddles, harness, ammunition wagons, forges, ordnance stores of every description. The fortunate find, shortly afterwards, of a thousand sets of mule harness made it possible to adopt mule draught for the wagons of batteries, and, although this decision was greeted with many indignant protests, the good qualities of the mules as campaigners soon justified their admission to the ranks of the Royal Artillery.

Before closing this account of the halt at Bloemfontein the writer may perhaps digress for a moment to record a striking feature of the stay of the army there, which was brought to the particular notice of the artillery staff owing to their being billetted in the Bishop's House. Sunday after Sunday, morning and evening, the cathedral was crowded by soldiers of all ranks. And not only that. Each service had to be repeated for the benefit of those who had been waiting in long queues for the first congregation to come out and make room for them.

CHAPTER
XXX.

1900.

Preparations
for Advance.

April.

CHAPTER XXXI.

THE INVASION OF THE TRANSVAAL.

The Advance to the Vaal—Vet River—Zand River—Kroonstad—Doorn
Kop—Johannesburg—Pretoria—Diamond Hill—The Magaliesberg—
Zilikat's Nek—Derdepoort—The West—Rooidam—Mafeking.

THE NATAL ARMY.

Helpmakaar—Botha's Pass—Alleman's Nek—Joining Hands.

MAPS 13 & 22.

CHAPTER
XXXI.

1900.

The
Advance to
the Vaal.'

May.

By the beginning of May all was ready[1]. The advance
was to be made on an immense front, its "far-flung
fighting line" extending 350 miles from Ladysmith on the
right to Kimberley on the left, its various portions moving
simultaneously under the supreme direction of Lord
Roberts. It will be convenient first to follow the fortunes
of the main body under the personal command of the Field-
Marshal, and then to see how the wings fared.

The main body consisted of the cavalry and mounted
infantry and three and a half infantry divisions. There
was now also a considerable force of heavy artillery, in
addition to the field artillery of the divisions, and the pom-
pom sections that were coming into the field. In the
centre there were assembled at Karee Siding, 20 miles
north of Bloemfontein, the 7th and 11th divisions with a
mounted infantry corps, and for heavy artillery, two 5″,
two 4·7″, two 12-prs. On the right Ian Hamilton at
Thabanchu had a cavalry brigade, some mounted infantry,
and two infantry brigades, with the Xth brigade, R.F.A.[2]

On the left the 10th division had taken the place of
the 1st on the Kimberley line, and the latter was moving
in towards the centre. Behind the line were left the 9th

---

[1] The Order of Battle will be found in Appendix E.
[2] Transferred from the 6th division—the 76th, 81st, and 82nd batteries, with, in addition, the 74th, lent by the 8th division.

division behind Ian Hamilton, the 6th division garrisoning Bloemfontein, the 3rd division guarding the railway from Bloemfontein to the Orange River, and the 8th and Colonial divisions in the east of the Free State.

CHAPTER
XXXI.

1900.

The
Advance to
the Vaal.
3rd May.

On the 1st May Ian Hamilton successfully cleared the Boers out of Holtnek ten miles north of Thabanchu, and on the 3rd the general advance began. The centre had not covered many miles before the enemy were encountered, and the IIIrd brigade, R.F.A., was seriously engaged. The Boer gunners had got the range to a nicety, but fortunately their shrapnel burst high in the air, and the 18th and 75th batteries soon silenced them. Next day Hamilton had another lively action, and on the 5th the main body encountered some resistance on the Vet River. The guns were all brought into action, but the mounted infantry turned the flank and the infantry had nothing to do. The next obstacle was the Zand River, and before that was reached French had brought up the cavalry division which had been left behind to complete its re-fitting. On the 9th he secured the crossings on the left, while Hamilton made good those on the right, and next morning the heads of all the divisions were across the river. Here and there, however, there was hard fighting. On the right, where the kopjes came down to the river, the infantry of Hamilton's force, supported by the 74th and 76th batteries, had to fight hard to clear the hills, while the naval 4·7″ and the 5″ of 56/R.G.A. dealt with the enemy's guns. In the centre the 7th division carried all before them, though the 18th battery, crossing the river in close support, came under a hot fire from our guns lost at Sanna's Post. On the left the cavalry got into some difficulties, but the battle may well be summed up in de Wet's own words— "All day we were driven relentlessly, the British herded us like sheep, to the incessant shriek of shells and the whizz of bullets, and by evening we were a demoralised rabble fleeing blindly across the veld".

President Steyn had no choice but to evacuate

CHAPTER
XXXI.

1900.

Kroonstad.

12th May.

24th May.

Kroonstad where he had established his Government after
being driven out of Bloemfontein, and the town surrendered to Lord Roberts on the 12th May.   But the rapid
advance had taken toll, especially of the horses, and the
enemy had blown up all the bridges on the railway, so
another halt was unavoidable.   It was only ten days this
time, however, and before leaving Lord Roberts issued a
proclamation annexing the Orange Free State.

No resistance was offered to the passage of the Vaal,
and French and Hamilton (who had been brought across
from the right to the left of the main body) moved direct
on Johannesburg, while the main body, crossing at
Vereeniging,[1] followed the line of the railway to Elandsfontein.   But Botha was not going to give up all the
riches of the Rand without a struggle, and he had gathered
a considerable force on the Doorn Kop, a ridge above the
Klip river which covered the town from the south.   Against
this position Ian Hamilton advanced on the afternoon of
the 29th May.   His two brigades were side by side, supported by the field batteries, while two guns of
56/R.G.A. dealt with a long-range gun firing from
behind the Boer line at 8,500 yards.   The field batteries
moved in to 3,200 yards as the infantry advanced, and
then, to cover the assault, the 82nd and a section of the
76th, hitherto held in reserve, were pushed in to two
thousand or less.   As twilight fell the infantry gained the
forward crest, but, as on so many occasions, had to bring
their bayonets into play before the summit was won.
French with the cavalry and mounted infantry on the left
had got right round their flank, the horse artillery
brushing aside such small opposition as was encountered.
The Royal Artillery may indeed claim a full share of the
success of the day, for in the words, once more, of de Wet

---

[1] It was the Sunday after Ascension Day, and there was an open-air
service on the banks of the river.   By a curious coincidence the lesson
was the account of the crossing of the Jordan and the entrance of the
Israelites under Joshua into the Promised Land.

"The British . . . ran no risks with their infantry con-
fining themselves to most unpleasant gun-fire. For the
first time for many days we too had guns in action and
there were several batteries of Creusots blazing away.
The gunners suffered terribly".

On this day the main column was marching up the
railway line, and the mounted infantry were sent ahead
with "J" battery to secure the junction at Elandsfontein
where the lines from the Free State to Pretoria and from
Natal to Johannesburg cross. A 155$^{m/m}$ Creusot on the
railway gave some trouble until it was withdrawn,
frightened of having its retreat along the railway cut off by
the fire of "J", directed upon the track behind it. There
was some skirmishing among the slag-heaps, but by three
o'clock the mounted infantry had gained full possession
of the railway junction with all its railway stock.

Next day, while the Johannesburg authorities were
given twenty-four hours to clear the Boer stragglers out
of the town, the cavalry were working round both sides
on the way to Pretoria, and there were clashes with the
enemy's rear-guards in which "G", "O", and "R" were
engaged, and a field gun captured. On the 31st Lord
Roberts made his formal entry, and the 7th and 11th
divisions and heavy artillery marched past him in the
centre of the town. A two days' halt followed, during
which the artillery staff was called upon to deal with all
the explosives in the magazines of the fort built after the
Jameson raid to dominate the town. Bonfires were built
all round the ramparts, and the stuff—dynamite, gun-
powder, detonators, fuzes, etc.—was all brought up and
thrown into the flames. As related in Chapter III there
had been a terrible explosion when carrying out a similar
task at Kabul in 1879, in which the artillery officer in
charge was killed, and Lord Roberts was insistent on
every precaution being taken, and frequent in his
inquiries.

On the 3rd June the army was set in motion again,

guided by a driver of the 75th battery, who had lived in
these parts. It was the last lap, for the capital was only
fifty miles away. No opposition was encountered until the
advanced guard of mounted infantry and "J" battery
had crossed Six Mile Spruit, when they were met by fire
from the ridge which hid the town from the south. "J"
was under heavy fire until the 7th and 11th divisions and
the heavy guns came up, but, thanks to a position skilfully "false-crested", suffered little. The 11th division
brought the naval 12-prs. and the 83rd and 85th field
batteries into action on a kopje to the right of the road;
the 84th with the 4·7″ and 5″ guns on Zwart Kop. The 7th
division sent forward at first only sections, but later the
whole of the IIIrd brigade, R.F.A., was pushing forward
when orders were received for the attack to be delayed until
Hamilton's column on the left could come in on the flank
of the Boers holding the ridge. The modern fort, conspicuous on the highest point was silent, and the big 9·5″
howitzers[1] were not brought into action, but the sailors
could not resist the temptation to have a shot at it, after
which their fire was directed over the ridge in the hope of
hitting the railway station just behind. That evening
emissaries from Pretoria came out to arrange for its

surrender, and on the 5th June Lord Roberts made his
ceremonial entry, the flag of the Republic was hauled down
and the Union Jack hoisted. Best of all, the prisoners
were released[2]—the senior being Lieut.-Colonel H. V.
Hunt who had commanded the IVth brigade, R.F.A., at
Colenso.

---

[1] The story of the purchase of the "Skoda" howitzers and their
shipping from Trieste has been told in Volume II, and need not be
repeated here. The express object was to deal with the Pretoria forts,
and as soon as they arrived in South Africa 92, R.G.A., the siege
company at Modder River, was sent back to de Aar to take over the
howitzers. From thence the half-company took them on to rail-head
which was then at Roodeval, between Kroonstad and the Vaal. Here
they were met by a great train of oxen to draw the howitzers, and a
battalion to escort them, and thence they marched under Colonel
Perrott to Pretoria, where they arrived in time, but were not called
upon. Their further adventures have been told on page 162.

[2] All except a thousand taken east by the Boers.

President Kruger had fled to Machadodorp on the railway towards the Portuguese frontier, and .the great majority of the Transvaalers had drifted home, as had the Free Staters after the fall of Bloemfontein. But Botha managed to gather seven thousand men, and with these and all the guns, took up a position astride the line some fifteen miles east of Pretoria. Such defiance could not be allowed, and on the 7th Lord Roberts sallied out to disperse the gathering, leaving the 7th division to garrison Pretoria.

The Boer position was on a line of hills that crossed the railway at Pienaar's Poort. North of the deep ravine through which the line passed, the range extended for a dozen miles to a cluster of kopjes at Krokodil Spruit and Kameelfontein. The centre lay between Pienaar's Poort and the similar gorge of Donker Poort through which the main road passed. South of this the ridge continued by Diamond Hill for another dozen miles, making the total length of the line occupied not less than twenty-five miles.

On the 8th Lord Roberts' camp came under fire from one of the 155$^{m/m}$ Creusots which the Boers had mounted on a railway truck. This was run out of the shelter of Pienaar's Poort, but 56/R.G.A. dropped a 5″ shell between the rails at nearly 10,000 yards, destroying the track within a few yards of the muzzle, and the gunners sensitive to the danger of the line being cut behind them, hastily withdrew, as they had done at Elandsfontein. There ensued a couple of days negotiation, which proved fruitless, and on the 11th Lord Roberts set his troops in motion. On the left French with the 1st and 4th cavalry brigades and mounted infantry; on the right Ian Hamilton with the 2nd and 3rd cavalry brigades and the 21st infantry brigade; in the centre the 11th division and heavy guns.

On the left the day began with the leading cavalry brigade, aiming to outflank the enemy, bumping into them in position. Coming under a hot fire the cavalry had

CHAPTER
XXXI.

1900.

Diamond
Hill.

11th June.

to fall back, and there followed further attempts to "get round", finishing up with a race for a commanding point on the extreme flank. The burghers counter-attacked time and again, and the fighting, in which the cavalry were vigorously supported by "G", "O" and "T" batteries, went on until dark.

At the other end of the line Ian Hamilton also struck at the flank with his mounted troops, and again stirred up a hornet's nest. For the rest of the day there was a continuous fire-fight, but little movement on either side. Particular interest, from an artillery point of view, attaches, however, to an incident of the fighting owing to the evidence it affords of the effect of case-shot. Fortunately a first-hand account is available for in a discussion at the Institution, Major Phipps-Hornby, who was in command of the battery, described the incident. "I had the good fortune", he said, "once during the war to fire case shot—at the battle of Diamond Hill. On that occasion the Boers were all round us, very close up, and we fired eight rounds of case shot. The Boers were between 250 and 300 yards from us, and I cannot describe it better than by saying that it was like a scythe—the whole of the Boers, the grass, and everything were absolutely wiped out as if you had taken a scythe. We had two guns and fired eight case shot. I was sitting on a horse and I saw it done".

On the inner flank of his mounted troops, Hamilton's infantry brigade, supported by the 76th and 82nd batteries and by two 5″ guns of the 56th, had fared better. The Boers were pushed off their advanced position on a lower ridge which covered their main position on Diamond Hill, but neither time nor troops admitted of further advance.

In the centre the fighting was confined to long-range artillery, in which the other half-battery of 56/R.G.A. bombarded the Boer position between the two "Poorts", and the naval 4·7″ engaged in a duel with the "Long Tom" on the line—greatly enjoyed by the spectators.

The Boer gun would be run out from the ravine, fired, and then hastily run back under shelter again—if possible before the sailors could get in a shot.

It was a disappointing day. Botha had taken note of Lord Roberts' invariable preference for "going round", and had extended his line and strengthened his flanks to such an extent that the mounted troops at both ends of the line had been blocked, and were no more than holding their own, while in the centre the infantry had never been put in. If any progress was to be made the burden must be shifted on to their shoulders.

We have seen that Ian Hamilton's infantry and artillery had established themselves on a lower ridge facing Diamond Hill. To them on the morning of the 12th were sent the Guards brigade from the 11th division, the 83rd battery, the other half-battery of 56/R.G.A., and two naval 12-prs. This combined force was launched against the ridge occupied by the enemy from Donker Poort to Diamond Hill, and the crest was won without much difficulty. The infantry then found themselves on an open plateau, faced by a furious fire from a further range, and enfiladed from the other side of the poort where the ridge ran up to a commanding point unapproachable across the gorge. There was a call for the guns, and prompt was the response. Major Connolly took the 82nd up the steep slope of Donker Hoek, to find on reaching the top, the infantry lining a rocky ridge faced by the Boers on a similar feature. He brought his battery into action at once on the only position from which he could clear the infantry, and opened fire at 1,850 yards. The infantry reported an almost immediate relief, but the battery soon drew the fire not only of the riflemen but of the guns. Major Guthrie-Smith followed with the 83rd, which he took almost into the firing line, opening fire at 900 yards, to keep down the fire from the left. And when the mounted infantry, with a couple of pom-poms, gained a point which covered the right, the position was won. During the night the

Boers evacuated the whole line of hills, and next day
the British marched back to Pretoria, and set to work
to repair the damage caused by their rapid advance from
Bloemfontein.

They were not to be left at leisure for long. The
Boers, driven from their position on the east of the capital,
were soon gathering again in the north, beyond the
Magaliesberg. This range, stretching east and west for
nearly a hundred miles forms a rocky barrier between
the bush-veld and the open country in which Pretoria lies.
Being impassable for troops except at half-a-dozen passes,
these "poorts" or "neks" became features of great tactical
importance, and will appear frequently in this narrative.
Taking them in order from east to west, Derdepoort, north-
east of Pretoria, comes first; then Wonderboom Poort,
due north, where the road and railway from Pretoria to
Pietersburg find a way through the range; next Zilikat's
Nek, eighteen miles north-west; Commando Nek (where
the Pretoria—Rustenburg road crosses); Olifant's Nek, and
finally Magato Nek. Wonderboom Poort was included in
the defences of the capital, and was protected by a modern
fort. Derdepoort and Zilikat's Nek formed part of the
outposts and were held by cavalry. Against these the
Boers developed an offensive on the 11th July, a month
to the day after their defeat at Diamond Hill.

Up to the 10th July Zilikat's Nek had only a squadron
and a section of "O" battery (Lieut. W. P. L. Davies) to
guard it, but on that afternoon, in consequence of a
warning received, a battalion arrived. The infantry,
did not, however, occupy the summit of the range, but
remained in the nek itself which was not only commanded by the hills on each side, but had no field of
fire to the front owing to the bush which ran right up to
it. It had only been possible to clear this for two or
three hundred yards in front of the guns.

The attack came at dawn on the 11th, the Boers
working round the flanks by the main ridge as well as

through the bush in front. The guns fired shrapnel, set
at 1, and then all their case, against those creeping
through the bush, but the burghers on the hills were too
high to be reached at that short range, and their plunging
fire dominated the sangars. It was an impossible situation, and guns and gunners were captured with the cavalry
and infantry.

At the same time a column from Krugersdorp, including a section of the 78th battery under Lieut. A. J.
Turner, coming up from the south, expecting the co-
operation of the cavalry and horse artillery, fell into
a trap at Dolverkrantz. The guns came under very
heavy rifle fire, and soon only three of the gunners were
left unwounded, the section commander still working his
gun though hit three times. The unwounded men and
infantry[1] did all they could to save the guns, but the
limbers had been placed under cover at some distance in
rear, and both teams were disabled as they made the
attempt. The casualties in the section, in addition to the
section commander, were sixteen out of nineteen.

There was still a third episode on this day. A cavalry
regiment got badly hustled in the bush-veld, and had some
difficulty in getting back to Derdepoort, where were the
headquarters of the brigade and another regiment. Lord
Roberts fully expected that a serious attempt would be
made to force the line of the Magaliesberg covering
Pretoria, and General Marshall, with the headquarter
staff of the artillery, was sent out to take command of the outposts. The force under his command,
was quickly made up to a whole brigade of cavalry
(with its horse artillery) and two brigades of infantry who
held the hills for some miles on both sides of the poort.
The only attack was at daybreak on the 12th when the
cavalry outposts were driven in, and the Boers brought
some guns into action. These latter evoked a reply,

Margin notes: CHAPTER XXXI. 1900. Zilikat's Nek. 11th July Derdepoort 12th June.

---

[1] Captains Gordon and Young of the Gordon Highlanders received
the V.C.

however, which they had little expected, for a 6″ gun had been among the reinforcements hurriedly despatched to Derdepoort on the day before. In face of this display of strength the Boer attempt was never pressed, but what makes this incident worthy of inclusion in regimental history is that it was the first attempt to take a gun into action with mechanical transport. The traction engine which brought it from Pretoria covered the six or seven miles satisfactorily, but next morning when required to take the gun into position it was found lying on its side in a spruit to which it had gone for water. And so recourse had to be made to the primitive ox-teams after all, and three of these dragged the gun into action just before dawn.

A few days later the cavalry and mounted infantry holding the outpost line on the east of Pretoria were seriously attacked, and it became clear that steps must be taken to clear the country to the north and east of Pretoria. More troops were ordered up from Bloemfontein; columns swept the country north of the Magaliesberg; the 11th division advanced along the railway to the east, with the cavalry and mounted infantry on the right. It was an imposing array on a front of 35 miles, under the Commander-in-Chief, but on the night of the 25th July such a storm burst as no one that experienced it will ever forget. The mortality among the transport animals was so great that the operations had to be suspended.

We may take the opportunity to glance at the operations on the wings.

Soon after the relief of Kimberley Lord Methuen had pushed some of his 1st division up the line to Warrenton, where the railway crosses the Vaal. The enemy were entrenched on the opposite bank at Fourteen Streams, with a "Long Tom" and other guns, and there was a good deal of artillery activity across the river, in which the 20th battery was alone until reinforced by the 37th (howitzers) at the beginning of April. On the arrival of Sir A. Hunter with the 10th division from Natal,[1] Lord Methuen moved the 1st division up-stream, first to Boshof, then to Hoopstad, and so to Kroonstad where he arrived on the 28th May, just in time to take a prominent part in the defence of the lines of communication, as we shall see in the next chapter.

While the 1st division[2] drew in towards the centre the 10th division took its place on the west, the 66th battery going up to Warrenton, where also were the Diamond Fields Artillery and the 6″ gun from Modder River its fire directed from a balloon. These occupied the attention of the Boers at Fourteen Streams, while Hunter with the main body of his division, and the 28th and 78th batteries, crossed the Vaal lower down and attacked—and defeated—the enemy at Rooidam on the 5th May. Thus threatened in flank, those at Fourteen Streams fell back, and the crossing was won, although the railway bridge had been destroyed.

In his dispositions Hunter's great object had been to cover the movement of the force destined for the relief of Mafeking, where a little British garrison had been holding out ever since the outbreak of hostilities. A half-way house on the railway from Kimberley to Bulawayo, just outside the Transvaal border, its capture had been one of the first objectives of the Boers, and a considerable commando—at first under the redoubtable Cronje—was sent to take it.

No share in its defence can be claimed by the Regiment, but the steps taken by the garrison to form an artillery are worthy of note in regimental history. The

CHAPTER
XXXI.

1900.

The West.
April.
28th May.

Rooidam.
5th May.

Mafeking.

1899.

---

[1] The 10th division was formed in Natal after the relief of Lady-smith of the 5th and 6th infantry brigades and a brigade of field artillery formed for it under Lieut.-Colonel W. A. Smith, consisting of the 28th, 66th and 78th batteries—all from different brigades.

[2] The artillery left with the 1st division consisted of the 4th, 20th, and 38th batteries, and a section of the 37th—under Major Butcher of the 4th battery.

only guns in the place were four 7-prs. and a 1-pr. Nordenfeldt, but, as at Kimberley, the local engineering works quickly rose to the occasion. An old smooth-bore ship's gun, a relic of some native war, was mounted upon wagon wheels, round shot cast for it, and even shell improvised. It threw a 16-lb. shot three thousand yards, and with the 7-prs. drove back out of their range the 155$^{m/m}$ Creusot which the Boers had brought from Kimberley. The great effort of the railway workshops, under Mr. Coghlen, was, however, the construction of a 5″ howitzer,[1] the barrel of steel tube strengthened by iron rings shrunk on. It was mounted on the wheels of a threshing machine. Major F. W. Panzera of the British South Africa Police commanded the artillery throughout, and the guns put up a wonderful show, sometimes distributed in defence, sometimes massed for attack.

At the same time that Colonel Baden-Powell had been sent out to Mafeking, Lieut.-Colonel Plumer had gone to Rhodesia. There he raised a local force of a thousand men, with, for artillery a Maxim-Nordenfeldt 12½-pr.[2] and two 2·5″ left behind by the mountain battery after the Matabele Rebellion. A retired officer of the Regiment, Lieut. H. de Montmorency, was fortunately available to take charge. With this force Plumer worked down the railway towards Mafeking, with plenty of fighting on the way. Eventually he found the enemy barring the direct road too strong for him, and struck out into the open country west of the line. There he established himself, and there he was joined by "C" battery of the Royal Canadian Artillery (Major Hudon) which had been sent round by sea from Cape Town to Beira in Portuguese East Africa, and thence made the long march through Rhodesia, relays of mules for coaches and guns laid along the road.

---

[1] Now in the museum of the Royal United Service Institution.

[2] This gun was similar to those used in the Jameson Raid and captured by the Boers, which it will be remembered, were re-captured at Elandslaagte, and used throughout the defence of Ladysmith.

It was not until the middle of April that Lord Roberts was able to do anything to help, but he then formed a special force of twelve hundred mounted men under Colonel B. Mahon to start from Kimberley. For artillery Mahon had "M" battery (with a special establishment for the occasion,[1]) and a pom-pom section. At the beginning of May, Mahon slipped away from Kimberley, covered by the offensive of the 10th division, and, keeping well out to the west, got into touch with Plumer on the 15th. Next morning the two forces moved together upon Mafeking, the combined artillery massed between them— the Canadian battery on the right, the Rhodesian guns in the centre, "M" on the left. The Boers were in position seven miles west of Mafeking and greeted the appearance of the relieving force with a brisk shell fire, but the British guns galloped into action, and before the day was over the way was clear.

CHAPTER
XXXI.

1900.

Mafeking.

April.

May.

16th May.

THE NATAL ARMY.

We may now turn to the other flank where, after the relief of Ladysmith, the Natal Field Force had been re-organized by the amalgamation of the garrison and the relieving force, less the 10th division sent round, as we have seen, to Kimberley, and a "Drakensburg Defence Force" under Downing which included the 19th and 73rd field batteries and No 4 mountain battery. The naval brigades from the *Powerful* and *Terrible* returned to their ships, handing over their guns to No. 10 mountain battery, and three companies of garrison artillery[2] which arrived from home under Colonel P. S. Saltmarshe. This

---

[1] See Appendix H.

[2] {100 R.G.A. — Major F. A. Curteis.
    {101 do   — Major G. J. F. Talbot.
    {102 do. — Major C. E. Jervois.

left the naval brigade with the field force with only two 4·7″ and four 12-prs.

The force available for offensive operations thus consisted of the 1st and second cavalry brigades and 3rd mounted brigade, the 2nd, 4th, and 5th divisions, and corps troops. The artillery with these formations was:—

C.R.A.—Colonel L. W. Parsons.
Staff Officer R.A.—Captain R. W. Boger.
3rd Mounted Brigade—A/R.H.A.
2nd Division   —7, 63, & 64/R.F.A.—
                               Major W. L. H. Paget.
4th   do.   —II/R.F.A. (21, 42, 53)—
                               Lieut.-Colonel J. A. Coxhead.
5th Division   —I/R.F.A. (13, 67, 69)—
                               Lieut.-Colonel E. H. Pickwoad.
                  ⎧—61 & 86 (howitzer) R.F.A.
Corps Troops   ⎨—No. 10 M.B., 57, 100, 101, and
                  ⎩   102 R.G.A.

There were also the first three pom-poms to reach Natal.

This force was disposed round the north of Ladysmith facing the Boers who had taken up their position on the Biggarsberg, a spur of the Drakensberg which stretched across northern Natal. Here, in compliance with Lord Robert's wishes, Buller lay inactive throughout Mar. & Apr. March and April, but early in May, in accordance with the scheme for a simultaneous advance, he set his troops in motion. The first thing to be done was to clear the enemy off the Biggarsberg, and this was effected without much difficulty by an encircling movement, which commenced on the 11th May.

11th May.   All the fighting was on the right, where the 2nd
Helpmakaar. division and 3rd mounted brigade were directed against Helpmakaar, their right covered by a force of irregulars under Colonel Bethune, to which were attached sections

of 100/R.G.A. under Lieut. N. Tandy, and of the Natal CHAPTER
field artillery under Lieut. Livingstone. On the 13th the XXXI.
2nd division advanced to the attack, the 4·7″ of 101/R.G.A. 1900.
commencing operations by silencing two Boer guns which Helpmakaar.
had opened fire on them from a distance of 7,000 yards, 13th May.
just as they were limbering up to file off. The mounted
brigade galloping forward on their right surprised and
captured some advanced positions, but the opposition
soon proved too strong for mounted troops, and "A"
battery, behind the firing line, came under so rapid and
accurate a bombardment that the guns had to be man-
handled out of action. The long range[1] of the 12-prs. with
Bethune now proved invaluable, for it enabled them to
subdue the Boer guns. The infantry and artillery of the
2nd division then came up, a pom-pom was brought into
action—the first on the British side in Natal—and during
the night the Boers evacuated the position. The mounted
brigade, with "A" battery, followed up, driving their
rear-guard from successive positions, and on the 15th 15th May.
Dundee was occupied. It was not until they had reached
the famous position of Laing's Nek that the Boers stayed
their retreat, and Buller halted his troops on the Ingogo
facing them. The strength of the position that had
defied Colley in 1881 forbade a direct attack, and Buller
determined on a turning movement through the Drakensberg passes.

The heavy guns of the naval brigade and the garrison Botha's Pass.
companies were distributed[2] between the lofty mountain
of Inkwelo on the left of the British position facing Laing's
Nek, and van Wyk's Hill commanding the mouth of
Botha's Pass, which was seized by a brigade and the

---

[1] The Boer guns were beyond the range to which even these naval
guns were sighted, and could only be reached by sinking their trails
and using clinometer elevation.

[2] Inkwelo—two 5″, two 4·7″, two 12-prs.—R.G.A.

Van Wyk $\begin{cases} \text{two } 4·7″, \text{ four 12-prs.—R.N.} \\ \text{two } 5″, \text{ two } 4·7″, \text{ two 12-prs.—R.G.A.} \end{cases}$

CHAPTER
XXXI.

1900.

Botha's Pass.
5th June.

8th June.

13th field battery on the 5th June. The precipitous side of Inkwelo, towering 700 ft. above the plain, were more than the ox-teams of 57/R.G.A. could tackle, and after one gun and its team had hurtled down thirty feet, man power was called in, and five hundred infantry on a $4\frac{1}{2}''$ rope walked the $5''$ guns to the summit by moonlight.

At ten o'clock on the morning of the 8th June all was ready, and two infantry brigades were launched straight up the pass, supported respectively by 7 and 63, 13 and 69, and a couple of pom-poms, the mounted brigade covering the right. The heavy guns on Van Wyk's Hill completely commanded the pass, and dominated the crest of the Drakensberg on each side, so that the advance met with little opposition, and in spite of the steep climb the crest was gained by 2 p.m. Here, however, the attack was checked by the fire from a line further back on the plateau, and now was seen the value of the heavy guns dragged up Inkwelo. Not only did they enfilade the Boer line on the plateau, but commanded the ground by which reinforcements from Laing's Nek could reach that line. Thus the Boer use of their mobility to transfer troops to a threatened point, so successful in meeting attacks on the Tugela, was frustrated. Two guns of "A" and a pom-pom hauled up the Drakensberg cliffs by the mounted brigade, also swept the plateau with fire, and by four o'clock all opposition was at an end.

Buller was now in the Free State, and next morning he turned north towards the Transvaal. A spur of the Drakensberg lay across his path, pierced only by a narrow defile—Alleman's Nek—the approach to which was over perfectly open rolling downland. But no time could be wasted, for the army was cut off from its base with only three days' supplies. The heavy guns—four $4.7''$ and six 12prs.—moved in at once to three thousand yards, and the two infantry brigades with their field artillery advanced over the open. Paget, leading his brigade, noted how enjoyable it was to hear the "jingle of the traces as they moved forward over a nice open plain instead of the bump—smash—crash of the wheels going over rocks and boulders to which so long accustomed in Natal''.

An outwork on the left was rushed by the infantry, and then all guns were concentrated on the nek and the cliffs on either side. The howitzers were brought up under Buller's orders to clear the Boers out of their last stand among the rocks at the summit, and the pass won. But the guns sent forward in pursuit could do nothing against the curtain of smoke from the grass fied by the Boers as they retreated.

On the right the mounted brigade and ''A'' battery had been fighting hard to gain the ridges from which the cross fire had taken heavy toll of the infantry, and the final crest was never won. But with the pass in our hands the position was no longer tenable, and during the night the burghers drew off. So also did the garrison of Laing's Nek as soon as they saw their retreat threatened. At last Natal was clear of invaders.

The railway tunnel under Laing's Nek had been blown in, however, so it was some time before a further advance into the Transvaal could be undertaken. Meanwhile Boer commandos hovering on the flanks received attention, but we need not pause over these operations. Suffice it to say that the important railway and communication centre at Standerton was occupied on the 23rd June, Sir F. Clery with a column of all arms, including ''A'' R.H.A., 63, 86, R.F.A., and 57/R.G.A., joined hands with troops sent by Ian Hamilton from Heidelberg thus opening up a second and shorter line of communication and drawing a barrier between the two Boer States. On the 6th July Buller took train to Pretoria—the junction was complete.

CHAPTER XXXII.

The Lines of Communication.

The Orange Free State—Baken Kop—The Brandwater Basin—The Escape of de Wet—The Western Transvaal.

Map 13.

Chapter
XXXII.

1900.

The Orange
Free State.

May.

Lord Roberts had let nothing interrupt his triumphal progress from capital to capital, but as soon as he had reached Pretoria he turned his attention to the everlengthening lines of communication through the Orange Free State. He had left the 6th division to garrison Bloemfontein, the 3rd division to guard the line from thence southwards,[1] while the 1st division was moving in from the west towards Kroonstad. The east was the direction from which danger was to be apprehended. To keep order there when Ian Hamilton swept northwards with the right wing of the grand army, the 9th division advanced to Heilbron, and Rundle, with the 8th and Colonial divisions, moved forward to the Zand river, fighting at the Biddulphsberg on the way.[2] But before the end of May misfortunes commenced.

Botha had written to de Wet—"What I desire from your Honour now that the great force of the enemy is here, is to get in behind him and break or interrupt his communications. We have already delayed too long in destroying the railway behind him". No time was lost

---

[1] The divisional artillery of the 6th division had been transferred to Ian Hamilton's force, and of the 3rd division to the 8th, which, as we have seen, had been sent out from England without artillery.

[2] At the occupation of Senekal on the 25th May, a very widely known officer of the regiment—Major H. S. Dalbiac—was killed. He had retired some years before, but joined the yeomanry on the outbreak of war.

in carrying out these instructions. A whole battalion of yeomanry was captured at Lindley, then a convoy on its way to Heilbron.

Rail-head at this time was at Roodeval, thirty miles north of Kroonstad, and there a great mass of supplies and clothing required by the army at Pretoria was collected, pending the reconstruction of the bridge over the Rhenoster River. On this, and on the neighbouring stations—Rhenoster, and Vredefort Road—de Wet swooped down on the 7th and 8th June, putting a thousand men *hors de combat*, tearing up the line, burning the bridges, and destroying an immense quantity of stores. It was fortunate that the siege battery had got away in time, for the capture of the great 9·45″ howitzers would indeed have been a feather in his cap.

The blow at the line was a serious matter. Kroonstad with its vast accumulation of stores might be the next object of attack. A brigade of infantry and the 17th battery were hastily despatched from Bloemfontein, but this left the capital with only one battalion for garrison. Colonel Long, R.A.—happily recovered from his Colenso wound and in command there—enrolled all the convalescents and details in a provisional battalion. Lord Methuen with the 1st division drove the raiders from the scene of their exploit and attacked them again near Heilbron. A raid on the line at Honing Spruit a few days later was repulsed with loss. The danger was over, and Lord Roberts set about organising measures for clearing up the situation in the north-east of the Free State into which the raiders had withdrawn.

These measures involved the breaking up of the 6th and 9th divisions, and of many brigades (both of infantry and artillery), and even of batteries. Flying columns, convoy escorts, garrisons of posts, all clamoured for artillery. The treks were long and hard, and when there was fighting the guns were apt to be the target for the enemy's fire. But to follow the fortunes of sections and single guns is beyond the purpose of this history. All that can be done is to indicate the general nature of the operations in which they were engaged, and to record some of the most noteworthy incidents.

Moving on Bethlehem four guns of the 38th and two of the C.I.V. battery were with the mounted troops covering the left flank of Paget's brigade when they found the road barred by the enemy about Baken Kop. The guns were brought into action by sections along a ridge, with an interval of a hundred yards or so between sections, and the shape of the ground was such that the sections could not even see each other. An artillery duel ensued, during which the other troops were withdrawn behind the ridge, and after some time Major Oldfield, thinking of his ammunition supply, ordered the guns to cease firing, and the gunners to lie down, twenty yards or so behind them. Quickly grasping their opportunity the Boers crept up through a mealie field that lay in front of the right section and suddenly poured in a rapid fire at very close range. The gunners rushed to their guns, but the section commander, Lieut. W. G. Belcher, dropped dead, and they were only able to get off a single round of case before the enemy were on them, and all not killed or wounded were captured with the guns. Oldfield, hurrying up from the other flank, fell mortally wounded. The limbers of the next section had been ordered up as soon as the alarm was given, and one gun was got away. But the drivers and horses of the others were shot down, with the Captain (G. A. Fitzgerald) and all the gunners. Fortunately the Boers, elated by their success, delayed to attempt the removal of the three captured guns, and Captain Budworth, R.A. (of the C.I.V. battery) seized the opportunity to gallop to the nearest troops for assistance and lead them up. Before their charge the burghers broke and fled. Meanwhile the section of that battery on the left under Major McMicking, had been fighting hard, and at one time the guns were firing trail to trail, one towards the raiders on their right, one against an attempt to get round their left. But with the flight of the raiders the recovered guns were brought into action again and the enemy driven off.

Seeing the danger of encirclement at Bethlehem, de Wet had abandoned his position there, and fallen back into the mountains surrounding the Brandwater basin, whither the Free State Government, and the bulk of their forces had already retired. This basin was a fertile valley, dotted with thriving farms, teeming with flocks and herds, which ran down to the Caledon river—the boundary of Basutoland. On the other three sides it was surrounded by the rugged ranges of the Wittebergen and Roodebergen, pierced by only five passes practicable for wheeled vehicles. Under the general direction of Sir A. Hunter the columns hastened to block these exits, but before the net could be drawn tight de Wet broke out with a couple of thousand men and the members of the Government. Leaving the cavalry to pursue him, the columns closed in. One pass after another was forced, one loophole for escape after another was closed—the great bulk of the Boer forces still in the toils.

On the 30th July over four thousand men, under Prinsloo, laid down their arms, and the capture included horses, wagons, cattle, arms and ammunition. Best of all to a gunner, two of the guns lost at Sanna's Post.

In these operations the guns were freely used, and their incessant bombardment did much to shake the resistance of the burghers, but as might be expected in such country, there was little opportunity for massed action. The batteries engaged were the VIth brigade (74, 77, 79), the Xth brigade (76, 81, 82), the 5th, 8th, 28th and 38th batteries, R.F.A., the C.I.V. battery, and 103/R.G.A.

In evasion of the terms of surrender agreed to by Prinsloo—the officer elected to the chief command after the departure of de Wet—Commandant Olivier had slipped away with a couple of thousand men. With a reorganized

CHAPTER
XXXII.

1900.

The Brand-
water Basin.
13th Aug.
force, including the 5th, 8th, and 82nd field batteries, and one 5″ gun of 103/R.G.A., Hunter marched north to cut him off from de Wet, and on the 13th August brought him to battle at Spitzkop on the way to Heilbron. Rundle meanwhile had reached Harrismith,[1] and established himself in that pleasant little town, whence he reopened communication with Natal through Van Reenen's Pass which had been seized by Major-General Downing, R.A., with the Drakensberg Defence Force. From Harrismith Rundle swept the country with movable columns, and by the end of August the Free State east of the line Frankfort—Bethlehem—Ficksburg was clear of any organized body of burghers.

After his escape from the Brandwater Basin, de Wet made off north-west with the cavalry and mounted infantry hot on his heels. There was some fighting, but he got across the main railway line, and then stood at bay with his back to the Vaal. Infantry were ordered up, and Lord Kitchener was sent down to direct operations. Facing de Wet in the Free State he ranged the 2nd and 3rd cavalry brigades, Ridley's mounted infantry, Hart's and C. Knox's infantry brigades, and the Colonial division, while Methuen and Smith-Dorrien were railed round into the Transvaal to block the drifts behind him. Yet somehow,

the Boers slipped across the river and made north, with Methuen dogging their footsteps. De Wet was always ready to fight, and to sacrifice his guns if necessary to get his laager away, but Methuen was not to be denied. Well backed by the gunners his infantry drove the Boer rear-guard out of position after position, until, deceived by a wily turn of direction, they flashed over the scent.

Leaving his infantry to guard his ox-wagons, Methuen cast forward with his guns and mounted men, and picked

---

[1] Rundle remained at Harrismith until the end of the war, and on his departure was presented with an illuminated address by the inhabitants bearing witness to the consideration he had shown in his application of martial law.

it up again.   The hounds were running to view now, and
with all their hackles up, for right across the line stretched
Magaliesberg, and Hamilton was holding the passes—
the earths were stopped.   Alas, not all.   It boots not how,
but Olifant's Nek was unguarded, and de Wet and Steyn,
with all their train, slipped through.   An eye-witness has
left us a picture of the scene.

"We got a splendid view of them, going like blazes,
just like a hunting crowd breaking away.   You could
hear the chaps all over the mountain whooping away as
they viewed them".

Chapter
XXXII.

1900.
The Escape
of de Wet.

14th Aug.

Nothing has been said so far regarding the lines of
communication through the western Transvaal.   After
the relief of Mafeking the 10th division and other troops
marched through the western Transvaal, the 10th division
to Krugersdorp, those from Mafeking to Rustenburg.   At
first there was little or no opposition, but it was not
long before the spirit of the burghers began to stir
again.   Roberts decided that his strength was not
sufficient to hold the Rustenburg—Mafeking line while
keeping enough troops for the field operations he contemplated.   To carry out the evacuation of the posts
Carrington with the "Rhodesian Field Force"[1] was ordered
from Mafeking and Ian Hamilton took a newly formed
force[2] from Pretoria.   Methuen in the west was clearing
the country of marauding bands, stocks, and supplies,
and there was a good deal of fighting, notably at Quaggafontein on the 31st August where the 3rd cavalry brigade
and Colonial division—with "M" and the C.M.R. battery
were hotly engaged.

The Western
Transvaal.

---

[1] A mixed force of Colonials, including a field artillery brigade
manned by Australians, and New Zealanders, with officers and N.C.O's
of the Royal Artillery.

[2] The force under Ian Hamilton consisted of Mahon's and
Hickman's mounted brigades, and Cunningham's infantry brigade.
With Mahon was "M" battery, with Cunningham the 75th and Elswick
batteries.   The two regiments in Hickman's brigade were both commanded by gunners—Rochfort and von Donop.

CHAPTER XXXIII.

THE ADVANCE TO THE PORTUGUESE FRONTIER.

The Advance on Belfast—Geluk—Bergendal—Lydenburg—Barberton—
Komati Poort—The Change of Command.

MAPS 13 & 23.

CHAPTER
XXXIII.

1900.

The
Advance on
Belfast.

July.

August.

IT will be remembered that Lord Roberts' move eastward from Pretoria at the end of July had been baulked by a storm of unprecedented violence. The movement had, however, cleared the neighbourhood of the capital of the commandos which had been threatening the outposts, and the 11th division occupied Middelburg on the Delagoa Bay railway. Buller held the Natal railway, but so long as large Boer forces were in the field in the north-east of the Free State, his army was fully occupied in guarding against any interruption of the line. The surrender of Prinsloo removed this anxiety, and all was ready for a combined movement against Botha, who had established himself with the main Boer army near Belfast.

Early in August, Lord Roberts set his troops in motion—Pole-Carew with the 11th division and Hutton's mounted infantry along the railway from Middelburg directly on Belfast; French with the 1st and 4th cavalry brigades and Mahon's mounted brigade over the highveld between that railway and the Natal line; Buller with the 4th division and cavalry and mounted brigades, striking due north from Standerton.[1] The latter found

---

[1] The artillery taking part in this movement consisted of:—

11th Division.	...	XI Brigade R.F.A. (83, 84, 85) and 56/R.G.A.
Hutton's M.I.	...	"J"/R.H.A.
French	...	"M". "O", and "T"/R.H.A.

Natal Army.

Mounted Brigade.	...	"A"/R.H.A.
4th Division	...	II Brigade R.F.A. (21, 42, 53).
Corps Artillery	...	61/R.F.A., No. 10 M.B., 57, 100 and 101
		R.G.A.

the Boers at once, and there was skirmishing all the
way; but no real opposition until Van Wyk's Vlei was
reached.

The resistance stiffened again at Geluk on the 23rd,
and the mounted troops were rather severely handled.
The fighting at Geluk showed that the Natal army had
come up against the flank of a position which was held in
force, but the interest of the action from the regimental
point of view lies in a personal incident.

Colonel R. W. Rainsford-Hannay, R.A., had been
invalided home, but before leaving South Africa he was
given leave by Lord Roberts to visit his two sons at the
front. One of these—Lieut. F. Rainsford-Hannay—was
a subaltern in the 21st battery, and his father reached
the battery the day before the action at Geluk. Next
morning a galloper came in to say that the cavalry advance
was held up by some Boer guns, and young Rainsford-
Hannay's section was sent up to turn them out. In action
the guns came under a sharp fire of shrapnel and pom-
pom shells, and there were several casualties. Seeing
the supply of ammunition beginning to flag, the Colonel
gave a lead by shouldering a portable magazine. The
fact of a full colonel in the Regiment carrying up the
ammunition to the guns commanded by his subaltern son
must be unique in its history.

On the 24th the 11th division occupied Belfast without opposition, and on the 25th Lord Roberts arrived
there. His plan was for Buller to attack the Boer flank,
while the 11th division occupied the centre, covered on
its left by French with the cavalry. On the 26th the
Boer artillery were active while the troops were closing
in to striking distance, the British heavy guns replying
with some success, though one "Long Tom", at some-
where about nine thousand yards, defied all their efforts.

Botha had taken post astride the railway in order
to protect the seat of government, which had been
established by President Kruger at Machadodorp, against

CHAPTER
XXXIII.
1900.

Geluk.
23rd Aug.

24th Aug.

26th Aug

Bergendal.

the advance of Lord Roberts along the line from Middelburg. But the movement of the Natal army obliged him to throw back his left parallel to the railway, so that his line formed an elbow, whence the position facing the Natal army ran along a ridge thickly strewn with boulders. The most prominent point, where the elbow lay, was crowned with an outcrop of rock that must have reminded many of the "tors" on Dartmoor. Near it stood the farm buildings of Bergendal surrounded with trees. At between three and four thousand yards was a parallel ridge, along which, at dawn on the 27th, the heavy batteries of Buller's force were deployed in the following order :—No. 10 M.B., 61st R.F.A., 100, 57, and 101, R.G.A.—two 12-prs., six 5″ howitzers, two 12-prs., two 5″ guns, two 4·7″.

The 21st and 42nd field batteries were rather further forward down the slopes on the left and right flanks respectively, the 53rd still more to the left, linking up with "A" which had gone forward with the mounted brigade to a position on the flank.

For three hours the Boer position was subjected to a fierce bombardment in which the heavies with the 11th division joined. The Boer artillery on the heights in rear was soon reduced to impotence, and the fire of the heavy guns was then concentrated on the cluster of rocks near Bergendal farm, while the shrapnel of the field batteries swept the hollow in rear so as to prevent any reinforcements reaching the defence.

It seemed that no living thing could be left alive in such an inferno. And yet when the infantry breasted the slope they were met by a burst of musketry which laid many low. The howitzers were brought to bear, dropping their shells into the crannies impenetrable to gun fire, and continuing their support until the infantry were within fifty yards of their target. Many of the defenders then bolted, chased by shrapnel, and the rest of the survivors threw up their hands. But that cluster of rocks was a veritable shambles. The "Zarps" had once again proved their worth.

The capture of Bergendal broke the whole Boer position, and during the night Botha fell back through Machadodorp. Buller's troops, pushing forward in pursuit, were checked by the fire of a 155$^{m/m}$ Creusot and a Krupp 120$^{m/m}$ howitzer, and there was a call for the heavy guns. Unfortunately they had been put far back in the line of march and 57/R.G.A. had to take its 5″ guns forward at a trot,[1] and eventually gallop into action over an exposed crest, losing some horses from the enemy's fire and a gun overturned. But the infantry soon righted the gun, and as soon as the battery was in action the Boers made off : their guns had held up the British long enough to enable the non-combatants to get out of range. The 11th division advanced along the line practically un-opposed, releasing nearly two thousand prisoners. It was the end of Botha's force as an army. The President, for whose protection it had been fighting, slipped across the frontier : Botha, with a couple of thousand men. struck north into the mountains towards Lydenburg : the same number, under Smuts, made south for the Barberton goldfield : the remainder, probably three thousand under Viljoen, still falling back towards the frontier. Lord Roberts sent Buller after Botha, French against Barberton, Pole-Carew to follow Viljoen along the railway.

On the 1st September Buller struck north into the tangle of ridges and ravines into which the long range of the Drakensberg dissolves. Taking the road to Lydenburg he found the Boers facing him in a formidable position on the Crocodile river. Fortunately Ian Hamilton had come up to Belfast,[2] and was immediately despatched north to turn the Boer flank. Under the menace of

---

[1] This battery had exchanged its ox-teams for horses previous to the advance from Standerton—see Chapter XXXVII.

[2] His artillery consisted of 20/R.F.A., D/R.C.A., and two 5″ guns.

encirclement Botha vacated his position, and Lydenburg was occupied on the 7th September without further opposition.

Buller's task was, however, by no means finished. Further east another road led from the railway at Nels Spruit to Pilgrim's Rest, where, according to report, Kruger contemplated establishing his government. It was of primary importance to block this avenue, and on the 8th he took the road again, taking with him "A" R.H.A., the 53rd and 61st batteries R.F.A., No. 10 mountain, and 57/R.G.A. The Boers were in position again within gun-shot of Lydenburg, and the 5″ were brought into action within the town, but the burghers slipped away as the infantry toiled up the steep slopes. So it went on, the Transvaal artillery taking every opportunity to check the advance by the fire of their Creusot guns and Krupp howitzers at ranges too long for the field guns. Unfortunately the heavy guns were too often placed far back in the order of march, and by the time they had been brought up the Boer guns had got away. There were occasions when the latter waited too long, as when from the top of the Mauchberg (9,000 feet) the whole of the enemy's convoy could be seen trekking along the valley below. The 5″ shelled the road beyond to check the flight, the 155$^{m/m}$ shelled the pass to check the pursuit. The advanced guard pushed on, and the Boer guns were only saved by the coolness of the Boer gunners. "When the cavalry were barely half-a-mile behind the rear gun, and we regarded its capture as certain", says an eyewitness, "the leading Long Tom deliberately turned to bay and opened with case shot at the pursuers over the head of his brother gun. It was a magnificent coup, and perfectly successful".

On the 15th September Buller occupied Spitz Kop, thus closing the last avenue to anything but small bands unencumbered by guns or transport. Ten days later he pushed on to Pilgrim's Rest, and from there turned back again to Lydenburg which he reached on the 2nd October.

There he handed over his command and left for home, receiving an enthusiastic send-off from the troops he had led so far. His final operations had been carried out under almost incredible difficulties of ground, and the strain on the artillery horses, and especially those of the heavy battery, had been tremendous. But where the State Artillery could get their guns the Royal Artillery were not to be outdone. The following extract from a battery commander's diary shows what they had to compete with :—

Sept. 19th. "We began the descent from the Mauchberg to the next camp—over 2,000 feet down. The road was awful, much of it on solid blasted rock with no surface trimming at all, winding in and out of the heads of ravines hundreds of feet deep. . . . . It wanders along until, in desperation at finding itself approaching an absolute precipice, it plunges down a little cleft worn by water out of the solid rock, more like a giant's stair-case, with steps two or three feet high (rounded off by constant wear) than a road. With a desperate struggle to avoid the precipice it reaches a narrow nek, not more than twenty yards wide, which carries it safely on to a lower spur. . . . We had to let our guns and wagons down one by one, all the gunners holding on ropes and even then we nearly killed some horses".

While Buller was struggling through the mountains in the north, French was sweeping forward south of the railway, with Hutton's mounted infantry between him and the line. The combined forces had a fight on the 9th when the Boers held the Roodelvogte defile, but, attacked in front by the mounted infantry and in flank by the cavalry, their resistance crumbled. By the 12th, in spite of a stiff climb of 1,500 feet, and the difficulty of the ground, Hutton gained the ridge which commanded all the ground to Barberton, and that place was occupied by French next day.

CHAPTER XXXIII.
1900.
Lydenburg.
4th Oct.

Barberton.

9th Sept.

12th Sept.

CHAPTER
XXXIII.

1900.

Komati
Poort.

24th Sept.

Between the two wings whose progress we have been following, Pole-Carew, supported by Ian Hamilton, had been continuing his advance along the line of the railway, practically unopposed. On the 24th September he occupied the frontier station, Komati Poort, and there made a great haul of supplies and rolling stock, as had French at Barberton. Hemmed in against the frontier, the Boers were forced to destroy the guns which could no longer be taken with the bands making their way through the mountains. And so there were found at Komati Poort and Hector Spruit great masses of guns and ammunition— the former mostly in fragments :—

1 155$^{m/m}$ Creusot.
3 75$^{m/m}$ do.
2 do. Krupp.
1 do. Vickers-Maxim.
4 old mountain guns.
2 Pom-poms, and
2 12-prs. captured at Sanna's Post.

On the 25th October a ceremonial parade was held in Pretoria for the Proclamation of the Annexation of the Transvaal. After cheers for the Queen, and the presentation of decorations, the troops marched past, headed by the Chestnut Troop. All branches of the Regiment were represented, the R.H.A. by "A" and "J"; the R.F.A. by the 18th battery; the R.G.A. by the 56th. Five days later Lord Roberts left Pretoria, though his departure from South Africa was delayed by the serious illness of his daughter.

On his arrival in England Lord Roberts was welcomed by his brother officers at a dinner at the Mess, at which the Colonel-in-Chief presided. In his speech (after giving some particulars of the unprecedented development of the artillery in South Africa) Lord Roberts explained a matter which had caused considerable comment, namely his recent transfer from Colonel-Commandant of the Royal

Artillery to Colonel of the Irish Guards.   He assured his brother-officers that when offered the latter appointment he had no idea that its acceptance would entail his removal from the Regiment; that he had immediately raised the question; and that he hoped, ere long, to find himself re-instated.

With Lord Roberts went home General Marshall to resume his post as G.O.C., R.A., at Aldershot.   On his retirement in 1904, he was entertained to dinner by fifty of his old brother-officers, and after his death, a memorial tablet was placed in All Saints Garrison Church at Aldershot—where he had commanded in succession the Chestnut Troop, the Horse Artillery Brigade, and the whole of the Artillery.

Other senior officers whose services were required for the training of the artillery at home were Colonels Eustace and Parsons.   The former became Commandant of the School of Gunnery (Horse and Field), the latter took up the command of the artillery on the newly acquired Salisbury Plain.   The invaluable services they rendered in those key positions, as did General Knox later on in command of the artillery in Ireland, have been described in Vol. II.

CHAPTER XXXIV.

Guerilla Warfare—The First Phase.

1900.

The Advent of Guerilla Warfare—The High-Veld—The Bush-Veld—
Rhenoster Kop—The Western Transvaal—Nooitgedacht—The Free
State—Bothaville—de Wet's attempt to enter Cape Colony.

Map 13.

CHAPTER
XXXIV.

1900.

The
Advent of
Guerilla
Warfare.

25th Oct.

The Annexation Parade at Pretoria, with its bestowal of rewards for gallantry in action, formed a fitting conclusion to the period of regular warfare. It had ended in victory for the regular army, but it was very soon evident that the enemy had no intention of accepting that result as final.

The Boer army had broken up on the Portuguese frontier, and the burghers had made their way home by devious ways. But the military organization by "commando" still existed, and under the inspiration of Botha and the other leaders the stalwarts soon recovered their *moral*. The commandos were perhaps all the more formidable for the weeding out of the weaklings, while the destruction of the greater part of the artillery increased their mobility. If small individually they were soon found to be capable of combining with extraordinary secrecy and rapidity when an incautious move presented an opportunity—to scatter as quickly when retribution threatened.

To meet the changed conditions a reorganization of the British forces was obviously called for. The Natal Army ceased to exist as a separate entity, divisions disappeared as fighting formations, their place taken by "columns" ranging in strength from five to fifteen

CHAPTER
XXXIV.

1900.

The
Advent of
Guerilla
Warfare.

hundred men. Artillery brigades no longer functioned as tactical units, lieut.-colonels taking charge of artillery interests in districts, with their headquarters at convenient centres on the lines of communication, and their ammunition columns—immobilised and renamed "local" —as depôts for men, horses, ammunition and everything that the artillery could require. Batteries, whether with columns or in posts, were for the most part distributed by sections, or, in the heavier natures, by single guns : those that still figured as "batteries" in the constitution of columns rarely took more than four guns into the field. The last of the Naval Brigades returned to their ships, the Chestnut Troop went home with the Household Cavalry, the H.A.C. Battery with the City Imperial Volunteers.

The forces were scattered to an extent almost impossible to realise. Columns ranged far and wide over an immense area : garrisons held innumerable posts along hundreds of miles of railway. It was weary work sitting month after month in garrison, or trekking hither and thither across the veld in bitter storm or tropical sunshine, with little to keep the interest alive. And yet every post and column, every train and convoy, must be ever on the alert, for a ubiquitous foe was ever ready to punish the least mistake. Here are a couple of pictures of the gunner's part, in column and garrison respectively, during those early days :—

With a column.

"We start off each morning in some direction till we meet Boers, we then engage them with our guns, and after a bit they clear out and we follow on after them until it is time to dump ourselves down for the night, and next day the same sort of thing goes on. . . It is weary, harassing work, and I don't see that we are doing much good. They are so mobile that they get away long before our infantry can get up and have a go at them. . . . We are not mobile enough to

CHAPTER
XXXIV.

1900.

The
Advent of
Guerilla
Warfare.

attack them unless they wait for us, which at present they don't seem inclined to do, and so we go wandering over the veld in a most aimless way".

In Garrison.

"I have been up at 2.30 every morning for a fortnight. But yesterday the attack was delivered at midnight to try and catch us asleep. I was woken by hearing two shots, and rushed off to rouse the men. By that time the alarm was sounding all round, and heavy firing had begun on our most outlying work. We harnessed up and stood ready to move if required, the gunners having their carbines ready in case we were rushed. About 2 a.m. the firing got heavier just in front of us, so we lobbed a shell at 1,000 yards range over our outpost works. It was a lucky shot in the dark, bursting a few hundred yards in front of our infantry, and among the attacking Boers. It frightened them a good bit, I think, as they had not thought we should use our guns in the dark. Later on we fired another shot in another direction, but the attack had been repulsed all round, and I think we were too late. After daylight we fired half-a-dozen shells at the retreating Boers".

Any attempt to follow the fortunes of individual batteries, or even of those columns commanded by gunners, would be wearisome to the reader, and of little value for regimental history. And yet to ignore their interminable wanderings and minor engagements would be to do a grievous injustice to those who "stuck it out" so patiently, month after month, and year after year. All that will be attempted, therefore, in dealing with this period will be to give such general account of the operation as is necessary to follow the methods of warfare gradually evolved, and to illustrate it by descriptions of the incidents most noteworthy from a regimental point of view. It will, it is thought, be least confusing if the three years—1900,

1901, 1902—are treated in successive chapters, and if the main geographical divisions of the country are considered separately in each—High-veld, Bush-veld, Western Transvaal, Free State, Cape Colony.

CHAPTER XXXIV.

1900.

The Advent of Guerilla Warfare.

Owing to the fact that the section gradually became the ordinary unit for the artillery the work of the guns fell more and more into the hands of the subalterns, as it had done in the later stages of the war in Burma. Time and again they fell at their guns, with all their gunners round them. But not before they had shown the value of the system of section command which had for so long differentiated the artillery from the other arms.

If the subalterns monopolised the work with the guns, many of their seniors found opportunities for distinction as commanders of columns; and, before the victory was won, we shall find Lord Kitchener calling upon the Regiment to leave their guns and prove their prowess as riflemen. But that is another story, to be told later on. In this chapter we are concerned only with the first phase of the guerilla warfare, i.e. up to the end of the year 1900.

The cavalry were the first to experience the new tactics of the enemy. Returning from the Portuguese frontier between the Delagoa Bay and Natal railway lines, with three brigades and "M", "O", and "T" batteries, French had to fight his way over the high-veld in drenching rain, continually beset by aggressive bands. It was a distressing contrast to the brilliant exploit which had relieved Kimberley eight months before, and on arriving at Heidelberg the division was broken up.

The High-Veld.

September.

North of the high-veld stretched a wild, sparsely populated country—the bush-veld. Out of this remote region had sprung the commandos which attacked the passes of the Magaliesberg in July. Since then Paget and Plumer with the 38th, C.I.V., and "D" Canadian batteries, two 5″ guns, and two pom-poms, all under Colonel L. J. A. Chapman, had driven the Boers deep into the

The Bush-Veld.

wilderness. But after the debacle at Komati Poort Botha and the remainder of the Transvaal Government made their way through the mountains to the north, and were followed by the burghers in driblets until gradually sufficient were collected to reconstitute a fighting force.

Rhenoster
Kop.

29th Nov.

Against this force, in position at Rhenoster Kop, Plumer and Paget advanced on the 29th November. Their artillery was now reduced to the 7th and 38th batteries, R.F.A., with two 12-prs. of 103/R.G.A., and, unfortunately, the latter could find no position from which to deal with the Boer guns at less than 7,000 yards. On the other hand the field batteries, whose task was to support the attacking troops, had the greatest difficulty in finding ground which would allow them to do so except within musketry range. The 7th at 900 yards, and a section of the 38th at even less, fought the riflemen successfully with shrapnel, but the attackers could make no real headway against the overwhelming fire from the main position. It was indeed all that they could do to hold on to the ground gained until darkness put an end to the fight. Fortunately the approach of another British force, threatening to cut off the enemy's retreat, caused them to withdraw during the night.

The
Western
Transvaal.

3rd Dec.

Under Botha's plan for the conduct of the guerilla war, de la Rey was to take charge of operations in the western Transvaal. But many British columns were in those parts, and so peaceful did the country seem that convoys passed regularly on the Pretoria—Rustenburg road. De la Rey was only biding his time, and on the 3rd December he swooped down upon the long train of ox-wagons "winding slowly o'er the lea". The escort—a couple of companies and a section of the 75th battery—held their own all day, though they lost half their number —the gunners nine out of fifteen. We may well quote the Official History:—

"The guns, finely commanded by Captain H. J. Farrell, an intrepid officer, who, when many of his

men were down, armed the rest with rifles taken from
the slain, and laid the field-pieces himself—were run
trail to trail, and with depressed muzzles shattered the
front of the charge at only forty yards distance with
case shot and shrapnel fuzed at zero. The infantry
around showed equal valour".

The gallant stand of the escort deprived de la Rey of
the full reward for his enterprise, but he captured over a
hundred wagons, and drove off or destroyed ten times
that number of oxen. It was a rude shock to British
complacency, and worse was to follow.

Clements and Broadwood had been clearing the
Magaliesberg in co-operation, but Clements did not think
himself strong enough to attack de la Rey until many of
his men who had been detained on garrison duty rejoined.
While awaiting their arrival he went into camp at Nooit-
gedacht, and de la Rey, marking his isolation, called
Beyers from the north to his assistance, and lured
Broadwood away.

Clements had 650 mounted men, six companies of
infantry, P/R.H.A., 8/R.F.A., 66/R.G.A., and a
pom-pom. The horse and field batteries had each four
guns, the garrison only one 4.7". The camp was pitched
at the foot of the Magaliesberg, which rose almost sheer
behind it for a thousand feet. Four companies were
posted on the summit.

At dawn on the 13th the outposts at the foot of the hill
were attacked by de la Rey, and it was only after severe
fighting, in which "P" and the pom-pom were prominent,
that he was driven off. Then came an outburst of
musketry from the summit. Beyers was playing his part,
and soon the bullets were kicking up the dust in the
camp. The infantry left to guard the summit had been
swept away, the Boers were lining the crest: the camp
was at their mercy. Under such a plunging fire Clements
had no choice but to abandon it, and during the time that
the transport was being collected (to quote again the

CHAPTER
XXXIV.

1900.

The
Western
Transvaal.

Nooitged-
acht.

13th Dec.

Official History) "nothing but the admirable practice of the guns kept the enemy from pouring down the mountain side". Then came their turn to withdraw. Sir Godfrey Thomas got away the section of his battery that was in the outpost line, but when he turned to the other section he found its commander and all one detachment killed or wounded. Any attempt to bring the teams up to the guns would have been fatal, so a picquetting rope was made fast to the trails, and the guns pulled back to where the horses were sheltering behind some scrub.[1] The 8th battery, which was in action in the centre of the camp, was withdrawn with less difficulty, but what was to be done with the 4.7″ gun that was firing shrapnel at only 900 yards? Fortunately it was somewhat screened by bushwood, and was safely limbered up. When, however, the No. 1 looked for the team, he found the driver and leader and half the oxen had been shot. Drag-ropes (with wheel purchases) were put on, but there were few to man them, and the weight was too much. Major Inglefield got some volunteers from the infantry to help, and the big gun began to move. Once under weigh it rolled rapidly down into shelter, where half a team of oxen had been collected. But the No. 1 and four of the men on the ropes had fallen.

Having withdrawn his force for a couple of miles, Clements took his stand on some rising ground, and brought his batteries into action. A couple of guns which de la Rey brought up were soon silenced, but the position of the little force was critical, for their losses had been heavy, their supplies had been lost, and ammunition was short. Fortunately the Boers were busy looting the camp, and Clements seized the opportunity to slip away. It was 23 miles to Rietfontein, the nearest post on the road to Pretoria, but Alderson's mounted infantry and "J" battery were sent to meet him, and at dawn the survivors marched in, tired and hungry,

---

[1] This was a practical application of the old principle of the "prolong".

but with their confidence in their general undiminished. In his orders he wrote : "The manner in which the artillery —Horse, Field, and Garrison—assisted the mounted infantry and dismounted troops, and eventually secured the retirement of their guns, under the heaviest musketry fire, is beyond praise".

Within a week, as part of a gathering of columns under French, the survivors of Nooitgedacht were striking back, and at Hekpoort the burghers were hunted off, "pursued", as a Boer spectator described it, "hot-foot by the English, the white puffs of the shrapnel breaking over them as they went, the horsemen hard on their heels".

It is time to turn our attention to the Free State to which de Wet had managed to find his way back. At first he lay quiet, gathering his followers, but Lord Roberts telegraphed to Sir A. Hunter at Kroonstad that he was to be given no rest but "closely followed up by every available man in whichever direction he may go". Once more he sought refuge in the Transvaal, but Barton,[1] in a position covering the passes of the Gatsrand, repulsed all his efforts to break through. On the 25th October de Wet gave up the attempt and fell back towards the Vaal, but before he could cross C. E. Knox was on him, and he was glad to get back into the Free State with the loss of a couple of guns. In darkness and storm he disappeared, his last wagon, with its store of dynamite, blown up by a lucky shell. But Hunter would allow him no respite. We need not follow his dodgings, or the marches and counter-marches of the columns, except to note the loss in one of the engagements of one of the best-known officers of the Regiment, Major J. Hanwell.[2]

---

[1] His artillery consisted of the 78th battery, a section of the 28th a 4·7″ of 66/ R.G.A., and a pom-pom.

[2] Hanwell had won the Bugle at the "Shop", and the Kadir Cup in India. He had been best man-at-arms at the Military Tournament, and Master of the R.A. Drag. He came of a gunner family, being third in a succession from father to son, covering the whole of the XIXth century.

The climax came when le Gallais, with a thousand mounted men and "U" battery, stumbled upon de Wet's trail. Following the tracks of the gun-wheels, he topped a low ridge, as the sun rose, and there saw at his feet the whole of the long-sought laager—the burghers busy collecting their animals. Those who could catch their horses made off—saddle or not—among them Steyn and de Wet,[1] while those left behind put up a desperate fight. The State Artillery brought three guns into action in the open to cover their efforts to get the others away, but the fire of "U" battery prevented all such movement, and the Boers took refuge in a kraal, the stone walls of which defied the 12-pr. shrapnel. There were no signs of yielding, and indeed the little British column was in some danger, for the fugitives were returning to the field, and, as they closed in, the gunners of "U" began to suffer from their fire. Le Gallais was killed, leaving Major P. B. Taylor of "U" in command, but he could do no more than hold the enemy to their ground. There were anxious hours until de Lisle galloped on to the field, and then the defenders of the kraal gave in. The spoil included six guns—a 12-pr. from Sanna's Post, a 15-pr. from Colenso, four Krupps and a pom-pom. It was fitting that Major Taylor should be in command when the loss of his battery at Sanna's Post was so amply avenged.

De Wet was not the man to be discouraged even by such a blow as this. It had been arranged among the Boer leaders that the next move should be a descent on Cape Colony in order to rouse the Dutch there to rebellion. De Wet lost no time in setting about it.

All along the Basuto border lay a series of detached kopjes, forming a covered way to the upper crossings of the Orange river. To reach this de Wet had first to cross the main line of railway, and then run the gauntlet of

---

[1] "I had heard a good deal about panics—I was now to see one with my own eyes . . . . all those who had already up-saddled were riding away at break-neck speed, many even were galloping off bareback". (de Wet).

the line of forts which stretched from Bloemfontein by
Thabanchu to the Basuto border. Both of these he
accomplished, and then fell upon Dewetsdorp, held by a
small British force under Major W. G Massey, a section
of whose battery (the 68th) was included in the garrison.
After three days of continuous fighting, in which the
section lost 16 men killed and wounded out of 18,[1] the
place surrendered, and de Wet was well on his way with
his prisoners before the relieving force arrived.

Every effort was made to head him off, and for
once the weather was against him. The Orange river,
previously fordable almost everywhere, became impassable
except at the bridges and drifts, and these were guarded.
Columns gathered behind him, and C. J. Long was sent
across the river with Pilcher's and Herbert's to meet him
should he get over. Shut in between the Orange and the
Caledon, both in flood, de Wet was in a parlous plight.
All chance of reaching the colony was gone, and there was
but slight hope of saving his commando. He thought
that the English had caught him at last. Then the rain
stopped, the Caledon ran down as rapidly as it had risen,
and he was across—though with the loss of a gun, the
whole of his laager, and many hundred horses. But
columns commanded by gunners were still hot in pursuit.
J. S. S. Barker (with a section of the 86th battery), W. H.
Williams (with the 85th), and W. L. White (with the
76th) were often in sight of him, if rarely within range;
Long was close behind; Sir C. Parsons (with the 44th)
was coming in on his left; Natal shut off his right; the
Thabanchu line stretched right across his front. In it,
however, there was a gap between two of the forts, a
thousand yards wide, and through this he galloped—like
French at Klip Drift—and vanished into the mountains.

CHAPTER
XXXIV.

1900.

De Wet's
attempt to
enter Cape
Colony.
23rd Nov.

14th Dec.

---

[1] The Farrier fired the last round.

CHAPTER XXXV.

Guerilla Warfare—Second Phase.

1901.

Botha Attacks the Railway—The First Drive—Botha Strikes at Natal
  —Itala—Bakenlaagte—The Bush-veld—The Western Transvaal—
  De Wet in Cape Colony—Smuts in Cape Colony—The Block-
  house Lines—Tafel Kop—Tweefontein.

Map 13.

Chapter
XXXV.

1900.

Botha
Attacks the
Railway.

29th—30th
Dec. 1900.

1901.

7th—8th Jan.

With the year 1900, the first phase of guerilla warfare closed. De Wet had been baffled in his attempt to
enter Cape Colony, but was still at large in the Free State :
in the Western Transvaal de la Rey had gained a startling
success at Nooitgedacht : it was now Botha's turn, and he
waited until the last hours of the old year. During the
night of the 29th—30th December he captured three out
of the four enclosed works which formed the post of
Helvetia, and nearly three hundred British soldiers were
made prisoners. In one of the works was a 4·7″ gun
and 21 gunners of 101/R.G.A.—the carrying off of the
gun was deeply felt by the army, and especially by the
artillery. Encouraged by this success Botha staged a
general attack on the line. During the night of the
7th—8th January, Machadodorp, Dalmanutha, Wonder-
fontein, Wildfontein, Nooitgedacht,[1] and Pan were all
assailed, and not only these smaller posts but Belfast
itself, the headquarters of this section of the line. Its
garrison was large—a couple of thousand men—but
the perimeter was long, and the attackers were commanded by Botha in person. The fighting was fierce,
but the defenders stood firm, and before dawn the burghers
had drawn off. At the time of the attacks the artillery

---

[1] Not to be confused with the scene of de la Rey's victory.

on this line was distributed as follows—the batteries split up in many cases by sections in the outlying posts :—

CHAPTER
XXXV.
1901.
Botha
Attacks the
Railway.

42nd Battery R.F.A.			— Machadodorp.
61st	do.	do.	— do.
66th	do.	do.	— Pan.
84th	do.	do.	— Belfast.
97th Company R.G.A.			— do.
101st	do.	do.	— Dalmanutha.
100th	do.	do.	— Machadodorp.

Such a challenge must be met, and Lord Kitchener determined to clear the country between the Delagoa Bay and the Natal railway.

During the first phase of guerilla warfare each of the newly-formed "columns" had been allotted to a district in which it was supposed to break up any hostile gatherings. The result had been disappointing—for commandos simply moved on when threatened, and a new system was now to be tried in which several columns were to co-operate under one direction. Extended in line across the tract of country to be driven, they were to sweep everything before them until brought up by some more or less impenetrable barrier such as a mountain range or fortified railway.

For the drive that was to inaugurate the new system, French was given seven columns of an average strength of about 1,500 men and half-a-dozen guns and pompoms. In addition to these Smith-Dorrien was to strike south from the Delagoa Bay railway with 3,000 men and 12 guns.

The long line of horsemen, stretching right across the high-veld from railway line to railway line, started moving east on the 28th January. After some fighting on the way, the columns were approaching Ermelo, where Botha was reported to be in force, when the latter got in the first blow. An hour or two before dawn on the 6th

CHAPTER
XXXV.

1901.

The First
Drive.

6th Feb.

February, Smith-Dorrien's camp at Lake Chrissie was fiercely assailed. A tremendous fusillade stampeded the horses, and the burghers, driving these back before them, got through the outposts. There was much confusion in the camp, but the picquets stood firm, and the attack was handsomely repulsed with heavy loss to the enemy. Under cover of the attack Botha had, however, got across the railway to the north, out of the path of the drive, with a couple of thousand men. French continued his sweep eastward as far as the mountains which edge the high-veld, with a final beat right down to the Zulu border. There was a daily tale of men and guns, wagons and cattle captured, but the weather was atrocious, the roads became impassable, and no convoys could get to the columns. Men and horses were half-starved, and, owing to Botha's evasion, the "bag" was only thirteen hundred men, though with seven guns, a couple of pom-poms, and countless horses and cattle. Botha's main fighting force had escaped, but the new move undoubtedly shook the confidence of the burghers, and all through the next few months the high-veld was harried. A new technique was being developed, very different from the somewhat aimless wanderings of columns described in the last chapter. Boer laagers which had settled down to rest in fancied security were apt to find themselves rushed at dawn by a column that had marched twenty miles or more across the veld during the night. As de Wet wrote :—"From May 1901 the British began to make night attacks upon us : at last they had found out a way of inflicting severe losses upon us. These night attacks were the most difficult of the enemy's tactics with which we had to deal". But to be successful these lightning strokes required the right combination of column commander and intelligence officer, and prominent among these were Colonels G. E. Benson, R.A. and Wools Sampson.

So serious was the effect of these new tactics upon the burghers that Botha determined upon a bold move to counteract them. In September he struck at Natal without warning. It was only by good fortune that his scheme was detected in time.

Some mounted infantry, with a section of the 69th battery, on a reconnaissance from Dundee, came across a band of Boers at Blood River Poort near Vryheid on the 17th September, and immediately attacked them. The tables were soon turned. Boers appeared on all sides, galloping at the guns in great numbers, and after a few minutes mêlée the whole force was captured. But the encounter disclosed the danger. Columns hastened to the rescue, and the Boers, ridden off from Natal, edged away into Zululand. There was nothing to stop them but two small posts—Itala and Fort Prospect—which dated from the Zulu War of 1879. At Itala, the garrison under Major Chapman consisted of three hundred men and a section of the 69th battery, and they were attacked by greatly superior numbers during the night of the 25th September. An outwork was taken, but the main work showed so bold a face, though closely pressed all through the 26th, that the enemy gave up the attempt and drew off at nightfall. The section of the 69th upheld worthily the honour of the Regiment. Lieut. C. Hebert was severely wounded, but when Chapman called for volunteers to carry ammunition across a space swept by cross-fire, gunners and drivers stepped forward at once. For his gallantry on this occasion Driver F. G. Bradley received the ꝩ℃., and four others of the section the D.C.M.

The garrison of Fort Prospect showed similar resolution, and the check inflicted by these two repulses was sufficient . British columns were gathering in, and Botha only extricated himself with difficulty. On his way back, however, he determined to take the opportunity to strike a blow at the most troublesome of his enemies on the highveld, and issued orders to "attack with all their force, whenever possible, Benson's restless columns".

At this time Benson's strength was close on a couple of thousand men, with the 84th battery and two pom-poms. Unfortunately he had been obliged to change his well-tried infantry for fresh troops who had been for some months doing garrison duty in posts on the line. The blow came on the 30th October, when the column was just going into camp at Bakenlaagte, forty miles south of Middelburg. Seeing his rear-guard attacked, Benson at once returned with such troops as he could gather, to find the rear-guard falling back. In a storm of rain Botha launched a thousand mounted men upon them. The infantry were over-ridden, but Major E. Guinness, commanding the 84th battery, brought a section into action, and here he was joined by Benson and some of the mounted troops, while another gun of the battery, coming out of camp, came into action, on a knoll between. But Botha was not to be denied. From every dip in the undulating ground mounted men sprung up as if by magic, and a couple of thousand galloped down on Benson and those with him. The fire of the guns stopped the mounted attack, but the burghers were soon closing in on foot. The last rounds of case were fired, and then Guinness ordered up the teams. The section commander (Lieut. J. M. Maclean) had fallen, but the drivers dashed out of their cover without hesitation. It was hopeless: men and horses were shot down, and the team from the gun on the knoll fell beside the others. The little band on the ridge had no thought but of selling their lives as dearly as possible, and Guinness died amidst his men and horses. When silence told the Boers that resistance was over, and they swarmed on to the ground, Benson, mortally wounded as he was, found strength to send a soldier into camp with orders to shell the ridge so as to prevent the Boers carrying off the guns. His orders were carried out, and the Boers ventured no more. When the relieving troops arrived they had disappeared.

Of the 280 men with Benson and the guns on the ridge 66 were killed and 165 wounded. Of the 32 with the section of the 84th battery there fell 28.

Of all the leaders of small columns none had gained a

higher reputation than Benson. His death was a
grievous loss to the Regiment.[1]

In the early months of 1901 Lord Kitchener was too
busy with matters elsewhere to attend to affairs in the
north. But in March a couple of thousand men, with
the 18th battery and three pom-poms, were assembled
under Plumer at Pienaar's river, forty miles north of
Pretoria, and on the 8th of April they rode into Pietersburg, 130 miles further on, without serious opposition.
The place itself, and the railway between it and Pretoria,
were fortified and garrisoned, under Lt.-Col. F. H. Hall,
R.A., and, with it in his possession, Lord Kitchener was
able to see to the clearing of the bush-veld between
Pietersburg and Lydenburg, in which the Transvaal
Government was lurking. Columns working under the
direction of Sir Bindon Blood surprised their refuge at
Roos Senekal, and captured over a thousand men, with
all the state papers and treasure, but the most important
members of the Government escaped. The remains of
the 4·7″ from Helvetia were found, however, and one of
the 15-prs. lost at Colenso, a Krupp, and a pom-pom.
The last of the "Long Toms"[2] also met its fate in these
operations, blown to pieces by its devoted gunners when
escape was impossible.

In the west the early months of 1901 saw many British
columns on the move, and many skirmishes in which the
artillery took their full share. Methuen was always
ready to fight, and in February brought off a spirited
action at Hartebeestefontein. Next month Babington
recovered two 15-prs. and a pom-pom at Taailbosch Spruit,
and in April Rawlinson got a 12-pr. and another pom-

---

[1] His command of Arab levies in the Eastern Sudan will be remembered. (Chapter XVIII).

[2] This was the gun which opened fire on Ladysmith from Pepworth,
was blown up in the sortie, repaired at Pretoria, and then sent to
Kimberley.

CHAPTER
XXXV.

1901.

The
Western
Transvaal.

pom at Brakpan. After a longish night march he had rushed the laager at dawn, and the section of "P" was in action on a narrow ridge, when rifle fire was suddenly opened upon them from behind. Some burghers had crept up through the rough ground, and had surprised and captured the column commander. The guns were turned right round, and their fire soon drove the intruders away, and freed the captives.

In May another of the Zilikat's Nek 12-prs. was recovered, but at the end of the month Dixon was very roughly handled at Vlakfontein, just south of the Magaliesberg. After hunting for some guns reported to be concealed in the neighbourhood, the column was returning to camp when the rear-guard, under Major Chance, R.A., was rushed by several hundred Boers under cover of the smoke from a veld-fire. They galloped down on the guns (a section of the 28th battery), shooting down men and horses, and scattering the escort. The few gunners who did not fall to the first volley attempted to open fire with case, but the portable magazines caught fire. Lieut. J. P. MacDougall refused to surrender and was killed with nine of his men: the drivers attempted to extricate the guns, but every horse was shot down. When all resistance was crushed the Boers turned the captured guns on· to the camp, but a counter-attack was launched, and they lost heart and decamped, leaving the guns they had captured where they had found them.

In September, Kekewich, whose artillery consisted of three guns of the 28th battery and a pom-pom, was attacked by de la Rey when in camp at Moedwil. In the darkness before dawn the outpost line was broken, and the horses stampeded. For a time the situation seemed serious, but a firing line was formed, the guns got into action, and the attack was checked. Counter-attack followed, and, as at Vlakfontein, the enemy gave way.

Next month it was Methuen's turn at Kleinfontein. He was in two columns, one under his personal command, the other under Major von Donop, who had a

CHAPTER
XXXV.

1901.

The
Western
Transvaal.
24th Oct.

thousand men, the 4th battery, a howitzer of the 37th, and a couple of pom-poms. The columns were crossing the worst part of a very difficult bit of country when de la Rey seized the opportunity to fall upon the weaker of the two. While the wagons were winding of necessity along the road, six hundred men swept out from a wood in which they had been concealed, and charged the flank and rear. Lieut. H. N. Hill, commanding the section of the 4th battery with the rearguard, brought his guns promptly into action, but had only fired three rounds of case when he, and all but two of his men, were shot. With these two men, however, Lieut. R. F. A. Hobbs, R.E., kept the section in action; the escort fought valiantly; and the burghers never reached the guns. Meanwhile von Donop, having collected his mounted men, and posted the other three guns, came to the rescue of the rear-guard, and the Boers drew off, leaving fifty of their number dead on the field.

We may now leave the Transvaal and return to the Orange Free State. Here de Wet had scattered his followers to rest and recuperate after his abortive attempt to enter Cape Colony at the end of 1900. In January he led them south again, but the columns were soon on his trail. In spite of their efforts, however, he eluded them, and crossed the Orange River into Cape Colony. The columns were close on his heels, but the river was rising and the gun teams had to be brought back to get the wagons over. "It was fine to see them humping themselves up the steep bank out of the river bed" wrote an onlooker. Once across the river de Wet turned westward, but the rain fell in torrents, and he was encumbered[1] with supply wagons as well as guns, for his force was too big to live upon the country. The rain fell in torrents, and soon his movement west-ward was blocked by

---

[1] "We inspanned 30 oxen to each gun, but if it got stuck fast in the mud 50 were sometimes not sufficient to move it". (de Wet).

CHAPTER
XXXV.
1901.
De Wet in
Colony.
Cape
28th Feb.
morasses and swollen rivers. Doubling back along the left bank of the Orange River, looking for a place to cross, he was attacked opposite Hope Town by Plumer, and lost a 15-pr. and a pom-pom, besides wagons and prisoners. But on the last day of February a little-known drift near Colesberg Bridge was found to be negotiable by mounted men. A way of escape was open at last, and all who still had horses got across. Of the commando that a fortnight before had defied all efforts to deny it entrance to the Colony, there struggled back into the Free State only a broken rabble—their track marked by abandoned guns, horses, wagons, ragged and foot-sore men. Lord Kitchener was determined that they should have no peace at home, but set about the organization of a series of big drives to sweep the country. We need not follow these as they rode up and down the Free State. The results were rather disappointing in that, though the captures of wagons and cattle were many, those of fighting men were few.

Among those who escaped from the above drives was Mr. Smuts, the State Attorney of the Transvaal. He was bound for Cape Colony, and, in spite of every effort to stop him, he rode right through the Free State to the Basuto border, where he crossed the Orange river with three hundred men[1]. Once in Cape Colony he had no difficulty in finding guides and supplies, and in getting into touch with the other bands of Boers that were keeping alight the embers of rebellion.

Direct pursuit had no chance of catching small parties with no guns or wagons, picking up at every farm food and fresh horses, and news of the British. But the railways enabled the latter to bring troops round and establish barriers across their line of advance. In spite of all, however, Smuts pressed southwards until the eyes of his

---

[1] One of the small detachments guarding the drifts was surprised at Quaggafontein, and its one gun (of 38th battery) captured—but not until a dozen of the little band of artillerymen had fallen.

followers were gladdened by a distant view of the Indian Ocean, gaining on the way a sensational success against a cavalry outpost at Tarkastad, in which the subaltern of the one gun with the squadron[1] was killed. Headed off from Algoa Bay Smuts turned west, and his patrols were in sight of Table Mountain before their path was barred again. But down the Olifants river their way to the sea was open, and the burghers, many of whom had never seen the sea before, rode into the Atlantic surf. We must leave them there and return to the Free Sate through which Smuts had made his way so skilfully.

The small results of the "drives" in the Free State had shown the weak point in the system—the difficulty of preventing determined men breaking out of the net. The country was all open : the only barriers the fortified railways and the few lines of posts which had been established. But the areas between these were so vast that it was impossible to guard every avenue of escape. The need once recognised, however, ingenious minds soon found a way of limiting the enemy's freedom of movement.

The new system was developed from the invention of a type of block-house susceptible of mass production at a trifling cost, which could be carried in a wagon, and erected by half-a-dozen men in a few hours.[2] With these it became possible to run out lines of block-houses from the railways, cutting up the whole country into manageable districts, and thus to strike at the roots of the enemy's policy of constant evasion.

De Wet, who had dispersed his men after the big drives, began to gather them together again in order to check this extension of the block-house lines. After a repulse with loss by Dartnell at Tiger Kloof Spruit, he surprised Damant at Tafel Kop.

---

[1] No. 4 Mountain Battery, Lieut. Hay-Coghlan.

[2] A small blockhouse made of a double skin of corrugated iron filled in with gravel between.

CHAPTER
XXXV.

1901.

The Block-
House
Lines.

Tafel Kop.

20th Dec.

Dressed in British uniform, and imitating the movements of regular troops, the burghers got within a couple
of hundred yards of the hill on which Damant had posted
his guns, and from which he was watching the movements
of his scattered force, before their identity was discovered.
By that time they were lining the crest of the kopje, and
after four rounds of case the guns were in their hands.
But every man of the gunners, and their escort, from
the column commander downwards, fought it out individually, and before the burghers could take the guns
away they were chased off the scene by the rest of the
column.

Out of 80 men on the hill 77 were killed or wounded.
Captain H. J. P. Jeffcoat, terribly wounded, lived only
an hour, but before dying wrote a will bequeathing £50
each to two unwounded men of his pom-pom section;
Shoeing-Smith A. E. Ind of the same section received
the 𝖁𝕮; 2/Lieut. R. G. Maturin of the section 39th
battery was awarded the D.S.O.

Still bent upon cutting the tentacles which threatened
to strangle his activities, de Wet noted that the head of the
block-house line that was under construction from
Harrismith to Bethlehem was covered only by some five
hundred yeomanry, with one gun of the 79th battery and
a pom-pom. Their camp was on the slopes of a kopje,
the other sides of which were almost precipitous. Up the
steepest of these cliffs the burghers crept in the small
hours of Christmas morning, and rushed the defenceless

camp. After firing a couple of rounds, the gunners fell
to a man round their officer, Lieut. S. T. Hardwick, and
as dawn broke de Wet marched off with the guns and two
hundred prisoners.

It was a sad ending to a year that had been notable
for the large number of artillery officers selected for the
command of columns:—Colonels Barker, Jeffreys,
Parsons, and Rochfort, Lieut.-Colonels White, Wing,
Williams,[1] and Dawkins.

---

[1] Struck by lightning on Chistmas eve, and invalided.

CHAPTER XXXVI.

Guerilla Warfare—The Final Phase.

1902.

The Artillery Mounted Rifles—The Storm Centre—The Drives—The Western Transvaal—Tweebosch—Boschbult—Rooiwal—The Last Drive—Cape Colony—The Treaty of Vereeniging.

Map 13.

After the failure of the peace negotiations in 1901, Lord Kitchener realised that the only way to bring the war to an end was by relentless "hunting" of all burghers still under arms.   To carry out this policy he required mounted men in large numbers, and these were provided.   But many of them could neither shoot nor ride, and lacked both discipline and self-confidence.   The Commander-in-Chief's eye noted the numbers of artillerymen redundant to requirements at this stage of the war—some indeed already under orders for home or India.   Here were officers and men, horsemen and horsemasters, who had proved their steadfastness and discipline on many occasions.   And, if they had done little musketry, it presented no mystery to gunners—practice only was required.

And so, when Colonel Sclater presented himself for the usual morning's interview with the Chief on the 12th December 1901, he was told to set about the conversion into mounted riflemen of all the gunners that could be spared.   With Lord Kitchener procrastination was not encouraged, and the orders went out that afternoon for the conversion of the two howitzer brigades (VIII & XII). Hardly had these been assembled at Pretoria for conversion than that of two more brigades (V & IX) was decided upon, and this was followed by that of six batteries of horse artillery.[1]   The officers selected for the command of

CHAPTER XXXVI.

1901.

The Artillery Mounted Rifles.

12th Dec.

---

[1] Less one section of "J" retained in General Rimington's column at his special request.   A copy of the order for the formation of the first corps will be found in Appendix I.

CHAPTER
XXXVI.
1901.
The
Artillery
Mounted
Rifles.
24th Dec.
1902.
January.
these three corps were Lieut.-Colonels J. W. Dunlop, Sir J. H. Jervis-White-Jervis, and J. L. Keir.

The first corps was complete on Christmas Eve, and marched off on Boxing Day, being inspected by the Commander-in-Chief a few miles out of Pretoria, and gaining a highly complimentary order.   Before the end of the first week in January 1902 the first two corps had crossed the Vaal and entered the storm-centre in the Free State, and the third was forming.   But before considering their work it will be well to add to the very brief sketch given in Chapter XXV a few further details regarding the country over which they were to operate.

Broadly speaking this debatable land lay between the railway lines from Bloemfontein to Johannesburg and from Johannesburg to Natal.   On the east it was bounded by the Drakensberg range, from which to the Bloemfontein railway ran two block-house lines.   The northern of these led from Botha's Pass to Heilbron by Tafel Kop and Frankfort, the southern from van Reenen's Pass to Kroonstad by Harrismith, Bethlehem and Lindley.   The western part of this area is of the usual high-veld type— long, rolling, ill-defined ridges, with here and there a flat-topped kopje.   East of Heilbron a range of hills forms the watershed between the Rhenoster River and the Wilge (with its tributary the Liebenberg Vlei) and beyond this watershed the ground becomes broken.   At first there are isolated blocks of hills, such as the Bothaberg, and some big solitary ones, like Tafel Kop, rising precipitously hundreds of feet above the plain.   But gradually these merge into the foot-hills of the Drakensberg, seamed with deep ravines cut by the mountain torrents.   Here the country is an intricate maze of ridge and valley, in the recesses of which the farms are hidden, while the precipitous sides of the kloofs, honeycombed with caves, provide endless secret storehouses.

The drives of 1901 had lost half their effect through the lack of "stops".   By the beginning of 1902 this want had been made good by the block-house lines which cut up the veld into manageable areas. And with this came many improvements in the technique of the drives. In front rode a continuous line of scouts, stretching right across the ground to be covered—perhaps fifty miles, often more—the flanks touching block-house lines that enclosed the area to be driven. Behind the scouts came their supports in compact formation, and behind them again the main bodies of the columns. Even when the veld was open, the control of an extended line of such a length was difficult enough. When the movement was interrupted by bush or kloof or enclosure it was almost impossible to prevent loss of direction, overcrowding, and gaps—of which the Boers were quick to take advantage.

At night a chain of picquets, dug in a hundred yards apart, formed a line of resistance. The block-house lines were strengthened with battalions of infantry, and sometimes further stiffened by columns moving forward in advance of the flanks of the line. When railways were available as stops they were patrolled by armoured trains, provided with guns and search-lights.

There was no rest. Each day, in rain or sleet, icy blast or scorching sunshine, the line moved forward from dawn to dusk : each night entrenchments had to be dug, and miles of entanglements laid out, before the few who could be spared from guard snatched a few minutes slumber. There was always the anticipation of a violent effort to break through, heralded by a storm of musketry. And, for recompense, continual small captures—sometimes the surrender of a bunch, more often the discovery of individuals hiding in the reeds, or up a tree, or in an ant-bear hole. Sometimes a gun was fished up out of a pool, sometimes a *cache* of ammunition was found in a cave. Always there was the rounding up of the enemy's apparently inexhaustible supply of stock and wagons. Here are a few words from a pom-pom officer's letter :

"We left here on the 26th, no tents, no baggage, two blankets per man. A short march and an early start the following morning. Saw at least 250 Boers, fired 88 rounds, and got 5 wagons. Next day awful cold sleet, got in at mid-day, but started again at 11.30 p.m. and marched all night. Halted at 5.30 a.m. for an hour to feed. Three wagons reported, trotted after them and caught 18 and 2,000 cattle. Killed 2 and captured 3 Boers, fired 22 rounds. Got into camp at 2 p.m. quite tired. Camp sniped at night. Started 5 a.m. on advanced guard. Fired 29 rounds, 4 dead picked up. I am very lucky getting here".

For the first of the great series of drives through the north-east of the Free State in 1902 the columns were ranged along the Liebenberg Vlei, and drove westward against the railway between Kroonstad and Heilbron. For this Dunlop's corps of artillery mounted rifles were in Byng's column, Jervis's in Rimington's. On the 6th February the long chain moved forward on a front of 60 to 70 miles, and during the afternoon several hundred Boers were sighted in front of Jervis driving a huge herd of cattle. This turned out to be de Wet himself, and during the night he broke through the block-house line on the left. There were still, however, some hundreds within the net, who spent the night riding up and down the picquet line looking for a gap, and making occasional rushes for what seemed a weak place, only to be driven back by the fire of the picquets. An exciting night for the gunners with these bursts of musketry, lighting up first one part of the line and then another, the glare of searchlights from the armoured trains reflected on the clouds, and occasionally the bright flash and loud report of a field gun.

On the 8th the line drew in to the railway and parties of burghers began to give themselves up, huge herds of gemsbok and springbok galloping about distractedly between the columns and the line. The total number of Boers captured was only about 300, but the guns lost at Tweefontein on Christmas morning were recovered.

As de Wet wrote "The guns had not been of much benefit to us for the English kept us so constantly on the move that it had been impossible to use them".

Two or three days only were allowed before the columns took up their places again. This time the line started from the railway between Heilbron and Elandsfontein and moved eastward, between the Natal railway and the Heilbron—Botha's Pass block-house line, as far as Standerton. There it made a great wheel to the right, and drove south between the Drakensberg Range and the Wilge River. De Wet had been caught in the toils again, and this time he determined to try his luck with the columns. While one party came up against Jervis's mounted rifles, and were driven back in confusion, de Wet himself, with the indomitable President, rushed the picquets at another part of the line with five hundred men.

The adjoining picquets wheeled back so as to form a fresh front and prevent the line being rolled up. Lieut. A. R. G. Begbie (writer of the letter quoted above) brought up his pom-pom and became the pivot of defence. He was cheering on the defenders when he fell shot through the heart. It had been a desperate fight, and more than 80 men lay among the carcases of horses and a mob of cattle. 12 burghers and 70 horses were lying within fifty yards of the pom-pom.

The drive swept on, and two further efforts to break through were handsomely repulsed. On the 27th February, now a fateful day for the Boers, all those left in the net surrendered. The bag was 800 men, 2,000 horses, 200 wagons, and countless cattle, but once again the real object of the drive had fought his way out.

The third drive was a repetition of the first, but Keir dropped out, having been given a couple of small columns in addition to his mounted rifles, and moved further north, while Dunlop's corps was increased by the addition of the 79th battery, converted into riflemen. The pace was hurried, and consequently there were break-outs, so

CHAPTER
XXXVI.

1902.

4th Drive.
20th Mar.
—4th Apr.

that the result was poor except for the findng of De Wet's main magazine in a cave.

No time was lost before the columns faced about and drove east again, but heavy rain made the spruits impassable, and the whole line was hung up between the Liebenberg Vlei and the Wilge River. Garratt had got a small portion of his column across the Wilge before it rose, and they were now starving on the far side, and at the mercy of the enemy. Dunlop, recalling "bridging" days on the pond at Shoeburyness, set to work to construct a raft, and got his mounted riflemen across. Baldock, temporarily in command of Rimington's column, decided to go back across the Liebenberg Vlei so as to get to the bridge over the Wilge at Frankfort. On the 30th March Gubbins with two batteries swam the Vlei, carrying nothing but a day's food for horse and man, and got round in time to stop the gap. By the 1st April the waters had abated, and the other columns were able to cross, so that the drive could be resumed on the 2nd. In spite of the delay there were still burghers in front, and that night they swept down on the picquet line, some hundreds breaking through. On the 3rd, Baldock drove across Garratt's front and a hundred Boers tried to break through Jervis's line. About half succeeded. Next day the Drakensberg was reached and the drive terminated with poor results except for the discovery in a deep pool of three 75$^{m/m}$ Krupps, the last of the Free State armament. The columns concentrated on the passes through which supplies and remounts could be drawn from Natal.

After a couple of days halt they were off again, sweeping northwards this time, with their right on the Drakensberg. Working down the valley of the Commando Spruit, after a long night march, Dunlop's column, which was leading, was forced to fall back before superior numbers. The rear-guard was sharply pressed, the Boers coming on in constantly increasing numbers, and on fresh horses. Reinforcements from the main body finally drove them off, after more than one gallant stand to save a wounded comrade on the part of the 86th and 87th batteries. On the 15th the Natal-Johannesburg railway was reached, and the drive came to an end.

Sir John Jervis was at last obliged to take a rest. For a year and half he had been marching and fighting without leave, and had been specially selected to command the second corps of mounted rifles. Colonel Baldock, commanding the Vth brigade in this corps, was senior to Jervis, but at the time of the conversion had not had any experience of work with a column. He obtained permission, however, to serve under Jervis, and now obtained the command of the corps in his place, increased about this time by the 53rd battery.

Advantage was taken of the presence of the columns from the Free State on the Natal railway for a drive thence across the high-veld to the Delagoa Bay line, and back again. Meanwhile in the Free State, the Boers, getting desperate at the extension of the block-house lines, started a series of systematic attacks upon them, and two block-houses were captured. The situation was serious, and Bruce Hamilton brought all his columns from the Transvaal to help. By the 4th May a line was formed from the Bloemfontein railway through Heilbron to the Liebenberg Vlei, the drifts of which were held. It was decided to cover the whole distance to the Kroonstad— Lindley line, forty miles, in one day, taking no vehicles except pom-poms. It was an impossible task for such an extended line over such a country. There were no halts in which to fill gaps; and many of the Boers escaped in consequence.

A return drive over the same ground brought little more result, and then came the truce while the terms of peace were being discussed at Vereeniging.

This series of "drives" in the Free State has been given in some detail, both as an example of what the drives became at their greatest development, and also because

CHAPTER
XXXVI.

1902.

7th Drive.
9th—15th
May.
the Artillery Mounted Rifles took such a prominent part in them.

Among the column commanders active in the above-mentioned operations were the following artillery officers :—Colonel J. S. S. Barker, Lieut.-Colonels Baldock, Dawkins, Dunlop, du Cane, Sir J. Jervis-White-Jervis, Keir, and Wing.

We must turn now to the Western Transvaal where de la Rey had been fighting his finest campaign.

At the beginning of 1902 Lord Methuen decided to establish his own head-quarters at Vryburg, and to hand over the command of his column to von Donop, who had been his lieutenant in all the fighting in the west. Throughout February von Donop was constantly on the war-path, harrying the burghers, and capturing convoys, but by the 23rd his wagons were empty, and he had to despatch them to Klerksdorp to refill, sending an escort of 500 men, with a section of the 4th battery and a pom-pom with them. But de la Rey swooped down in full strength. Attacked in front, flank, and rear, the troops put up a fine fight, but the numbers against them were overwhelming, and the convoy was lost, though the enemy paid dearly for their victory.

Lord Methuen was determined to recover the guns, and came himself from Vryburg, with what troops he could collect. On the 6th March he camped at Twee-bosch, starting off again next morning before dawn. In front went the convoy, with an escort which included a section of the 4th battery and a pom-pom. The main body followed, and then came the rear-guard, with a section of the 38th battery and a pom-pom. While it was still dark heavy fire was opened upon the rear-guard, and a large body of burghers charged, firing from the saddle. The medley of untrained and ill-disciplined yeomanry and oddments from colonial corps broke and fled, leaving the guns isolated in the midst of the enemy. Here we may well quote the official history :—

"Now, not for the first time, were training and tradition to illumine the blackness of disaster. Lt. T. P. W. Nesham and his artillerymen of the 38th battery were men of the same blood as they who had given way, but to them flight was not even a last resort it was an impossibility. Until every man had fallen the guns were served with case, and even when the pieces were actually captured and lost to sight amidst the surging crowd of Boers, the young officer in command, the only unwounded member of the *personnel*, refused to surrender and suffered death for his gallantry".

For two more hours Lord Methuen, with the infantry and the section of the 4th battery, kept off all attacks on the convoy, though surrounded by marksmen. The two guns of the 4th battery were fought as nobly as those of the 38th had been. Even after Lt. G. R. Venning, their commander, was killed the gunners remained at their work until all were down. The infantry fought as stubbornly, and Lord Methuen was, himself, the central figure of the defence until he fell with a fractured thigh. Then, five hours after the first attack, the column disappeared. It was long since so complete a catastrophe had befallen the British arms. "Nothing was saved except the honour of the infantry and the gunners". Lord Methuen's despatch bears official witness to their heroism :—

" . . . All mounted troops then in rear broke and galloped past the left flank. The section 38th battery was thus left unprotected but continued in action until every man except Lt. Nesham was hit. I am informed that this officer was called upon to surrender, and on his refusing to do so, was killed. . . . I remained with the guns of the 4th battery and infantry until wounded. They held out in a most splendid manner until about 9.30 when all the men round the guns had been shot down, and Lt. Venning, commanding the section, had been killed. . . . I would also call attention to the gallant manner in which Lts. Nesham and Venning stuck to their guns".

In his own despatch Lord Kitchener bore further witness to the gallantry of these officers and of the Royal Artillery under their command. Lord Methuen related afterwards how he was lying wounded close to the trail of one of the guns of which all the detachment had been placed *hors de combat* save two—Lieut. Venning and a bombardier. The subaltern layed the gun and then fell dead across Lord Methuen's legs. The bombardier turned round, said "There's another good man gone" and went on serving the gun. Directly afterwards he himself fell dead.[1]

The position was serious. Troops poured into Klerksdorp, and among them Keir[2] with, in addition to his horse artillery mounted rifles, mounted infantry and a colonial corps. Lord Kitchener came down himself to inaugurate in the West the system of drives which was proving so successful in the East.

For the first the columns trotted out through the night for forty miles passing right through the Boers. At dawn they turned about, extended, and swept back again. There was some hard fighting, and though many escaped a good number of burghers were caught in the toils. The three guns and two pom-poms lost in the recent disasters were recovered, but eighty to a hundred miles in twenty-four hours was a high trial, and many of the horses were so exhausted that they collapsed when a heavy hail-storm caught them.

About the end of March Cookson and Keir set out on a reconnaissance in force. At dawn on the 31st March a Boer convoy was sighted, and at once pursued by Keir's M.I. who were in advanced guard. They were suddenly brought up by heavy fire from in front, while numbers

---

[1] Letter from Lord Cromer to Major J. H. Leslie.
[2] Major Mercer had succeeded him in command of his corps of Mounted Rifles.

of burghers were seen galloping round the flanks; they had stumbled upon de la Rey's main body. A defensive position was hastily taken up along the line of the Brak Spruit, while the screen was being driven in, fighting hard —its withdrawal covered by the fire of Keir's four guns and pom-pom.

The Boers advanced mounted in a long line, firing from the saddle, and gradually got up to about a thousand yards from the position, at which range a heavy fire-fight ensued without movement on either side. Nearly three-quarters of a mile on the flank of the position was a detached farm which was held by "P" and "T" batteries under Major Lecky, and about 4 o'clock a large party charged down upon it. The horse artillery reserved their fire until the burghers halted to dismount at about 600 yards, and then poured in such a stream of bullets that their further advance was stopped. It was their last effort, and after its failure the whole attack collapsed. By half-past five the enemy had drawn off.

It was evident that the large number of columns now collected in the western Transvaal required central direction, and early in April Sir Ian Hamilton arrived to take command. Disposing his columns round the Brak Spruit he was moving south on the 11th April when Kekewich's mounted screen (commanded by von Donop) was surprised by some fifteen hundred Boers, whose khaki uniforms and formal movement caused them to be taken for men of one of the co-operating columns. Though surprised, and temporarily thrown into confusion, von Donop's men maintained a series of detached stands. Grenfell who was close behind was at first deceived, but, as soon as warned, dashed to the front and seized a ridge, bringing a couple of guns into action. This formed a rallying point, and a line was formed in a rough semi-circle. By this time, however, the burghers were within 600 yards, cantering steadily forward shooting from the saddle. On they came in spite of the fire—unfortunately

CHAPTER
XXXVI.

1902.

Boschbult
or
Brak Spruit.

31st Mar.

Rooiwal.

24th Feb.

mostly ineffective—of the untrained irregulars until they were within a hundred yards—and then they turned and galloped away. In the pursuit which followed two 15-prs. and a pom-pom were recovered.

Ian Hamilton ranged his columns in line under Kekewich, Rawlinson, and Walter Kitchener, for a final drive, with Rochfort south of the Vaal.[1] This time the sweep was to be right through the western Transvaal, and across the border to the Mafeking—Buluwayo railway which was strongly held. Expecting a turn-about, as on previous occasions, the burghers kept ahead, and on the 15th the line reached the Transvaal border. Only fifteen miles ahead ran the railway, heavily fortified and garrisoned, with half-a-dozen armoured trains steaming up and down. That night the burghers made a desperate attempt to break back through the line, but silhouetted against the beams of the search-lights they had no chance against the watchers in their entrenchments. Next morning the white flag went up, and the columns marched in to the railway with nearly four hundred prisoners.

Among the column commanders the Royal Artillery were well represented by :—Colonel Rochfort and Lieut.-Colonels Dawkins, Keir, Scott, and von Donop.

The situation in Cape Colony at the beginning of 1902 was disquieting, for the rebels were still in possession of large districts, and the mass of the population were disaffected. Fortunately the Boer Governments do not seem to have appreciated the possibilities, for, though they appointed Smuts Commander-in-Chief in the Colony, they sent him little assistance. He could do nothing beyond keeping the British troops busy by threats here and there, and the occasional capture of a convoy. Their chief attention was, however, devoted to covering the construc-

---

[1] Previous to this Rochfort had been working independently with five columns under him, sweeping the banks of the Vaal between Christiana and Bloemhof after capturing Schweitzer Reneke.

tion of a block-house line, commenced in December 1901,
from Victoria West by Carnarvon and Clanwilliam to the
sea at Lamberts Bay, known sometimes as the "Lines of
Torres Vedras", sometimes as the "Great Wall of China".
In the eastern districts Colonel Lukin, with a mounted
division formed by the Cape Government, gradually rounded
up the marauders, and the command of the railways by
General French enabled his columns to head off any serious
threat by Smuts. But in the far west there were no railways, and the country was difficult for large bodies of
troops. The Prieska district commanded the drifts of the
Orange River by which might come reinforcements from
the Transvaal, and the million square miles of Namaqualand were patrolled only by a single British column,
which had been formed at Ookiep in August 1901 by Major
W. L. White, R.A. By the rapidity of his strokes he had
established his hold on the country, and Smuts now set to
work to reduce the posts which he had established in the
important copper-mining district of Ookiep. Some fell,
but Ookiep itself held out against investment and assault,
although an attempt to relieve it by a "Namaqualand
Field Force", brought round by sea to Port Nulloth, failed.
Smuts was called away to attend the Peace Conference and
on the 4th May the investing force broke up, and White
marching up from Garries re-opened the road.
In these operations in Cape Colony the columns
commanded by Colonel Sir C. Parsons, Lieut.-Colonel
White, and Major Jeudwine, were active.

Any account of the various moves in the negotiations
which eventually resulted in the Treaty of Vereeniging
would be outside the scope of this history. Suffice it to
say that, after a meeting with Lords Kitchener and Milner
at Pretoria, the leaders went off on a ride round the
commandos to arrange for the election of delegates, thirty
from each state, to meet in council. On the 15th May
these delegates met at Vereeniging, where there was much
heated debate, and very nearly a breach between the Free

CHAPTER
XXXVI.

1902.

The
Treaty of
Vereeniging.

31st May.

State and the Transvaal. Eventually a deputation went to Pretoria, and returned with definite terms to be answered "yes" or "no". Steyn, indomitable to the last, denounced them *in toto* and resigned the Presidency which he had maintained with such undaunted determination. There were three more days of debate, and the issue was still in the balance when de Wet, seeing the inevitable ruin of his country if the struggle were prolonged, moved the acceptance of the terms. His lead was accepted almost unanimously, and on the stroke of midnight of the 31st May the Treaty was signed.

The block-house system had slowly strangled the freedom of movement which alone had enabled the Boers to carry on their guerilla warfare. They had no longer any escape from the relentless drives of the columns : there was nothing left but to bow to the inevitable. But while the delegates were debating Lord Kitchener left nothing undone to make a resumption of hostilities—if it should be forced on him—decisive. Rested and refreshed, all deficiencies in men, horses, or equipment, made good, the artillery was never fitter for war than on the day on which peace was declared.

CHAPTER XXXVII.

COMMENTS.

ORGANIZATION—ARMAMENT—EMPLOYMENT—CONCLUSION.

MANY a hard, practical, lesson in all branches of the CHAPTER XXXVII military art was learnt upon the veld, and it has been well said that if it had not been for the South African War the British Army would never have been able to stand up to the Germans as it did in 1914. Be this as it may, the experiences of 1899-1902 undoubtedly exercised a profound influence upon the organization, armament, and training of the Royal Artillery in the years that followed. They are therefore of exceptional importance in the history of regimental development. But to view them aright it is necessary to see them in the light of the time. And in this chapter an endeavour has, therefore, been made to present the problems as nearly as possible as they appeared, and to show how they were dealt with, during the course of operations.

ORGANIZATION.

The Army Corps—The Reinforcements—The Naval Brigades—Mobilization—Command and Staff—The Work of the Artillery Staff—The Artillery Mounted Rifles—The Pom-pom Sections and Galloping Maxims—Ammunition Supply—The Boer Artillery.

The Field Force ordered to South Africa in the autumn The Army Corps. of 1899 consisted of an army corps and a cavalry division. It sailed complete although circumstances compelled its breaking up on arrival, and so affords a good example of the principles of war organization accepted at the time.

The cavalry division consisted of two cavalry brigades and a brigade of horse artillery. Each division consisted of two infantry brigades and a brigade of field artillery. The army corps was composed of three such divisions and "corps troops", the latter including a "corps artillery" of one brigade of horse and two of field artillery—one of the latter being a howitzer brigade.

Brigades of horse artillery had two batteries, of field artillery three, and each brigade included an ammunition column. The following table shows the number of brigades and batteries and of guns of each nature.

Formation.	Branch.	Bde.	Bty.	12-pr.	15-pr.	5″ how
Cavalry Division	R.H.A.	1	2	12	—	—
Three Divisions	R.F.A.	3	9	—	54	—
Corps Artillery	R.H.A.	1	2	12	—	—
	R.F.A.	2	6	—	18	18
		7	19	24	72	18

The weak spot in the organization was the temporary nature of the brigades. In peace they were simply the batteries which happened to be quartered together for the time being, or which were at the top of the roster for foreign service,[1] and the lieut.-colonel appointed to command them was not even provided with a full-time adjutant. The war had not been two months in progress before the necessity of giving permanence to the brigades in South Africa was being considered, and this measure was adopted in 1900. Even then the administrative powers of their lieut.-colonels were small compared with those of the commanders of regiments and battalions,

---

[1] In the field artillery brigade sent from home to Ladysmith just before the war the lieut.-colonel came from Aldershot, the adjutant from Hilsea, and the batteries from Dorchester, Woolwich, and Weedon. The brigade from India was similarly collected—the lieut.-colonel from Lucknow, the adjutant from Rawalpindi, and the batteries from Secunderabad, Ahmednagar, and Deesa.

thus imposing a mass of administrative work on the artillery staff at headquarters quite unsuitable for active service.

It soon became evident that further troops would be required, and during the winter of 1899—1900 there were sent out four divisions and a cavalry brigade. They took their horse and field artillery with them on the usual scale, except the 8th division which had to go without its divisional artillery owing to the need for retaining some trained batteries for home defence. There were, however, large numbers of mounted and dismounted troops outside the regular formations—Yeomanry, Militia, Volunteers, Colonial Contingents—for which guns were required. And the enemy had introduced a novel feature by their use of heavy guns in the field, which must be met by the garrison artillery since the naval brigades could not be kept on shore indefinitely. The reinforcements included, therefore, artillery units of all branches :—

R.H.A.—A brigade (3 batteries) from home, and two batteries from India.

R.F.A.—Four brigades—one of howitzers.

R.G.A.—A mountain battery to replace that lost at Nicholson's Nek, a siege train, and twelve heavy batteries.

In addition to these regular units there were other artillery reinforcements drawn from various sources. From England came two volunteer batteries, the "C.I.V." and the "Elswick" The former was part of the "City Imperial Volunteers" raised by the City of London, and its *personnel* was, to a large extent, provided by the Honourable Artillery Company : the Elswick battery was formed from the 1st Northumberland Volunteer Artillery. The Antrim, Edinburgh, Donegal, and Durham Artillery Militia each sent out a company. From Australia came a field battery and Canada sent a brigade. In South Africa there were among the local forces many artillery corps, and these were all called out. Much assistance was also received from individual officers of auxiliary and colonial artillery. A complete list of all artillery units which took part in the war will be found in Appendix A, where also is given a table showing the total strength of the artillery in South Africa at various periods.

Any account of the organization of the naval brigades would be out of place in this history, but a few words may be added to explain how they came into the field.

During the anxious days when Cape Colony lay defenceless, a naval brigade, with a couple of 12-prs. of 8-cwt. formed a welcome reinforcement to the little garrison of Stormberg. When this garrison was withdrawn to Queenstown the naval brigade, reformed and re-armed, joined Lord Methuen in time to bear a hand in all the fighting from Belmont to Magersfontein. Further strengthened in men and guns, with Captain Bearcroft in command, vice Captain Prothero wounded at Graspan, it formed part of the army under Lord Roberts during the invasion of the Orange Free State and Transvaal, until all the sailors were recalled to their ships in October 1900.

In Natal we have seen how Sir George White's appeal for "detachments of bluejackets with long-range guns firing heavy projectiles" brought Captain Lambton's naval brigade from the "Powerful" to Ladysmith in the nick of time for its defence. A week later Captain Scott reached Durban in the "Terrible" and organized its defence, sending up a brigade under Captain Jones to join Buller. This latter took part in all the fighting on the Tugela from Colenso to Pieter's Hill.[1]

Throughout their service in the field all the naval

---

[1] The *personnel* of both naval brigades in Natal was augmented by the Natal Naval Volunteers.

brigades worked in the closest co-operation with the Royal Artillery, and the relationship was of the happiest.[1]

Special "Regulations for the mobilization of a Field Force for service in South Africa" were issued in June 1899 and worked smoothly enough on the whole. But, in common with all such regulations until the advent of Mr. Haldane, they made no provision for ammunition columns. In consequence batteries which did not form part of the army corps had to be drawn upon to supply officers, men, and horses for the ammunition columns. Then, when the mobilization of further divisions was ordered, and these batteries were required, they were found to have been already depleted. Thus the shortage increased progressively, and under this drain the artillery reservists fell short almost at once, though those of the cavalry were not all called up until the end of 1901.

For siege artillery the mobilization scheme provided for the expansion of each siege company into two siege batteries. Only one company was, however, so expanded, taking out eight 6″ howitzers. Another went as a single battery with four 4·7″ guns.

From the arrival of Sir Redvers Buller to the departure of Lord Roberts the whole of the artillery in South Africa was under the command of Major-General Marshall. He had come out from Aldershot as G.O.C., R.A., of the army corps, and remained at Cape Town when Buller went on to Natal. For staff he had his Brigade-Major from Aldershot (Major Sclater), who was appointed D.A.A.G. on mobilization, and Captain Kirby, A.D.C. On the arrival of Lord Roberts, and the large increase in the size of the army, Major Sclater was promoted to A.A.G.—with

---

[1] Nearly forty years afterwards, the writer well remembers the regret of all at the deaths of so many of the midshipmen who used to come to Headquarters for orders during the investment of Cronje at Paardeberg, and fell victims to the epidemic of typhoid at Bloemfontein.

the rank of lieut.-colonel, and Captain Headlam joined the staff as D.A.A.G.

In Natal Colonel Downing was C.R.A., with Captain Russell as staff officer. As soon as they were shut up in Ladysmith, Colonel Long was sent out to command the artillery of the relieving force, with Captain G. F. Herbert as staff officer; and when Colonel Long was wounded, he was succeeded as C.R.A. of the Natal Army by Colonel Parsons, who took his adjutant, Captain Boger, as staff officer.

On the arrival in South Africa of a large number of garrison companies it was decided to send out a senior officer to take command of the Royal Garrison Artillery. Colonel Rainsford-Hannay was selected for the appointment with Captain von Donop as staff officer, but by the time they arrived the companies had become "heavy batteries", and were scattered all over the country, some with columns, some in posts on the lines of communication. Any separate command of that branch was therefore impracticable, and Colonel Rainsford-Hannay was given command of the line from Bloemfontein to railhead, while von Donop was soon in the field in command of the mounted troops with Lord Methuen.

The only other artillery staff officer[1] during the first year of the war was Lieut.-Colonel Eustace who was appointed A.A.G., R.H.A., of the cavalry division on its reconstitution and the increase of its horse artillery to three brigades.

Until Lord Kitchener succeeded to the command, the artillery staff formed a separate and distinct branch of headquarters under the G.O.C., R.A. But Lord Kitchener made an entire change in principle. The position of G.O.C., R.A., was abolished, and the A.A.G. and D.A.A.G. remained as members of the Headquarter

---

[1] This only refers to officers on the artillery staff. A notable feature of the war was the large number of artillery officers who commanded columns of all arms or served on the army staff. A list of these will be found in Appendix K.

Staff, working with the A.G's. and Q.M.G's. branches in all matters affecting the artillery. During the greater part of this period there was no officer in a position analogous to that of Chief of the General Staff, and Colonel Sclater received his instructions on general staff questions direct from the Commander-in-Chief. It was soon evident that he had gained Lord Kitchener's entire confidence—to the great advantage of the artillery. And "those who served with the batteries in the field would be the first to acknowledge what they owed to his wise foresight, constant care for their interest, and personal sympathy with their difficulties".

To assist him Colonel R. Bannatine-Allason was appointed artillery staff officer in the Orange Free State, with headquarters at Bloemfontein, and, later, Major J. A. Tyler in Natal.

The normal duties of the artillery staff at the time of the South African War may be summed up as :—

i. Maintaining the efficiency of all artillery units.

ii. Supplying ammunition in the field to all arms.

In actual practice a great difference soon became apparent between the artillery and the other arms. When the latter were reduced in strength of men or horses from any cause they turned out with so many sabres or rifles the less, but when a battery could not put all its guns into the field there was an immediate outcry. A notable case was the state of the cavalry division after the occupation of Bloemfontein. Although regiments were parading with a strength of less than the establishment of a squadron, strong objection was taken to batteries turning out with only four guns. Similarly ammunition was in a different category from other supplies. It could not be found in the country, nor improvised, nor could troops "do without" even temporarily. Moreover it soon became of inestimable value to the enemy.

For the above reasons special provision had to be

made by the artillery staff for the upkeep of batteries
and ammunition columns. When columns were roaming
the country no opportunity for filling them up with
men, horses, and ammunition, must be missed. At the
same time no precaution for the safety of these consign-
ments in transit could be neglected. Further, owing to
the depletion of the artillery reserve mentioned above,
a continual watch had to be kept over men in hospital or
otherwise detached from their units, and precautions taken
to prevent their straying from the regimental fold.

At first there was an attempt to continue the posting
of officers, and the promotion of senior N.C.O.'s. by the
War Office, and it was common for officers on promotion
to be despatched, in accordance with the traditional "right
of fall", from batteries at the front to stations in India or
the Colonies just as they were becoming seasoned campaigners, while their places were left vacant until their
successors could be brought from, it might be, the most
distant part of the Empire. But it was soon found that
this system was utterly impossible if units were to be
kept in fighting efficiency, and it was arranged that while
promotions remained on one list, appointments and
postings in South Africa were confined to officers in that
country.

So far, we have been considering the normal duties
of the artillery staff, as understood at the time. But
Lord Kitchener was not the man to be bound by any
official distribution of staff duties. If he trusted a man
he piled work upon him, and he soon showed his confidence in Sclater by entrusting him with multifarious
responsibilities, the more important features of which will
be indicated in the following paragraphs.

When Lord Kitchener first decided to make use of
artillerymen as mounted rifles the project only envisaged
the conversion of the two howitzer brigades. This was
found to be such a convenient unit, both as regards size
and organization, that it became the model for subsequent

conversions. Each of the "corps"—or "columns" as they were generally styled—was self-contained with its pom-pom section, scouts, and signallers, and was commanded by a selected lieut.-colonel, with a staff officer, the brigades and batteries retaining their organization and nomenclature. The strength of a column might be taken as 750, with 20 mule wagons for baggage. A supply column was attached.[1]

Lord Kitchener's experiment proved an outstanding success. The warfare in which the artillery mounted rifles played their part was of the most trying description, for it entailed hard marching and hard living, with un-remitting watchfulness. But officers and men devoted themselves to maintaining the prestige of the Royal Ar-tillery undiminished by the change in their rôle, and showed how quickly good soldiers can become efficient with any weapon. Their efforts did not go unnoticed by the Commander-in-Chief. Directly after the declaration of peace he telegraphed home asking that "to mark his satisfaction with the conduct of the Artillery Mounted Rifles" those under orders for India should be sent home first, and those remaining in South Africa should be allowed home on furlough. Perhaps an even more con-vincing proof of his appreciation of their value had, however, been given before the termination of hostilities by his application for a thousand horse and field artillery-men from India to add to their numbers, and—when India refused—by a similar application to the War Office, at once agreed to.

The Boers introduced a new weapon in their $37^{m/m}$ Vickers-Maxim guns—soon universally known as "pom-poms". There was an immediate demand for similar armament, and fifty were despatched to Cape Town, with twenty-five artillery officers, under Major Crampton as "organiser" and "inspector". Men were obtained from

CHAPTER XXXVII.

The Artillery Mounted Rifles.

The Pom-pom Sections and Galloping Maxims.

---

[1] A copy of the order for the formation of the first corps is given in Appendix I.

CHAPTER
XXXVII.

The
Pom-pom
Sections
and
Galloping
Maxims.

the artillery "excess numbers",[1] and the first three pom-poms to take the field joined the army under Lord Roberts at Paardeberg. Soon afterwards it was decided that the unit should be the section of two guns, but, with the multiplication of columns, most of these had to be divided into two single guns, each under an officer—A/1, A/2 and so forth.

In the Egyptian Artillery every battery included a section of Maxim guns on galloping carriages, with mounted detachments as in horse artillery. These had done good service in the re-conquest of the Sudan, and it is not surprising, therefore, that the employment of similar weapons in South Africa should be suggested. Eventually guns were received from Egypt and four sections—"A" to "D"—were formed.

The South African War was the first in which the subject of ammunition supply assumed a position of importance, and it has therefore been thought advisable to describe here in some detail the system in force.

The supply of ammunition to all arms in the field, i.e., in advance of the Ordnance Depôts, was looked upon as a function of the fighting troops, and was entrusted to the artillery. An ammunition column, carrying small-arm as well as gun ammunition, formed part of every horse and field artillery brigade, and behind them came an "ammunition park" directly under the G.O.C., R.A. These were all regimental units, and, in addition to supplying ammunition to all arms, formed a reserve to the batteries for the immediate replacement of casualties in officers and men, horses and stores.

In the great re-organization of the transport[2] prepara-

---

[1] All units coming to South Africa brought 10% extra as "Excess Numbers". Those of the artillery were concentrated at the Cape.

[2] In this re-organization all regimental transport was pooled under the Director of Transport, who appointed "transport officers" to take charge. In the case of artillery this was, of course, ludicrous, and, on arrival in Bloemfontein, all artillery transport was returned to regimental charge, and administered by the artillery staff.

tory to the invasion of the Orange Free State considerable changes were made in the constitution of these units. The columns were divided into two portions, the first consisting of horsed ammunition wagons and mule "buck" wagons, the second of ox wagons, the number of rounds per gun or rifle carried in each echelon being:—

CHAPTER XXXVII.

Ammunition Supply.

	Amtn. Wagons.	Buck Wagons.	Ox Wagons.	Total.
12 and 15-pr.	50	100	100	250
5" howitzer	23	27	100	150
Small-arm	—	25	50	75

This organization worked well, the first portion being able to keep in close touch with batteries, while the ox wagons followed a march behind. But, as the character of the war changed, the organization was modified to suit local conditions, until by the beginning of 1901 most of the columns had been broken up or immobilised as "local" columns—really depôts—at convenient centres on the lines of communication.

The "record" expenditure was at Magersfontein where the four batteries of horse and field guns fired, respectively, 1,179, 940, 1,000, and 721 rounds. This gives an average of 960, the highest not only of any action in the South African War, but in any war since the introduction of rifled guns. The previous record was that of the German artillery at Mars-la-Tour.

The Ammunition Park never functioned as a mobile unit. In the reorganization of January 1900 it was decided that it should work along and from the railway in advance of the Ordnance Depôts. In addition to ammunition it held stocks of useful articles on which batteries could draw at any time, took care of baggage and spare vehicles left behind, and received and despatched to their units artillerymen discharged from hospital. On several occasions of threats to the line the Park manned guns for the defence of posts, and an advanced section was captured at Roodeval by de Wet in June 1900. At Pretoria the Park mobilised sections of released prisoners which were afterwards absorbed into batteries. In July 1901 it ceased to exist as a regular unit. During its life it had made a generous contribution to the efficiency and comfort of the artillery.

A special feature of the later stages of the war was the incalculable value of our ammunition to the enemy—it was their only source of supply. On assuming command Lord Kitchener therefore extended the responsibility of the artillery staff at headquarters to the control of all ammunition in South Africa. This included not only the fixing of the positions of all ammunition depôts, and of the amounts to be held in each, but also precautions for safety in transit, and supervision of all issues—especially to irregular corps.

Subsequently, on account of the use of explosives by the Boers, there was added the control of *all explosives* in the Transvaal and Free State, with the administration of the Dynamite Factory, the military charge of which was entrusted to the Ammunition Park.

Before concluding these remarks upon the organization of the British Artillery a few words may be added upon that of the enemy.

After the Jameson Raid the Transvaal artillery was augmented and reorganized, the field and fortress being separated. Very fine barracks were built at Pretoria, and the Royal Artillery who occupied them were lost in admiration at the contrast between their sumptuous accommodation and anything experienced at home. There was electric light and hot and cold water everywhere : the officers had villa residences with gardens; and every horse had a separate stall. The avowed object was to make the artillery the nucleus of an army. Its peace establishment was :—21 officers, and 545 other ranks.

In war the strength was brought up to 800 with reservists, and there were also many volunteers from the commandos. The service was for three years extendable to six.

The Free State artillery was not so well organised as that of the Transvaal, and there was no division into batteries. Its strength was about half that of the sister Republic.

In both cases the officers were mostly German or Dutch, and a large number of the men were Germans who had been trained by the Artillery of the Guard at Potsdam. The smart appearance of the *personnel* was noted by visitors before the war.

ARMAMENT.

General Considerations—Horse and Field Guns—Field Howitzers—Mountain Guns—Heavy Guns and Howitzers—Pom-poms and Galloping Maxims—Sights and Telescopes—Mechanization—Railway Mountings and Armoured Trains—The Boer Artillery—Maintenance—Personal Equipment—Harness and Saddlery—Vehicles—Draught animals.

In the many wars in which the Royal Artillery had been engaged in the past the efficiency of its weapons had never been seriously questioned. They had always been at least as good as those—if any—possessed by the enemy. But in 1899 the position was very different. The term "quick-firer" was hypnotising men's minds, and the Boers had just become the possessors of some field guns which were labelled "Q.F." They were few in number, and their claim to the title was questionable, but it was sufficient to cause anxiety, and the less responsible papers took occasion to predict for the British batteries under orders for South Africa "nothing less than positive annihilation should they be so foolhardy as to try conclusions with their powerful adversaries".

Thus the efficiency of our artillery armament had become a matter of popular interest before hostilities commenced, and when the first encounter disclosed the use by the enemy of other novel weapons there was very general concern. The question, therefore, demands

General considerations.

examination here, and in the following paragraphs some particulars will be given of the various natures brought into use, with the opinions formed at the time as to their merits and defects.

During the decade immediately preceding the war the horse and field artillery had been re-armed—the horse artillery with a specially designed, simple, light, 12-pr.: the field artillery with the old 12-pr. converted into a 15-pr. on the introduction of cordite. Both horse and field artillery fired only shrapnel and case.

A great nation cannot hurriedly rearm without due experiment, and both these guns were "breech-loaders" and not "quick-firers". The problem of keeping the carriage of a field gun undisturbed when the gun was fired —the essential attribute of a true quick-firer—had not yet been solved, except, perhaps, by the French. Many improvements had, however, been made with the object of increasing the rate of fire, and in England they took the form of a spade suspended from the axle. Added at the last moment, this worked well in minimising the recoil, thereby reducing the exposure and fatigue of the detachment, besides making possible a quicker fire when required.

The shrapnel shell had been specially designed to give the deepest possible zone to the bullets, so as to obtain the maximum effect against deep formations, or successive lines in movement. This meant a narrow cone of dispersion, which was the reverse of what was wanted for dealing with the widely extended lines of the Boer defensive positions. A further handicap was the want of a longer-burning time fuze. That in use (No. 56) only ranged to about 4,000 yards, but a modification, generally known as the "Blue Fuze" (No. 57), which increased the range by nearly 2,000 yards, was received early in 1900. Experience again showed that shell had often to be carried fuzed on service, but even in the dry climate of South Africa they were found to deteriorate rapidly when so exposed to the weather. Another serious defect of the ammunition was that instead of being packed in complete rounds, each component was in a different package.[1]

At first there was a cry for the re-introduction of common shell for field guns, but the demand was found to come entirely from officers of the other arms who had never seen the effect of our shrapnel from the target end of the range, and judged it from the insignificant effect of the Boer shell. Later experience, and especially that of being fired on by our captured 15-prs., corrected the error, and, for horse and field artillery, shrapnel held the field as a man-killer.

There was also a great difference of opinion as to the value of case shot, and this among gunners themselves. As will have been noticed, it was used on many occasions, but on most of these, there were, alas, no survivors to tell the tale. We have, however, clear testimony as to its value at Diamond Hill, though, on the other hand, the evidence of several officers who fired case against flocks of sheep and cattle during the "drives" was that its effect was trifling.

The necessity for any reduction of weight behind the splinter-bar from the old-established rule of 5-cwt. per horse in horse artillery, and 7-cwt. in field artillery, was not established, but it was generally agreed that ammunition wagons must be kept down to the same weight as the guns.

In the above we have been dealing with the service equipments : a word must now be said regarding those brought out by the two volunteer field batteries from England. It had been intended that these should have 15-prs. (with which they had been trained), but, by the generosity of patriotic individuals, both were provided with the latest productions of armament firms before sailing.

The C.I.V. battery had 12½-pr. Vickers-Maxims, presented by the Lord Mayor. These had single-motion

---

[1] The same defect had been found just a century before by General Lawson when organising the artillery of Sir Ralph Abercromby's force for landing at Alexandria.

breech-mechanism, recoil buffers, and traversing gear, and used fixed ammunition with Krupp fuzes ranging to 6,000 yards. Their long-range shrapnel fire was an undoubted advantage, but great difficulty was found in keeping the battery supplied with its special ammunition wherever it might happen to be.

The Elswick battery was armed with 12-pr. Q.F. guns of 12-cwt., as used by the navy and in coast defences, but mounted on special field carriages. They fired 15-pr. shrapnel as well as 12-pr. common shell, and with the latter a range of 8,000 yards could be obtained. These guns were presented to Lord Roberts by Lady Meux.

The Australian artillery brought 15-prs. the Canadian 12-prs. of 6-cwt. The South African Corps mostly started with 2·5″ but gradually acquired 15-prs. These were all of service pattern, but the three batteries of the Rhodesian Field Force were supplied with 15-prs. of various patterns requiring special spare parts, which were a constant source of anxiety.

The 5″ howitzer equipment introduced in 1897 on the formation of field howitzer batteries was unfortunately considerably heavier than that of the field gun. During the invasion of the Free State this proved a serious drawback, and the conditions on the Modder river, and in the Free State generally, were not such as to afford many opportunities for howitzers to show their value. In Natal, on the other hand, their peculiar powers proved invaluable, and the extra weight did not distress the horses in negotiating the rough ground as it had in the long marches over the waterless veld of the Free State.

Before the South African War high-explosive shell had never been used in the field, the only experience with them being in the bombardment of the Mahdi's tomb and the breaching of the walls at Omdurman. Absurdly exaggerated ideas of the capabilities of lyddite were widespread, and the disappointment when these failed to materialize was equally great, although the distinctive yellow fumes which too often accompanied their burst[1] gave the troops immense gratification. In the open the effect of lyddite shells was exceedingly local, but they were very effective in searching deep trenches, and worked tremendous havoc among men sheltering among rocks.

CHAPTER XXXVII.

Field Howitzers.

The 2·5″ R.M.L. jointed gun, with which the mountain batteries were armed, was hopelessly out-of-date before the war began. As a sufferer put it: "Our smoky powder gave away every position we took up, evoked the curses of our neighbours, and earned the sobriquet of 'the scrap-iron battery' for the unit". No more need be said about it here.

Mountain Guns.

The artillery normally included in a field army at this period consisted only of horse, field, and (in certain circumstances) mountain batteries. But the Boers brought heavy guns and pom-poms into the field, and immediate steps had to be taken to counter them. The ordnance stores of Cape Colony were ransacked for relics of the Kafir wars, war-ships and coast defences were stripped of their modern armament.

Heavy Guns and Howitzers.

The naval brigades which first filled the place of heavy artillery were armed with 4·7″ and 12-pr. guns on improvised mountings due to the ingenuity of Captain Scott of H.M.S. "Terrible". The first 4·7s were on "platform" mountings, which consisted simply of central-pivot upper-deck mountings bolted on to a transportable platform. They were thus very heavy and cumbersome to move and laborious to mount. The later pattern, on a wheeled carriage, was much more mobile, but still weighed 6 tons, and required a team of 32 oxen. The first two garrison companies to bring out 4·7″ guns had 6″ howitzer carriages, the remainder 40-pr. R.M.L. carriages, which proved very satisfactory with both 4·7″ and 5″ guns, the equipment weighing under 5 tons and

---

[1] The yellow fumes showed incomplete detonation. It was estimated that in only about 50% was this complete.

requiring only 20 oxen. The heavier shell of the 5″ guns, and the fact that the carriage provided a travelling position were distinct advantages, and, for land service, nothing was gained by the higher velocity of the 4·7″—rather the reverse indeed owing to its loss of searching power. The brass cartridges also added unnecessary weight.

The 6″ siege howitzer proved generally serviceable, and for use in the field it could be lightened by discarding the platform, top carriage, and buffer. Its range was, however, inadequate. The two 6·3″ muzzle-loading howitzers in Ladysmith—known as "Castor" and "Pollux"—afford an extraordinary example of the value of indirect fire. In spite of their antiquity, and the scratch lot of ammunition available, no Boer gun dared venture within their reach.

For the attack of such forts as had been erected round Pretoria the 6″ howitzers would have possessed neither the requisite power nor range, and a battery of 9·45″ howitzers was obtained from the Skoda works in Austria. The story of their acquisition and despatch to South Africa has been told in Volume II and need not be repeated here : suffice it to say that two of these weapons accompanied the army to Pretoria, though, owing to its surrender, they had no chance of proving their capabilities there. Their presence with the army did, however, show how it was possible, by dividing the loads, to combine great shell power with sufficient mobility for use in the field.

The relation of power to weight in the different heavy and siege equipments taken into the field is perhaps worthy of note.

Equipment.	Wt. of shell.	Total weight.	Number of loads	Heaviest.
4·7″ gun Naval (platform)	45 lb.	11 tons.	three	4 tons.
,, ,, (wheeled)	,,	6 ,,	one	
,, R.G.A.	,,	4¾ ,,	one	
5″ gun do.	50 lb.	4 ,,	one	
6″ how. do.	120 lb.	4½ ,,	one	
9·45″ how. do.	280 lb.	8½ ,,	two	

The general conclusion was that no heavier weapons were required as part of a field army than a gun and a howitzer weighing not more than four tons, and drawn by teams of twelve horses. The gun to be on the lines of the 5″ gun, but with a heavier shell, to range to 10,000 yards. The howitzer to be on the lines of the 6″ howitzer but to range to 8,000 yards. Heavier guns and howitzers on travelling carriages or railway mountings to be added as required.

The pom-pom was a 37$^{m/m}$ gun firing a 1-lb. common shell with a percussion fuze. The cartridges were in belts, and the gun was in fact a magnified machine gun. Before the war the only purchasers of these weapons had been the Chinese and the Boers, but their use by the latter caused an immediate demand for similar weapons, in compliance with which fifty equipments and one-and-a-half million rounds were bought. But in view of the decline of the pom-pom in popularity after the war it is not necessary to labour the subject. Rejected by one arm after the other the guns were returned to store and the ammunition scrapped.

The galloping maxims were of ·450″ calibre. Useful as they had proved in the Sudan, they suffered from fatal defects for service in South Africa—they offered a large target, and they were inferior in range to the enemy's small-arms.

Their work alongside the sailors brought home to the gunners how much they lost by the absence of telescopes and telescopic sights. Telescopes were provided for heavy and siege batteries, but not for their brigade headquarters, and horse, field and mountain artillery had none. The powerful telescopes possessed by the naval brigade were of immense value to them in locating the enemy's guns, but what the field artillery required them for, even more urgently, was to distinguish between friend and foe when supporting an attack.

The only telescopic sights in the land service were of the "Scott" pattern, which had been introduced with the B.L. guns in the 80's. But this inverted the image, a feature which seriously restricted its value for use in the field.

The only provision for indirect fire consisted of clinometers and aiming posts. The improvisation of gun-arcs by Major Gordon commanding the 61st battery has been mentioned in Chapter XXIX.

The addition of heavy batteries to the field army naturally turned attention to the possibility of using mechanical draught. But the only engines in South Africa were of the commercial traction engine type. Trials with these were carried out at Cape Town by Colonel Johnson, R.A., as early as February 1900 with two batteries armed with 4·7″ guns, and about the same time Major Callwell made a similar experiment with his 5″ battery in Natal. In neither case was the result such as to encourage the gunners to discard their oxen.

Somewhat later two 6″ guns on field carriages, weighing 12 tons, were railed to Pretoria, and one of these was taken out by a traction engine to the threatened outpost line at Derdepoort. This was the only occasion on which any attempt was made to take a gun into action with mechanical draught, and it proved a failure as described in Chapter XXXI.

The use of engine-draught for ammunition columns was ruled out by the fact that it had only proved reliable for ordinary supply purposes when worked "continuously, in its own time, in fine weather, between two points, on specially selected roads"—conditions obviously inapplicable to ammunition service. The eyes of artillerymen had, however, been opened to the possibilities of mechanization as soon as a more practical form of tractor had been evolved.

The first artillery to be mounted on railway trucks

CHAPTER
XXXVII

Railway
Mountings
and
Armoured
Tranis.

were 6″ and 4·7″ guns that were sent up from Durban and came into action near Chieveley during the last week's fighting on the Tugela. Somewhat more elaborate railway mountings for 6″ guns were constructed at the Cape. The first of these arrived at Modder River just in time to fire a few rounds against the Magersfontein position before Cronje evacuated it. Later on it was sent on to Fourteen Streams where it was in action for some time covering the movement for the relief of Mafeking.

Armoured trains were employed early in the campaign but were at first unprovided with artillery. With the inauguration of the system of "drives" under Lord Kitchener their number was greatly increased, until there were a score or more regularly organized as fighting units, including an artillery detachment. They usually carried a 12-pr. Q.F. on a pedestal mounting, but a few had 6-prs. or even 3-prs.

A complete list of the guns and howitzers in the possession of the two Republics is given in Appendix D, but the only ones to which any special attention need be directed are the Creusot fortress and field guns and the Krupp field howitzers.

The Creusot 155$^{m/m}$ fortress guns owe their chief interest to the fact that they were the first guns of this size to be taken into the field. Their design had no special features, but the carriages had travelling as well as firing trunnion holes, and this no doubt contributed largely to their mobility. The total weight was about $5\frac{1}{2}$ tons; the shell weighed 88 lbs.; the cartridge was black powder, and the nominal range 11,000 yards.

The Creusot 75$^{m/m}$ field guns were of 1894-5 model, and used fixed ammunition with smokeless powder. The carriage had a hydraulic buffer, spring recuperator, and traversing gear, but its most interesting feature was that the gun, instead of being mounted above the axletree, actually passed through the centre of it, thus firing in its axis. It was an ingenious attempt to attain stability, and the equipment generally is perhaps the most interesting specimen of the class which were being put upon the market in the 90's by all gun-makers in their attempt to solve the problem of the true quick-firing field gun. The shell weighed 14½ lbs., and the nominal range was 8,500 yards.

The file containing the official correspondence concerning the purchase of these guns was found at Pretoria. It included particulars of the ingenious arrangements by which the money filtered through to the makers without the Transvaal Government appearing openly in the transaction, but the most interesting point was that three batteries *had been paid for*, though only one had been delivered. It was suggested that we should present the receipted bill—which was in the file—and demand the other two batteries. Presumably it was not considered expedient.

The peculiar property of the *120ᵐ/ᵐ Krupp howitzers* was their extreme accuracy—otherwise there was nothing remarkable about the equipment. So marked was this feature that, when it came to finding a successor to our 5″ howitzers, it was some time before any could be found to approach the accuracy of one of the captured Krupps, sent home for comparison.

The weak point about all the Boer artillery fire was the feeble effect of their shell. The segment, or ring, shell used by their Krupps had little effect. Except in the howitzers the time fuzes rarely acted properly, and on percussion the bursts were feeble. It was only after a considerable time that the Boers discovered, and rectified, an inherent defect in the fuzes for their 155ᵐ/ᵐ guns which caused them to be blind. The 15-prs. captured from us caused more annoyance than any other gun in the hands of the enemy. A battery commander's comment on Sanna's Post is illuminating: "I hate their taking our guns, our shell and fuzes act, theirs don't".

Before concluding this account of the weapons in use by the artillery in South Africa a word must be said regarding the steps taken for their maintenance in serviceable condition in a country entirely devoid of armament works.

The fact that so many of the guns in use were not of service pattern, and so required special "spare parts", greatly complicated the provision of these essentials for keeping the guns in action. Still more serious anxiety was caused by the gradual wearing out of the guns themselves. Early in the war Major Bushe, R.A., accompanied by an Arsenal artificer, was continually touring the front, examining guns and carriages even when in action. Later on, periodical examination by Ordnance officers was instituted, and a record of the life of every gun in the country[1] was kept by the artillery staff at Headquarters. Arrangements were then made for the exchange of guns from one battery to another, so as to keep those showing signs of wear in posts where they were not likely to be required for heavy firing—such unorthodox "sentences" as "not to go on trek" were frequently passed.

A striking feature of the war was the immense amount of manufacture and repair of war material carried out by the railway and other engineering works with the assistance of Inspectors of Ordnance Machinery. In Ladysmith one of the 6·3″ howitzers was hit through the embrasure by a 155$^{m/m}$ shell, the breast of the carriage badly damaged, the elevating gear smashed. Working night and day, the ordnance artificers and railway mechanics got it into action again in forty hours. At Bloemfontein two 12-prs. were actually "half-bushed" under the direction of Captain Paul, I.O.M., and Mr. Duvant of the Railway Works.

The only fire-arms carried by the horse and field artillery were a dozen carbines strapped on the limbers, and the revolvers which had been issued to the drivers

---

[1] The original records are now in the Institution.

only just before the war. The consequent helplessness of batteries in camp, on the march, and in action, was soon shown practically, and the number of carbines increased. But the frequent moves of batteries by rail during the guerilla period showed that something more was required. The inconvenience—and often more—of having to find an escort became intolerable, and Lord Kitchener brought the number of carbines (or rifles) up to 48.[1] He also recommended that the rifle should be recognised as the personal weapon of all artillerymen.

Pole-draught had succeeded shaft-draught shortly before the war, but the artillery still retained the neck-collar, in spite of persistent efforts on the part of Lord Wolseley to get breast-harness adopted. The disadvantages of neck-collars on active service, where horses lose condition rapidly, even if they have not to be replaced, were soon apparent; and when mule-teams were issued, complete with breast harness, its advantages were apparent to all.

Another product of experience was the discovery that for artillery the simplest way of picketting horses was to tie them to a rope stretched between vehicles, and so dispense with the built-up rope, heel-pegs, and heel-ropes.

In addition to the guns and ammunition wagons the equipment of a horse or field battery included a number of technical vehicles, such as "forge" and "store" wagons, and in India a "captain's cart". Most of these were soon abandoned, and the field forge, artificers' tools, and office box transferred to one of the baggage wagons. But these latter were of a special pattern in the artillery —the "Ammunition and Store Wagon, R.A.". As early as the arrival of Lord Roberts its defects had become apparent. Its weight was excessive for the load it carried, the small size of its front wheels increased its

---

[1] Approximately the strength of a company—the infantry escort usually provided.

heaviness in draught, and its general design (in which
the connection between the front and hind wheels was
through the body) rendered it unsuitable for rough ground.
It was recommended that all special technical vehicles
should be abolished in the artillery, and that all stores
should be carried in the general service baggage wagons.

The horses received by the artillery on mobilization
were generally satisfactory. If they lacked the quality
desirable for horse artillery, they were perhaps better
adapted for field artillery work than those in the service,
shorter in the leg and bigger in the girth. The remounts
from Australia were at first of poor quality and appeared
to have been grass-fed. Argentines were tried, and
a consignment which reached Bloemfontein during the
halt there were greatly admired. Their appearance—
short, chunky, cobs—seemed just the thing for the pom-
pom sections being formed, but they soon dropped out of
favour. Stupid and stubborn, their want of "heart" was
fatal for active service. The London bus-horses were the
best of all for artillery purposes and the handiest : next to
them those from North America.

The wastage—except in Natal—was enormous,
although in this respect the artillery compared favourably
with the other mounted troops.[1] The first cause was that
in many cases the horses were called upon for severe
exertions before they had recovered from their long sea-

---

[1] Many commanders could echo Lord Methuen's report—"the R.A.
were the best horse-masters in my force", and the following percent-
ages of replacements, extracted from the report of the committee on
the purchase of remounts are instructive :—

	1901	1902 (5 months).
Cavalry	3·51	2·40
Artillery	1·95	1·79
Mounted Infantry ...	4·00	2·40
Imperial Yeomanry	4·15	2·72
Colonial Corps ...	4·20	2·30

voyage followed immediately by a train-journey.[1] This
in great heat, when they were putting on their winter
coats. Not only were the marches long and over
unmetalled tracks, but water was always scarce and sometimes unprocurable. Then, the horses were not accustomed to mealies, the staple food of the country, and
could not pick up a living off the veld grass. So they
starved when their regular rations were not forthcoming
until enterprising artillery officers learnt how to make use
of whatever was to be found in the country.

There had also to be taken into account the deadly
diseases which afflict horses in South Africa, about
which little was known. The following extracts from a
battery commander's diary during active operations in
1900 bring home the suddenness with which these plagues
struck, and the helplessness of all concerned.

Aug. 26th. "Most of the horses are off their feed
this morning, and two died before we moved off. . . .
During the day the sickness developed and 13 more
died, and about 20 were very sick."

Aug. 27th. "We started soon after day-break with
what was left of our horses—23 are dead, and about 50
so weak that they can scarcely pull a pound. All our
riding horses are in draught and we have had to borrow
horses from the ammunition column."

Aug. 28th. "Alas, there is not much chasing for us.
29 horses died of this unhappy poisoning and most of
the others only able to walk slowly. Some say it is
tulip poisoning, but others say that horses are never
affected by that. It is poison of some sort, however,
and they must have got it while grazing."

---

[1] The story of the IIIrd Brigade, R.F.A., which suffered so severely
in Lord Methuen's advance is a good example. So badly found were
the vessels provided that the whole voyage was a tale of break-downs,
beginning with a fire in the bunkers which kept one in the Mersey for
a week after the troops had embarked. Compasses out of adjustment,
condensers which only worked half-time, horse-decks without drainage.
One of the two had to put in to Las Palmas for repairs, the other to
call at St. Helena for water, with the result that one took 30 and the
other 35 days over the voyage. And then the batteries were hurried
into trains on the day they landed, and spent 36 hours in them.

The most important step taken towards coping with the wastage in horse-flesh was the replacement by mules of those in ammunition columns, and to a certain extent in batteries also. This measure caused some consternation at first, but so popular did the mules become that they found their way into horse artillery detachments, and one very well-known horse artillery major is credibly reported to have habitually commanded his battery on a cream-coloured one. The increase of the establishments, so as to provide ample "spares" when special exertions were required, also proved a real economy.[1]

CHAPTER
XXXVII

Draught
Animals.

Later on a great improvement was effected when batteries doing garrison duty, and local ammunition columns, became in practice regimental remount depôts, conditioning and training remounts and resting the tired. Under this arrangement batteries with columns were able to exchange their sick and tired horses for fit ones, whenever they got into the line, instead of having to work them until past recovery. At the conclusion of hostilities these depôts contained over two thousand horses being trained or rested, and over 25,000 had passed through them since the beginning of 1901.

For the guns of the Naval Brigade and of the heavy and siege batteries of the Royal Artillery, ox-draught was generally used. The teams were 32 for a 4·7″ on naval mounting, 24 for one on a 6″ howitzer carriage, and 20 for either a 4·7″ or 5″ gun on a 40-pr. carriage. Ox-draught was also, as we have seen, largely used in ammunition columns. The great objections to oxen for fighting troops were the enormous road space they covered, their slow rate of march, and the necessity for giving them time to graze by day, which limited their length of march to about a dozen miles. On the other hand they could draw far heavier loads than mules, no forage had to be carried for them, and, when rain turned the tracks into sloughs, they could work where mules were helpless.

---

[1] See Appendix H.

An interesting trial of horse-draught for heavy guns was made in the Natal army for the advance on Belfast in August 1900 by the substitution of horses for oxen with the 5″ guns of 57/R.G.A. and the 12-prs. of the Naval Brigade. The horses were specially selected from the two field howitzer batteries, which had a rather heavier stamp than the gun batteries, and they were used in 12-horse teams, four abreast. The 57th was given mule transport, and did wonderful work in the exceptionally rugged mountain country beyond Lydenburg, where the possibility of crossing a dangerous space rapidly proved of great value on several occasions.

EMPLOYMENT.

The
Doctrine

The "Doctrine"—Doubts and Difficulties—Smokeless Powder—Long Range—Concealment—Guerilla Warfare—Field Howitzers—Mountain Artillery—Heavy and Siege Artillery—Pom-poms and Galloping Maxims—The Boer Artillery—Conclusion.

The Royal Horse and Royal Field Artillery—the only artillery normally allotted to a Field Force—took the field in 1899 with a definite "doctrine" to govern their employment. The building up of that doctrine has been traced in Volume I, and need not be repeated here, but in order to render the following comments intelligible to a later generation, its salient feature must be recalled.

The duty of the artillery was stated to consist in, first silencing the enemy's guns, and then, successively, "preparing" and "supporting" the infantry attack. In order to ensure success in the artillery duel it was laid down that all available batteries should be brought into action from the earliest possible moment, and, since—with the means then available—the fire of dispersed batteries could not be concentrated, that the batteries should themselves be concentrated. The enemy's guns silenced, the artillery turned its fire on to the part of the hostile position selected for attack, in order to "prepare" the way for the infantry. Finally, when the infantry advanced, the duty of the artillery was to give the closest possible "support" up to, and including, the assault.

In gunnery the "bracket system" of ranging had been developed into a methodical process, strict adherence to which was insisted upon. The system was simple and sure, but it was desperately slow. The keynote of the training was steadiness, to ensure which "casualties" were constantly practised. Its spirit was summed up in the aphorism—"The essentials for fire discipline are the quiet, orderly, and correct performance of all duties under hostile fire".

In the early encounters the artillery was employed in strict accordance with these principles. But it was not long before people began asking themselves what was to be done if the unlimbering of this imposing array of guns was either received in silence, or greeted with shells from an invisible foe? Even when the enemy's guns could be located, how were they to be "silenced" when they were dotted about at ranges beyond the reach of time shrapnel, and shifted their position as soon as a shell came near them? And what was the good of plastering a hill-side in preparation[1] for the infantry advance if the defenders lay snug and silent in deep trenches over which the shrapnel bullets shrieked harmlessly?

In the words of a foreign military observer with the

---

[1] "Batter the hill thoroughly before the infantry advance to the attack" is a specimen of the orders given to the artillery in the early battles.

Boers :—"Many times it happened that ammunition, and still more valuable energy, were wasted on a fruitless bombardment without any profit accruing thereby to the infantry. When it came to the turn of the latter to advance everything still remained to be done; with what discouragement cannot be told, nor with what loss, for if discouragement is injurious in any state of life as a cause of helplessness, in war it is the first and surest factor of defeat".

The Boers were not playing the game in accordance with Aldershot rules, and the blame was cast upon the gunners. There was much irresponsible criticism, but there is fortunately available a more authoritative pronouncement. Before proceeding to the front in February 1900 Lord Roberts issued *Notes for guidance in South African Warfare based on the experience gained during the past three months*. The following are the paragraphs devoted to the artillery :—

"As a general rule the artillery appear to have adapted themselves to the situation, and to the special conditions, but the following points require to be noticed.

(i). At the commencement the artillery should not be ordered to take up a position until it has been ascertained by scouts to be clear of the enemy and out of range of infantry fire.

(ii) Preparation for attack should be thorough and not spasmodic. Unless the infantry are pushed to within nine hundred yards the enemy will not occupy their trenches, and guns will have no target. It is a waste of ammunition to bombard entrenchments when the infantry are likely to be delayed. To be of real value the fire of guns should be continuous until the assault is about to be delivered".

So far for general principles. A word must now be said regarding certain specific points which excited much comment at the time—smokeless powder, long range, and covered positions.

Smokeless powder for guns and rifles had come into general use shortly before the war, but all attempts to find a smokeless blank for use during manœuvres had failed. In consequence the training of the army had not prepared it for the change, and the "void of the battle-field" came as a real surprise, with far-reaching effects both moral and material.

In addition to the difficulty of finding a target in the early stages of the battle, the artillery were faced with that of distinguishing between friend and foe as the critical moment approached. There were cases in the early engagements—Talana Hill, Stormberg, Modder River—of the infantry suffering loss from the shells intended for their support. The batteries had no telescopes, and the khaki-clad men, scrambling among boulders and bushes, were hard to follow with the eye. Nor had they signallers to send back warning.[1] More than once, too, the batteries came under rifle fire from ground which they thought the other arms had made good. "It is better to have signallers of your own than to have to borrow them of the infantry, and scouts of your own than trust to the cavalry" was one of the first lessons of the war for the artillery. There were even demands for permanent escorts to be included in battery establishments. This was going beyond reason, for, as a general rule, the other arms were ready enough to provide all that was required. As an artillery staff officer wrote at the time:—"At peace manœuvres, if you ask an infantryman for an escort, he growls and asks if a serjeant and ten men will do. Now I go to a brigadier and say I am anxious to secure some ground from which the riflemen might pick off the gunners, and he replies 'Of course, my dear fellow, which will you have, a couple of companies or a battalion? Show me exactly where you want them and let me know if you want more later'."

[1] As has been previously mentioned, signallers had been abolished in the horse and field artillery.

The way in which the Boer artillery brought fire to bear from positions far beyond the reach of our field guns was the point on which the critics lavished their most scornful remarks about the British batteries. The gunners were not so much impressed with the value of this long-range fire, but there could be no doubt that being out-ranged shook the confidence of the other arms in the artillery. And even if the effect of the fire was generally insignificant, the delay it occasioned in the advance of the other arms was often of vital consequence. It was not unknown for so-called "pursuing" columns to be halted at the fall of the first shell, while the artillery were forced to waste invaluable time in the impossible task of "silencing" one or two invisible guns.

The scale of the country and the clearness of the atmosphere were very deceptive. The undulations of the veld especially, so much wider than they seemed, affected all artillery fire. The opinion of the Pretoria Committee was that one must be chary of drawing deductions from the abnormal atmospheric and topographical conditions which made it possible to see your enemy, and fire at him, at ranges of 10,000 yards. Owing to the impossibility of reliable observation in ordinary climates at a greater distance, and the difficulty of manufacturing longer-burning time fuzes, they considered it unnecessary for field artillery to strive for shrapnel fire at ranges over 5,500 yards.

Of greater real importance to the artillery was the range at which batteries in action were vulnerable to musketry fire. This had been a frequent subject of discussion during the last quarter of the XIXth century, and there were trials at artillery practice camps. For various reasons these failed to afford any reliable data, and artillery opinion generally supported the statement in the manual that musketry fire might be neglected at anything beyond a thousand yards. But many infantry officers claimed that they could inflict serious loss among the

gunners at a much greater distance. It was not long
before the burghers proved their case for them.

Another point in connection with the range of musketry
fire, which specially affected the horse artillery, was the
fact that the Boer mausers outranged our cavalry carbines.
The cavalry would not, therefore, dismount and engage
the enemy in a fire-fight, but depended entirely on the
horse artillery for fire effect. This was shown very clearly
in the relief of Kimberley, and on several occasions during
the invasion of the Transvaal. Later on the cavalry
exchanged their carbines for rifles.

The manual in use at the time of the war recognised
the "deliberate method" of occupying a position, unlimbering under cover and using aiming posts. But the
instruments provided were primitive, and it was distinctly
laid down that the first essential in an artillery position
was a clear view (over the sights) of the target and of
all ground on to which fire might have to be turned—the
last was cover. The inability to shoot—and hit—from
under cover inspired this distrust of covered positions.
Gunners felt that from such positions they would be
incapable of giving the other arms the support they
required. Rather than fail in this, they were prepared
to face the enemy in the open, and trust to their own fire
for their protection. Their maxim was that "The defeat
of the enemy, and not our own protection from his fire,
is the primary duty".

There could not be a better illustration of the pre-
vailing prejudice against covered positions than the
comment, in the Official History, on one of the occasions
when such a method had been adopted with success.—
"It might, however, be a misfortune if this example
were taken as one of general application under conditions
different from those of the particular day".

Undue loss could, it was thought, be avoided in the
occupation of a position in the open by the concealment of
any previous indication of the intention, the simultaneous

appearance of the whole line of guns, the pace of their movement, and their rapidity in opening fire. And so, time and again, batteries trotted out calmly on to the open veld, took up their positions quietly, and opened fire steadily, to the admiration of all. They lived up to their principles, but it could not be doubted that it was only the ineffectiveness of the enemy's fire which saved them from paying the penalty.

The adoption of guerilla warfare by the enemy, and the consequent break up of the divisional organization in the British Forces, brought many problems for the artillery. The formation of small mobile columns and the multiplication of fortified posts necessitated a complete re-allotment of the guns. Columns and posts clamoured for them, but with the change in the character of the war two new considerations came to the front—would the guns contribute to the object in view, and would they be safe from capture?

In posts guns were unquestionably of value in staving off attacks, and their inclusion in mobile columns was undoubtedly called for while the enemy were provided with artillery. But it was a different matter when they had lost all their guns. It had to be remembered that the object at this stage was to force the burghers to fight, and not just to move them on. De Wet has left on record how he took advantage of our habit of preliminary bombardments :—

"When the British saw two or three hundred burghers they would halt and bring their guns to the front. Then bombard the ridge and send out flanking parties, sometimes taking several hours before they could make sure that there were no Boers behind the rise. It was tactics such as these that gave my burghers time to retire".

In the final phase of the war the extension of the block-house system brought the safety of the guns from capture into prominence. For the block-houses were only bullet-proof, so that their existence depended upon the absence of guns with the enemy. It was, therefore, of the greatest importance that they should be given no opportunity of obtaining them, and the responsibility of ensuring that guns were not left in posts with insufficient garrisons, or taken with columns of inadequate strength, was placed upon the artillery staff. It was their duty also to keep touch with every gun that had originally belonged to the enemy's artillery, or had fallen into their hands. Whenever guns did fall into the enemy's hands the most strenuous measures were at once taken for their recovery. As de Wet writes, of the guns he had captured on one occasion "they had not been of much benefit to us for the English kept us so constantly on the move that it had been impossible to use them".

So far we have been speaking more particularly of the gun batteries, but in the last years of the XIXth century howitzers had re-appeared in the armament of the field artillery. There was at first some uncertainty as to their rôle, and it was only after considerable discussion and experiment that it was decided that they were to be kept as a special reserve for employment against entrenched positions, and shrapnel and case were withdrawn from their equipment. There was no time, however, for the army, or even for the artillery at large, to grasp the correct principles of their employment before the war, and, in consequence, they were, on many occasions, used for purposes for which they were neither intended nor equipped.

On arrival in South Africa the howitzer brigade[1] was split up like so many other formations. One battery went on to Natal, and of the two batteries which remained in Cape Colony, one was divided between the central and

---

[1] VIII Brigade (37, 61, 65). It is worthy of record here that among the officers of the brigade at this time were two who were to become Field-Marshals, and four more who were to reach the rank of General Officer and Colonel-Commandant. A second brigade—XII—was converted and sent out in January 1900.

western lines, and the other joined the army for the invasion of the Free State. Unfortunately this latter battery had to be allotted as divisional artillery to a division whose artillery had been delayed, and in consequence it was used for the most part as if it had been an ordinary field gun battery.[1] It was called upon for long and rapid movements with disastrous effect upon its subsequent mobility; its heavy lyddite shells were wasted on such targets as scattered burghers galloping over the open.

It was only in Natal that these new weapons were given an opportunity of showing their power. There, the positions of the Boers on the Tugela, provided exactly the class of target for which they were intended; the officer commanding the battery (Major A. H. Gordon) had realised the true functions of a howitzer; and he was given a free hand.

Before the war Major Gordon had come to the conclusion that the way to obtain the greatest advantage from the high angle of elevation which howitzers allowed was by the use of covered positions. He had accordingly trained his battery to the use of observing parties, and worked out a system of signals. This was further developed during the fighting in Natal, where covered positions became the rule for the battery, and proved of great practical value.

In the artillery duel the howitzers seldom got a chance owing to the distant positions occupied by the enemy's guns, although this was occasionally possible by moving in to a position defiladed from them, with the observing party left behind in the original position. But in the preparation of the infantry attack the steep angle of descent enabled their shells to enter the deep Boer trenches, and to search the reverse slopes and secluded kloofs, to the great discomforture of the burghers who relied so greatly on cover. Its greatest value was found, however, in the support of the

---

[1] Except during the investment of Paardeberg when the whole of the artillery was placed directly under the G.O.C., R.A.

assault. It was soon realised that the howitzers could maintain their fire until the infantry had got closer to the enemy than was safe with flat-trajectory guns, and they were in great demand by divisional commanders in consequence. The following extract is from a letter written shortly after the battle of Pieter's Hill, to a friend at the War Office: "On the 27th Gordon's howitzers were simply magnificent. My heart was in my mouth watching his shell dropping within 150 yards of the infantry. But the effect of reliable ordnance and nerve was that the Boers threw up their hands and surrendered a very nasty long trench which had twice before proved too much for our troops and caused attacks to fail". Another advantage of indirect fire, only incidentally discovered, was that the howitzers could continue their fire when heavy rain or darkness blotted out the target and arrested all gun-fire.

In the later stages of the war the short range of the howitzers, and their slow rate of fire, reduced their value for the defence of posts, while for work with mobile columns the extra weight of their equipment, and especially of their ammunition, were undoubted drawbacks. They were, however, much sought after by column commanders operating in broken country, where their lyddite shell were unequalled for bolting the burghers from rocky kopjes. There were instances of howitzer sections under resolute officers accompanying mounted troops for long and arduous periods, and rendering notable service in spite of the greater weight of the equipment. But when it came to converting artillerymen into riflemen the two howitzer brigades were the first selected.

South Africa, mountainous as much of it is, proved no country for mountain artillery. The inhabitants were past-masters in overcoming all the difficulties of the ground with wheeled transport, and no gun which could be carried on mule-back could be a match for the heavier metal that could be drawn by mules or oxen. Luck was against them too. Not only was their armament an anachronism, but

Mountain
Artillery.

the one battery which formed part of the pre-war garrison was involved in the disaster of Nicholson's Nek. The survivors did fine work throughout the defence of Lady-smith, and afterwards during the invasion of the Transvaal, but it was not with mountain guns. The battery sent out to replace it joined the army under Sir Redvers Buller just in time to be ordered up to Spion Kop, but had only got half-way when the defenders were met coming down. After surprising all arms by the way they negotiated the roughest and steepest hill-sides, during the fighting on the Tugela, it was allotted to the Drakensberg Defence Force, and later on it took part in the fighting in the eastern Free State and elsewhere, attached by sections to mobile columns, the gunners and ammunition carried in cape carts, the guns tied on behind. In the hills the guns were put again on to the backs of the mules, and the mountain gunners were able to show the way to the horse soldiers.

We have seen how the appearance of heavy guns with the Boers surprised the British in Natal, how Sir George White called upon the Navy for aid, and how prompt was their response to his appeal. The Royal Garrison Artillery lost no time, however, in showing their readiness to take their place alongside the Royal Horse and the Royal Field Artillery in the field army, and companies were soon on their way to South Africa as "Heavy Batteries".

The way the Boers brought their heavy guns into action, however difficult the ground, was a revelation to the British, but the burghers were in their own country, in which for generations they had been inseparable from their ponderous ox-wagons. What was far more surprising was the rapidity with which the garrison gunners from our coast-defences adapted themselves to their new rôle, and soon showed themselves as skilful as their opponents in surmounting the difficulties of what was to them an unknown land.

The heavy batteries were for the most part employed in dealing with the enemy's guns which the field batteries could rarely reach. Every column commander was anxious to get a heavy gun to clear the road for him, and was ready to put up with all the difficulties its addition entailed. There were grave doubts, however, as to whether their presence with columns at this stage was advisable, and a contemporary reflection may perhaps be quoted :—"Now that all the generals have got their desire and been provided with long-range guns they take good care never to get within range of the Boers, but content themselves with 'driving them off' with artillery fire". A more legitimate place for them was in the defence of posts, and by the middle of 1901 the majority were so employed.

Nothing need be said as to the employment of the siege train as such, for Pretoria surrendered without its services being called for, and, shortly afterwards, Colonel Perrott took the greater part to China. One of the siege batteries had, however, the opportunity of showing the value of siege howitzers with an army in the field. Arriving at Paardeberg on the 26th February 1900, it was hurried straight into action so as to give the Boers a foretaste of what bombardment with 120lb. lyddite shells would be like. Only twenty rounds were fired, but they were sufficient. At a meeting that evening the burghers decided to surrender next morning, in spite of Cronje's passionate pleading for postponement over Majuba Day. The gallant attack of Smith-Dorrien's brigade at dawn put the hour forward, but it was the howitzers which were responsible for the decision, as shown by the provision of a large white flag in every trench.

Like their use of heavy guns in the field, the unexpected appearance of $37^{m/m}$ Vickers-Maxim guns came as a disagreeable surprise to the British Army. Their little one-pound shells might do little actual damage, but the arrival in rapid succession of a belt-full was undoubtedly

Side notes:

CHAPTER XXXVII.

Heavy and Siege Artillery.

Pom-poms and Galloping Maxims.

CHAPTER
XXXVII.

Pom-poms
and
Galloping
Maxims.
disconcerting, and there was an immediate call for similar weapons. These were provided and organised as we have seen, the first joining the army at Paardeberg in February 1900. From that time until the very end of the war they were in constant demand for work with columns.

At first the sections were attached to batteries for convenience of administration. But, in practice, this was found to lead to their being treated as a part of the battery on the march, and even in action, while tactically their rôle was quite separate. The sections were, therefore, soon constituted as independent units, and as such—or preferably as single guns—found their best opportunities when used in free-lance fashion by officers with initiative and tactical vision.

"Galloping Maxims" were also, as we have seen, organised in sections under artillery officers in the Egyptian fashion. But experience showed that, good work as they had done in savage warfare, they offered too large a target to be usefully employed against an enemy armed with modern rifles. In any case the general opinion was that they were under no circumstances an artillery weapon.

In conclusion a word must be said about the employment of their artillery by the enemy. Generally speaking this seemed to be governed by the principles of "long-range, dispersion, and concealment", and might be summed up in the one expression "safety-first".

To obtain the longest possible range they overcharged their guns and sunk their trails, so that the muzzles sticking up in the air became a familiar feature of their positions. But the shell, arriving at a corresponding angle of descent, usually buried themselves before bursting, or, in the case of time shrapnel, were ineffective owing to their low remaining velocity. The dispersion of the guns precluded any systematic concentration or direction of their fire—time and again opportunities of obtaining serious results were missed through this cause. Only at Spion

Kop was any intelligent concentration effected, and there it could scarcely have been avoided. Cover for the detachments was always provided in the vicinity of the guns. "We can always keep down their fire by frightening their men away, but as soon as we stop out they come from their shelter and fire away", is the comment of a battery commander. Concealment was generally cunningly contrived, but often at the sacrifice of field of fire. Resort was frequently had to alternate positions, to which guns were skilfully transferred as soon as our fire became dangerous, but this only accentuated the draw-backs of dispersion. When things were going against them, the guns were withdrawn without a thought of covering the retreat of the infantry. It will be seen, therefore, that at first the Boer "doctrine" was the reverse of ours. A great change took place after the break-up of the Boer Army on the Portuguese Frontier, and the adoption of guerilla warfare. The few guns left to the Boers were used with much more freedom, and were frequently fought to the last with grim determination. In rearguards especially, they covered the retreat of the main body with skill and resolution on many occasions. The reminiscences of a mounted infantryman tell us how the private soldier used to watch what guns de Wet had. "You might be riding along on his track not taking much interest in things, when a fellow says to you 'de Wet called in at such and such a place and nabbed a couple of guns'. Then you take interest in things again, and hope to goodness some other column will work him till he loses those guns".

CHAPTER
XXXVII
Boer
Artillery.

CONCLUSION.

The South African War brought home to the army, and to the nation, the difference between fighting poorly armed savages, however numerous and brave, and waging war with a civilized people, provided with up-to-date weapons. For the first time for more than a generation all arms and all ranks—not forgetting the war correspondents—

experienced the sensation of being under artillery fire, and this caused a curious change in their attitude towards the artillery. Their first reaction was to lay the blame for this novel and disagreeable feature of the war on the shortcomings of the British batteries, and these had to suffer much criticism for their inability to relieve the other arms of the unexpected and unwelcome attentions of the Boer gunners. Later on it took the form of an almost embarrassing eagerness for the assistance of the guns and gunners they had disparaged.

Much of the early censure was cruelly unjust, but it may be admitted that the Royal Artillery—in common with the rest of the Army—had failed to visualise the effects of smokeless powder, and of the increase in the range, accuracy, and rapidity of fire of both guns and rifles. Judged, however, by a higher standard, the Regiment emerged from the ordeal by battle with no stain upon its escutcheon. If its methods and armament were in some particulars old fashioned, artillerymen showed time and again that they deserved, as fully as their predecessors of the Peninsula, the judgment passed on them by the French historian :—''les Canoniers Anglais se distinguent entre les autres soldats par le bon esprit qui les anime. En bataille leur activité est judicieux, leur coup d'oeil parfait, et leur bravoure stoique''. We may well conclude this acount of their services a century later with the equally generous tribute of *The Times*, on the occasion of the unveiling of the Regimental Memorial.

"The inspiring motto of the regiment, ''Ubique'', expresses the Alpha and the Omega of these great services. Each branch of the army, each arm, each department even, has come at some period under adverse criticism. The Artillery alone escaped scathless. The worst that the most severe critics could say of them was that they were equipped with an inferior weapon. Yet at the time when their services were at their greatest they often escaped the notice of those

chroniclers who were fascinated by the presence of
some tradition-bound regiment, some fashionable
battalion. The stereotyped expressions of 'a battery',
or 'the Artillery' or 'the guns' was deemed sufficient
for the units of the Royal Regiment. Even now, when
it is remembered that in the history of our Empire no
great battle has been won, no dire military disaster
averted, without the best efforts of the Artillery, it is
hard to peculiarise from among their services in South
Africa. . . . And their services were not even confined to their legitimate trade as gunners. When the
war was drawing to a close, and men, not guns, were
the essentials, they furnished the Royal Artillery
Mounted Rifles, which scoured the northern Free State
and Western Transvaal from end to end. A glance at
the despatches gives the only tangible evidence of the
Artillerymen's duty nobly done. Nine names are
marked as recipients of the Victoria Cross, while in the
collective despatch 516 officers and 640 non-commissioned officers and men find mention for conspicuous
service. But the performance of these great services
carried a heavy price—a price which the Royal Regiment has not forgotten, a price of which, we hope, the
nation is as sensible as are the comrades of those whose
names swell the heavy tale''.

—The Times.

"There have been incidents in this war which have
not increased our military reputation, but you might
search the classical records of valour and fail to find
anything finer than the consistent conduct of the
British Artillery''.

"It must be acknowledged that for personal gallantry
and for general efficiency they take the honours of the
campaign''.

—The Great Boer War—A. CONAN DOYLE.

The following letter from General Goodenough to Sir R. Biddulph came too late for inclusion in the account of Tel el Kebir given in Chapter XV. But, as the account by the G.O.C., R.A., of his leading, personally, into action the whole artillery of an army corps—three brigades in line —it is of too great historical interest to be omitted. So space has been found for it here, with the kind permission of Brig.-General H. Biddulph and the Society for Army Historical Research.

Tuesday, Sept. 19th, 1882.
Cairo.

"My Dear Biddulph,

. . . For the R.A. all depended on a good start, just to find the rendezvous! I took my Horse Battery, H/1 Indian Contingent, to the rendezvous myself; not so easy to find either. Then I put a marker (my acting Brigade-Major) square to the front of the line, with his back to the West, and dressed, myself, the Battery markers on the Pole Star, as a distant point. In the darkness I soon found it necessary to have a marker for each sub-division, and so I went on till at about ten o'clock I had the whole seven batteries, 42 guns, in one long line, at full intervals. One innocent subaltern said quite simply to me two days ago, 'It was a wonderful thing to see, when daylight came, our line in perfect order all through.' All night from 1.30 I steered the line from the front of the base (left) battery, guided by the stars, but keeping my interval from the 60th Rifles, which necessitated occasional inclines, but these did not upset our line. There was a good halt before we got near; then we went on and presently *felt* we were close up; it was quite dark; four signal shots were fired from some picket, and presently a rattling flash of musketry broke out from a long line in front of us. Almost simultaneously the shout of the Highlanders was heard,

and the firing of musketry was continued and was supplemented by guns, the shells from which we heard booming
away at a comfortable height of, I should think, 50 feet
over our heads. I halted the guns about 700 or 800 yards
from the works, and advanced a bit to get more into the
hollow. There was a moment of anxiety, when we thought
that, had the surprise failed! and should we have to fall
back and cover a re-formation! In a moment however
we saw that the flashes were as much towards the enemy,
as from him; and also that the guns had ceased. Then
we shoved ahead in echelon of Brigade-Divisions from the
centre, and trotted right up; I going on till—plump I found
myself atop of the trench; I could not see it in the dark
and smoke about ten yards off. There was a gap over
which I got; once over, it was suddenly daylight; there
was a column of Highlanders, the Black Watch, reforming
on my left front, and a number of stragglers from them
shooting and bayonetting what had been left in a bastion
on my right; from this line of trench to my right away to
the North, a heavy musketry fire was still proceeding. I
tried to get some Highlanders to flank it; but very soon I
found my good friend Schreiber had got his guns (I/2
Battery) on to the glacis thereof, and was firing case and
shrapnel in enfilade of the line beyond. He was excellent.
The two batteries lost 35 horses killed and wounded, but
only 1 man killed and 10 wounded, happily. I soon made
him stop, as the Abysinnians in front of him were bolting,
and I was afraid of hitting our own 1st Divn. attacking.
Meantime the centre batteries had got over, and one was
presently taken to the hill overlooking Tel el Kebir Station,
and fired into the trains going off, one of which was
taken. . . .

Yours sincerely,

(Sd.) W. H. GOODENOUGH."

SOUTH AFRICAN WAR—APPENDICES.

A. Units Engaged, and Total Strength.

B. Guns Used.

C. Ammunition Expended.

D. Boer Artillery.

E. Order of Battle, April 1900.

F. Distribution of Units, 1901.

G. Pretoria Committee.

H. Mafeking Relief—Special Establishment.

I. Mounted Rifles—Order for Formation.

J. Regimental Commands and Staff Appointments.

K. Army Commands and Staff Appointments.

L. Medals and clasps.

APPENDIX A.

UNITS ENGAGED.

Royal Horse Artillery	... ...	10 Batteries
Royal Field Artillery	... ...	39 „
„ „ „ (Howitzers)	... ...	6 „
Royal Garrison Artillery (Mountain)	... ...	2 „
„ „ „ (Heavy)	... ...	12 „
„ „ „ (Siege)	... ...	8 „
(Militia)	... ...	4 Companies
Royal Artillery (Pom-poms)	... ...	50 Sections
„ „ (Galloping Maxims)	... ...	4 „
Volunteer Artillery	... ...	2 Batteries
Australian Field Artillery	... ...	1 „
Canadian „ „	... ...	3 „
Rhodesian „ „	... ...	3 „
Prince Alferd's Own Cape Artillery	... ...	—
Cape Garrison Artillery	... ...	—
Cape Mounted Rifles (Artillery Troop)	... ...	1 „
Natal Field Artillery	... ...	1 „
Diamond Fields Artillery	... ...	1 „
Ammunition Columns (Cavalry Brigade)	... ...	4 „
„ „ (Divisional)	... ..	13 „
„ „ (Local)	... ...	9 „
„ „ (Artillery Brigade)	... ...	4 „
Ammunition Park	... ...	3 Sections

TOTAL STRENGTH AT VARIOUS PERIODS.

Periods	Regulars		Militia	Volunteers	Colonials	Total
	Home	India				
Garrison on 1st August, 1899	1,035	—	—	—	—	1,035
Reinforcements previous to declaration of war	743	653	—	—	—	1,396
Reinforcements during war						
11/10/99 — 31/ 7/00	14,145	376	617	358	692	16,188
1/8/00 — 31/12/01	2,244	—	289	—	—	2,533
1/6/02 — 31/ 5/02	1,294	—	—	—	—	1,294
Totals	19,561	1029	906	358	692	22,446
	20,590	[1]				

NOTE.—The maximum strength of the artillery in the field was reached about October, 1900, when the *personnel* numbered about 15,000 of all ranks. Over a thousand officers of the Regiment took part in the war.

[1] Units of R.H. & R.F.A. ... ... 12,230 ⎫
    „ „ R.G.A. ... ... 3,490 ⎬ 20,590 — including 600 A.S.C. drivers
   Drafts ... ... 4,870 ⎭   out for duty with Artillery

APPENDIX B.

Guns Used.

9·45″ B.L. Howitzer.

9·2″ B.L. Gun.

6·3″ R.M.L. Howitzer.

6″ Q.F. Gun.

6″ B.L. Howitzer.

5″ B.L. Gun.

5″ B.L. Howitzer.

4·7″ Q.F. Gun—several marks with different breech-mechanisms.

15-pr. B.L. Gun—service pattern.

do. do. —E.O.C. pattern, single motion breech-mechanism.

do. do. — do. · do. steep-coned do. do.

14-pr. do. —Hotchkiss.

12½-pr. Q.F. Gun—Maxim Nordenfeldt.[1]

do. do. — do. C.I.V. Battery.

12-pr. Q.F. Gun—of 12 cwt. Naval pattern.

do. do. —of 8 cwt. do. do.

do. do. —of 12 cwt. Elswick Battery.

12-pr. B.L. Gun—of 6 cwt. R.H.A.

75ᵐ/ᵐ Q.F. Gun—Krupp, two models.[1]

9-pr. R.M.L. Gun.

2·5″ do. do.

7-pr. do. do.

6-pr. Q.F. do.

3-pr. do. do. —Hotchkiss.

37ᵐ/ᵐ Q.F. do. —Vickers-Maxim (Pom-pom).

do. do. —Krupp.[1]

1″ Nordenfeldt.

·450″ Maxim.

---

[1] Captured and taken into use.

APPENDIX C.

AMMUNITION EXPENDED.

NATURE	PERIODS				TOTAL
	12/10/99 to 31/8/00	1/9/00 to 31/12/00	1/1/01 to 31/12/01	1/1/02 to 31/5/02	
6.3" Howitzer	765	—	—	—	765
6" Gun	98	195	24	—	317
6" Howitzer	19	36	—	—	55
5" Gun	3,099	1,067	1,167	147	5,480
5" Howitzer	6,572	1,270	1,288	15	9,790
4.7" Gun	1,838	705	443	490	3,035
15-pr Gun	90,203	19,802	50,089	6,454	116,548
12½-pr Gun	1,495	—	—	—	1,495
12-pr Gun (Q.F.)	2,091	1,344	2,400	308	6,143
„ „ (R.H.A.)	19,347	4,102	12,498	214	36,161
2.5" „	2,897	347	631	—	3,875
Pom-pom	—	—	(1) 166,708	27,129	193,837

(1) This figure is from the first appearance of the pom-poms in February 1900.

APPENDIX E.

ORDER OF BATTLE, APRIL 1900.

	Divisions.	Brigades.	Batteries.	Pom-pom Sections.	Galloping Maxim Sections.
MAIN ARMY	Cavalry	1st Cavalry	T	D	—
		2nd do.	Q	E	—
		3rd do.	R	I	—
		4th do.	O	J	—
	Mounted	1st M.I.	G	C & K	C & D
	Infantry	2nd do.	P	A & B	A & B
		Mahon	M	F	—
	Corps Artillery	R.H.A.	J	—	—
		XII R.F.A.[1]	43, 65, 87.	—	—
		R.G.A.	56, 91.	—	—
	1st	VII R.F.A.	4, 20, 38.		
		VIII do.	37[1]		
			Diamond Fields		
		R.G.A.	31		
	3rd	XV R.F.A.	5, 9, 17.		
	7th	III do.	18, 62, 75.		
	8th	VI do.	74, 77, 79.		
	9th	X do.	76, 81, 82.		
	10th	IX do.	28, 66, 78.		
	11th	XI do.	83, 84, 85.		
	Colonial	—	C. M. R.		
NATAL ARMY		Mounted	A	No. 1.	
	Corps Artillery	R.F.A.	61' 86'	Nos. 2 & 3.	
		R.G.A.	No. 10 M.B.		
			57, 100, 101, 102.		
	2nd	V R.F.A.	7, 63, 64.		
	4th	II do.	21, 42, 53.		
	5th	I do.	13, 67, 69		
	Drakensberg	R.F.A.	19, 73.		
	Defence	R.G.A.	No. 4 M.B.		
LINES OF COMMUNICATION		R.F.A.	44, 88.		
		Volunteer	Elswick		
		do.	C.I.V.		
		Australian	N.S.W.		
		Canadian	C, D, E.		
		R.G.A.	31, 62, 66, 68		
			97, 98, 99,		
			102, 103.		

[1] Field Howitzers.

APPENDIX D.

BOER ARTILLERY.

TRANSVAAL	Field	Fortress	Total
155^{m}/^{m} Creusot guns	—	4	4
150^{m}/^{m} Mortars	—	1	1
,, Howitzers	—	1	1
120^{m}/^{m} Krupp Howitzers	4	—	4
75^{m}/^{m} Creusot guns	6	—	6
,, Krupp guns	8	—	8
,, Nordenfeldt guns	4	—	4
,, Vickers-Maxim guns	4	—	4
15-pr Armstrong guns	1	—	1
12-pr ,, ,,	1	—	1
7-pr ,, ,,	6	—	6
3-pr ,, ,,	3	—	3
65^{m}/^{m} Krupp ,,	4	—	4
37^{m}/^{m} Vickers-Maxim guns (Pom-poms)	22	—	22

ORANGE FREE STATE			
75^{m}/^{m} Krupp guns			14
9-pr Armstrong guns			5
3-pr ,, ,,			3
37^{m}/^{m} Krupp ,,			1

AMMUNITION.

The ammunition obtained with the modern guns was as shown below, but some of this had been expended before the war.

Transvaal.	155^{T} Creusot guns	8,700	rounds.
	120^{T} Krupp howitzers	4,000	,,
	75^{T} Creusot guns	11,000	,,
	do. Krupp do.	5,600	,,
	do. Nordenfeldt guns	2,500	,,
	35^{T} Vickers-Maxim do.	72,000	,,
Orange Free State.	75^{T} Krupp guns	9,000	,,
	9-pr. Armstrong	1,300	,,

APPENDIX G.

PRETORIA COMMITTEE.

Assembled immediately after the occupation of Pretoria to consider the experience gained in the war as regards the Organization, Armament, and Employment of the Artillery.

PRESIDENT :

Lieut.-Colonel F. J. W. Eustace, R.H.A.

MEMBERS :

Lieut.-Colonel T. Perrott, R.G.A.

,,   F. H. Hall, R.F.A.

,,   H. B. Jeffreys, R.H.A.

ASSOCIATE MEMBERS :

Lieut.-Colonel J. S. S. Barker, R.F.A., for Field Howitzers.

Major P. J. R. Crampton, R.G.A., for 1-pr. and Galloping Maxims.

SUB-COMMITTEES :

Under the various members there were formed a series of sub-committees, composed as follows :—

Horse Artillery.—Lieut.-Colonel H. B. Jeffreys, Major B. Burton, Major Sir J. H. Jervis-White-Jervis.

Field Artillery.—Lieut.-Colonel F. H. Hall, Major W. F. L. Lindsay, Major H. G. Smith.

Garrison Artillery.—Lieut.-Colonel T. Perrott, Major E. G. Nicolls, Major R. C. Foster.

Field Howitzers.—Lieut.-Colonel J. S. S. Barker, Major G. M. Yonge-Bateman, Major A. M. Balfour.

Maxims.—Major P. J. R. Crampton, Captain C. Stirling, Captain H. A. Simpson-Baikie.

APPENDIX F.

Distribution of Units, 1901.[1]

Brigade.	Battery.	Armament.	Columns.	Allotment. Posts.
			R.H.A.	
	T	12-pr. 6 cwt.	6 Pulteney	
	Q	do.	4 Knox, 2 Benson	
	G	15-pr.	4 Bethum, 2 Alderson	
	P	12-pr. 6 cwt.	4 Cunningham, 2 Benson	
	R	do.	2 Scobell, 2 deLisle, 2 Herbert	
	O	do.	4 Allenby, 2 Broadwood	
	J	do.	6 Alderson	
	M	do.	2 Henneker	
	U	do.	4 Pilcher	
	A	do.	Gone home.	2 Burghersdorp, 1 Aliwal North.
			R.F.A.	
I	13	15-pr.	2 Gore.	2 Ingogo, 2 Volkenst.
	67	do.	2 Burn-Murdoch	2 Newcastle, 2 Dundee.
	69	do.	—	4 de Jaeger's Drift, 2 Nqutu.
II	21	do.	2 Campbell	4 Middelburg.
	42	do.	—	2 Schoeman's Kloof, 2 Whitclip, 2 Mackadodorp.
	53	do.	—	6 Lydenburg.
III	18	do.	4 Plumer	2 Pienaar's River.
	62	do.	—	2 Norval's Pont, 2 Rhenoster, 1 Hoenig Spruit, 1 Roodeval.
IV	75	do.	—	2 Pretoria, 4 Rustenburg.
	7	do.	2 Cradock	2 Springfontein.
	14	do.	—	2 Bloemfontein, 2 Colesberg, 2 Carnarvon.
V	66	do.	2 Smith-Dorrien	2 Pan, 2 Nooitgedacht.
	63	do.	—	2 Heidelberg, 2 Standerton, 1 Waterval, 1 Greylingstad.
	64	do.	4 Colvile	2 Platrand.
VI	73	do.	—	2 Ladysmith, 2 van Reenan's, 2 Albertina.
	74	do.	4 Dartnell	1 Elandsfontein, 1 Klip River.
	77	do.	4 Harley	2 Vrede.
VII	79	do.	—	4 Harrismith, 2 Bethlehem.
	4	do.	6 Methuen.	—
	38	do.	2 Shekleton, 2 Codrington, 2	—
			Kimberley Column	—
	78	do.	4 Babington, 2 Benson	—

VIII	37	5″ How.	1 Babington, 1 Benson	2 Mafeking, 2 Graaf Reinet.
	61	do.	—	2 Machadodorp, 2 Lydenberg, 1 Taibosch, 1 Meyerton.
IX	65	15-pr.	4 Burn-Murdoch	2 Welgelegen, 2 Vet River, 2 Sanna's Post.
	19	do.	—	2 Koenigsberg.
X	20	do.	—	2 Barberton, 2 Komati Poort.
	28	do.	2 Knox	4 Krugersdorp. 1 Welverdiend, 1 Banks.
	76	do.	2 Thorneycroft, 2 White	2 Thabanchu.
	81	do.	—	2 Pretoria, 2 Bronkhorst Spruit, 2 Wilge River.
XI	82	do.	2 Smith-Dorrien	2 Brandfort, 4 Victoria West.
	83	do.	4 Smith-Dorrien	2 Balmoral, 2 Brugspruit.
	84	do.	2 Crabbe, 2 Williams, 2 Jeffreys	2 Belfast.
	85	5″ How.	2 Pilcher	2 Pretoria.
XII	43	do.	1 Hickman, 1 Byng	3 Bloemfontein.
	86	do.	1 Allenby, 1 Pulteney	2 Standerton, 2 Vredefort Road.
	87	15-pr.	4 Campbell, 2 Williams.	2 Springfontein, 2 Johannesburg.
XIII	2	do.	4 Cunningham	2 Rietfontein.
	8	do.	2 Crewe	2 Boshof, 2 Matjesfontein.
	44	do.	2 Maxwell, 2 Munro, 2 White	2 Bloemfontein.
XIV	39	do.	2 Garland, 2 Kelham	2 Ventersdorp, 2 Kuruman.
	68	do.	2 Benson	2 Ladybrand.
	88	do.	2 Gorringe, 2 Byng	4 Winburg, 2 Hoopstad.
XV	5	do.	—	1 America Siding.
	9	do.	—	1 Armoured Train.
	17	do.	2 Hickman, 3 Crewe	1 Wolvehok, 1 Holtfontein Siding.
Ammunition Park and Depots		12-pr. 12 cwt.	—	1 Witkop, 1 Vereeniging, 1 Kromellenburg, 1 Kroonstad, 1 Pretoria.³
		12-pr. 8 cwt.²	—	
		15-pr.	—	1 Taibosch.
				1 Rooipynt.
Volunteer	C.I.V.	7½″ Krupp	Gone Home.	
do.	Elswick	4·7″	1 Knox, 1 Allenby, 1 Babington	1 Springs, 1 Edenburg. 1 Ventersdorp.
Australian	N.S.W.	12½-pr.	2 Colville	2 Vryburg, 2 Prieska.
Canadian	C	12-pr. 12 cwt.	} Gone home.	
	D	15-pr.		
	E	12-pr. 6 cwt.		
Rhodesian	1	15-pr.*	2 Thorneycroft	2 Zeerust, 2 Lichtenburg, 1 Kokemoor, 1 Ottoshoep.
	2	do.	1 Grenfell, 1 Wood	2 Klerksdorp, 2 Brandfort.
P.A.O.	3	do.		3 Mafeking, 1 Ottoshoep.
C.M.R.		do.	3 Maxwell	2 Warrenton, 1 Koffeefontein, 1 Jacobsdal.
Cape	Field	do.	Gone home.	
Natal Diamond Fields Body Guard	Police	2·5″	2 Dartnell	2 Taungs, 2 Barkly We, 2 Boshof.
B.S.A. Co.		do.	—	, 2 Mafeking.
Provisional Battalion		15-pr. V.M. }	—	2 Van Reenen's.
Brabant's Horse		14-pr. Hotchkiss }	1 Brabant	
Royal Scots		7½″ Krupp	—	1 Smaldeel.

Appendix F.   Distribution of Units, 1901.—(*continued*).

Brigade.	Battery.	Armament.	Columns.	Allotment. Posts.
			R.G.A.	
Mountain	No. 4 No. 10	2·5" 12-pr. 12 cwt.'	1 Campbell	2 Van Reenan's, 2 Nelspruit. 1 Laing's Nek, 1 Ingogo, 1 Whitclip, 1 Laing's Nek.
Siege	91	6" Q.F. 6" How. 5" B.L.	— — —	1 Pretoria. 2 Pretoria, 1 Eland's River, 1 Balmoral. 1 Elandsfontein.
	92	4·7" Q.F. 9·45" How. 4·7" Q.F.	— — —	2 Pretoria. 2 Pretoria. 3 Pretoria.
Heavy	31	2·5" 12-pr. 8 cwt.' 15-pr.'	— — —	2 Vryburg, 1 Kimberley, 3 Orange River. 2 Piquetberg Road, 2 Beaufort West, 2 Barkly West. 1 De Aar.
	56	6" How. 5" B.L.	— —	1 De Aar. 1 Middelburg, 1 Bethulie, 1 Heilbron, 1 Harrismith.
	57	5" B.L. 4·7" Q.F.	1 Smith-Dorrien —	1 Lydenburg, 1 Eden Kop. 1 Greylingstad.
	62	12-pr. 12 cwt.	Gone to China.	1 Graskop.
	63		do.	
	66	4·7" Q.F. 12-pr. 12 cwt.' 15-pr.'	1 Dartnell — —	2 Bloemfontein, 1 Rietfontein. 1 Blomfontein, 1 Glen. 2 Bloemfontein.
	68	6" Q.F. 4·7" Q.F. 12-pr. 12 cwt.' 9-pr. R.M.L.	— — —	1 Pretoria, 1 Bloemfontein. 1 Bloemfontein. 1 Bloemfontein. 2 Aliwal North

No.	Gun	Smith-Dorrien / Burn-Murdock	Locations
97	2·5"	2 Smith-Dorrien	2 Aliwal North.
	5" B.L.		1 Komati Poort, 1 Pietersberg.
98	do.	—	1 Zand River, 1 Rhenoster, 2 Kroonstad.
99	4·7" Q.F.[²]	1 Burn-Murdock	1 Standerton, 1 Paardekop.
	12-pr. 12 cwt.		1 Volksrust, 1 Houtnek, 1 Opperman's Krad, 1 Newcastle, 1 Dublin Hill.
100	4·7" Q.F.[²]	—	1 Middelburg.
	12-pr. 12 cwt.		1 Machadodorp, 1 Dalmanutha, 1 Utrecht, 2 Vryheid, 1 Doornberg.[²]
101	4·7" Q.F.	—	1 Mackadodorp, 1 Wakkerstroom, 1 Standerton.
	12-pr. 12 cwt.		1 Heidelberg, 1 Paardekop, 1 Standerson.
102	4·7" Q.F.	—	1 Newcastle, 1 Dundee, 2 Talana Hill.
	12-pr. 12 cwt.		2 Wakkerstroom.
103	6" Q.F.[³]	—	1 Pretoria.
	4·7" Q.F.[³]		1 Krugersdorp, 1 Frederikstad.
Militia	12-pr. 12 cwt.[²]	—	2 Naauwpoort.
Antrim	9-pr. R.M.L.		2 Naauwpoort, 2 Aliwal North.
Donegal	2·5"		1 Vryburg, 1 Kimberley, 3 Orange River, 1 Windsorton Road.
Cape Garrison Artillery	4·7"	—	2 Capt Peninsula.
	12-pr. 12 cwt.		1 Cape Peninsula, 4 Armoured Trains.
	15-pr.[²]		2 Rist River, 2 Kaffir River.
	15-pr.[²]		2 Modder River.
	3-pr.		1 Bethulie, 1 Christiania, 1 Armoured Train.
In Local Charge	6" How.	—	1 Orange River, 1 De Aar, 1 Aliwal North.
	5"		1 Bethulie, 2 Cape Peninsula (special carriages).
	5" How.		1 Kimberley (Long Cecil).
	4·7"		1 Rooipynt.
	12-pr. 12 cwt.		1 Naauwpoort, 1 Graaf Reinet.
	9-pr.		1 Kimberley, 2 Naauwpoort, 2 Aliwal North.
	6-pr.		1 Taungs, 1 Vryburg, 2 Armoured Trains (Cape Peninsula).
	3-pr.		2 Christiania, 2 Orange River, 1 Bethulie.

[1] Pom-pom sections omitted for considerations of space.
[2] Horsed.
[3] Taken over from Navy.
[4] Left by Canadian Artillery.
[5] Brought from India by "J"/R.H.A.
[6] E.O.C. Pattern.
[7] On Railway mountings.

APPENDIX H.

Mafeking Relief—Special Establishment.

O.C.,
  R.H.A. Brigade-division,
    Corps Artillery.

1. "M" battery and one section of your ammunition column will proceed to Kimberley as soon as railway carriage can be provided.
2. The following will be the establishment:—

	BATTERY		COLUMN	
	Horses	Mules	Horses	Mules
Detachments	60	—	—	—
Serrefiles	10	—	12	—
Guns	36	—	—	—
Ammunition Wagons	18	24	—	24
A & S Wagons	—	24	—	—
Spare gun carriage	—	—	—	6
Forge	—	—	—	8
Water Cart	—	—	—	4
	124	48	12	42

            Spare   64 horses and 20 mules
            Total 200 horses and 110 mules

3. The horses will be made up in the brigade-division, the mules and mule harness from the 9th divisional artillery. The large proportion of spare horses is to provide relief teams. Only 3 wagons are to be horsed, and harness need only be carried on the usual proportion of spare horses.
4. The ammunition column will be made up to the following *personnel* in your brigade-division:—

    1 officer, 1 R.S.M., 1 B.S.M., 1 Sergt., 1 Farr. Sergt., 1 S.S., 1 Cr.-Mr., 1 Whr., 1 Trpt., 2 Corporals, 2 Bombrs., 10 Grs., 12 Drs. = 35.

5. The C.-in-C. has sanctioned the following establishment here for the ammunition column of a single battery. This, however, is on the assumption that operations would be in the vicinity of a railway and that further supply is on the rail in R.A. charge. For this latter purpose 400 rounds a gun has been ordered to be sent to Kimberley by the P.O.O. Cape Town.

AMMUNITION COLUMN.

3 ammunition wagons carrying 300 rounds or 50 per gun.
5 buck wagons   do.   600   do.   do. 100 do. do.
2 ox   do.   do.   600   do.   do. 100 do. do.
and one buck wagon for supplies.

The 200 rounds per gun for the buck and ox wagons must be taken with you—the G.O.C. Kimberley has been asked to provide the transport.

By order,

D.A.A.G., R.A.

18/4/00.

APPENDIX I.

MOUNTED RIFLES—ORDER FOR FORMATION.

G.O.C. ——

1. The General Commanding-in-Chief has decided to employ the personnel of howitzer batteries as Artillery Mounted Rifles.
2. Each battery will form a company under its own officers. It will be known by its battery number, will continue the present pay-list, ledger, &c., and will retain the present organization by sections and subdivisions. All ranks will be armed with rifles and bayonets.
3. The companies will be mounted on their own battery horses, except that the heavier horses will be exchanged locally as far as possible for any more suitable which there may be in contiguous batteries or in the local ammunition column.
4. The attached lists show the articles of kit, &c., required: any articles required to complete may be drawn from Ordnance on the authority of this memorandum.
5. The equipment, baggage, rations, and forage will be carried on the lines laid down for a squadron of cavalry in Appendix N of "Transport Organization, South Africa—February 1901" except that the regimental reserve of S.A.A. will be 200 instead of 130 rounds. For this purpose the three A & S wagons (or other mule wagons) and water cart which form part of each battery will be retained with each company, and the scotch cart will be provided by the local ammunition columns.

   If it is required to carry rations and forage for more than two days, buck wagons for that purpose must be obtained from the Transport.
6. Ammunition will be carried as follows:—

   On the man —in bandoliers   —150 rounds.
   Per section   —on a pack horse — 2 boxes.
   Per company—in a scotch cart — 14 do.

   Making a total of 350 rounds per rifle.
7. The howitzers and equipment[1] now in possession of batteries which will be no longer required on their conversion will be returned to Ordnance Store. Any horses, mules, or mule wagons which may become surplus will similarly be handed over to the local ammunition column.

8. On conversion each battery should establish a small depot, where those men, if any, who are quite unsuited for mounted work in the field should be left.
9. Will you kindly arrange the assembly of the howitzer batteries in your district; and issue orders for the commencement with least possible delay of mounted drill and rifle practice—these should not be delayed until the assembly of the battery.

---

[1] Range-finding instruments will be retained with the companies.

APPENDIX J.

Regimental Commands and Staff Appointments.

Note.—It has not been found possible to follow all the changes of rank which took place during the course of operations owing to the abnormal run of regimental promotion and the many brevets awarded. Nor has it been feasible to indicate the cases where batteries became separated from their proper brigades.

Headquarters.

G.O.C., R.A.	—	Major-General G. H. Marshall.
A.-D.-C.	—	Captain A. D. Kirby.
A.A.G.	—	Lieut.-Colonel H. C. Sclater.
D.A.A.G.	—	Major J. E. W. Headlam.
Attached	—	Captain R. A. Bright.
		Lieut. D. Smith.
A.A.G., R.H.A.	—	Lieut.-Colonel F. J. W. Eustace.
Staff Officer, O.F.S.	—	Colonel R. Bannatine-Allason.
do. do. Natal	—	Major J. A. Tyler.

Natal (Ladysmith).

Colonel-on-the-Staff	—	Colonel C. M. H. Downing.
Staff Officer	—	Captain E. S. E. W. Russell.
Orderly Officer	—	Captain R. A. Bright.

Natal Army.

Colonel-on-the-Staff	—	Colonel C. J. Long, J. A. F. Nutt, L. W. Parsons.
Staff Officer	—	Captain G. F. Herbert, R. W. Boger.

Corps Artillery.

Colonel-on-the-Staff	—	Colonel W. L. Davidson.
Staff Officer	—	Captain H. H. Tudor.

Royal Garrison Artillery.

Colonel-on-the-Staff	—	Colonel R. W. Rainsford-Hannay.
Staff Officer	—	Major S. B. von Donop.

Brigades and Batteries.

R.H.A.

Brigade Commander	— Lieut.-Colonel	F. J. W. Eustace, B. Burton, R. Bannatine-Allason.
Adjutant	— Captain	A. d'A. King, F. W. Heath, H. J. Brock.
"O" Battery	— Major	Sir J. H. Jervis-White-Jervis.
"R" do.	— ,,	B. Burton, H. S. Horne, B. F. Drake.
Ammunition Column	— ,,	C. E. Maberly, H. S. Horne, E. H. Armitage, M. B. G. Jackson.
do. do.	— ,,	S. Belfield, F. W. Heath, R. England.

Brigade Commander	— Lieut.-Colonel	W. L. Davidson, E. M. Flint.
Adjutant	— Captain	G. W. Biddulph, H. F. Askwith, A. T. Butler.
"G" Battery	— Major	R. Bannatine-Allason, H. F. Mercer.
"P" do.	— ,,	Sir G. V. Thomas, A. B. Scott.
Ammunition Column	— Captain	J. P. DuCane, G. Baillie, G. R. F. R. Talbot.

Brigade Commander	— Lieut.-Colonel	A. N. Rochfort, H. B. Jeffreys, Sir G. V. Thomas, A. H. C. Phillpotts.
Adjutant	— Captain	W. B. Norwood, J. C. Wray, J. F. N. Birch, H. H. Tudor.
"Q" Battery	— Major	E. J. Phipps-Hornby, G. Humphreys.
"T" do.	— ,,	F. B. Lecky.
"U" do.	— ,,	P. B. Taylor, H. M. Campbell.
Ammunition Column	— ,,	E. C. F. Holland, P. Wheatley, G. Humphreys.

"A" Battery	— Major	E. A. Burrows, W. L. H. Paget.
"J" do.	— ,,	P. H. Enthoven, L. H. Ducrôt.
"M" do.	— ,,	H. K. Jackson, E. H. Armitage.
Ammunition Column	— Captain	P. E. Gray, S. F. Gosling.

R.F.A.

I Brigade.

Brigade Commander	— Lieut.-Colonel	E. H. Pickwoad, E. S. May, E. A. Burrows.

Adjutant	— Captain	J. A. Tyler.
13th Battery	— Major	J. W. G. Dawkins.
67th do.	— ,,	J. F. Manifold.
69th do.	— ,,	F. D. V. Wing.
Ammunition Column	— ,,	E. S. May, H. W. A. Christie.

II Brigade.

Brigade Commander	— Lieut.-Colonel	J. A. Coxhead, W. E. Blewitt, R. A. G. Harrison, R. E. Boothby.
Adjutant	— Captain	A. L. Walker, H. L. Reed, C. Evans.
21st Battery	— Major	W. E. Blewitt, G. F. Herbert, H. Corbyn.
42nd do.	— ,,	C. E. Goulburn, R. G. Ouseley.
53rd do.	— ,,	A. J. Abdy, L. G. F. Gordon.
Ammunition Column	— Captain	R. G. Ouseley, H. Corbyn, G. J. C. Stapylton.

III Brigade.

Brigade Commander	— Lieut.-Colonel	F. H. Hall.
Adjutant	— Captain	F. B. Johnstone, L. M. Phillpotts.
18th Battery	— Major	A. B. Scott, G. Humphreys, H. E. Stockdale.
62nd do.	— ,,	E. J. Granet, W. H. O'Neill.
75th do.	— ,,	W. F. L. Lindsay, J. F. N. Birch, N. E. Young.
Ammunition Column	— ,,	N. E. Young, J. H. Johnstone.

IV Brigade.

Brigade Commander	— Lieut.-Colonel	H. V. Hunt, R. A. G. Harrison.
Adjutant	— Captain	H. D. White-Thomson.
7th Battery	— Major	C. G. Henshaw.
14th do.	— ,,	A. C. Bailward, F. W. Heath.
66th do.	— ,,	W. Y. Foster, C. C. Owen, A. G. Glanville.
Ammunition Column	— .,	W. A. Smith.

V Brigade.

Brigade Commander	— Lieut.-Colonel	L. W. Parsons, W. L. H. Paget, E. A. Burrows, T. S. Baldock, C. E. Coghill.
Adjutant	— Captain	R. W. Boger, H. G. Sandilands, J. S. Ollivant.
63rd Battery	— Major	W. L. H. Paget, R. F. Fox, C. H. de Rougemont, G. Campbell-Johnston.
64th do.	— ,,	C. E. Coghill.
73rd do.	— ,,	C. M. Barlow, N. E. Young.
Ammunition Column	— ,,	N. D. Findlay, G. H. W. Nicholson.

VI Brigade.

Brigade Commander	— Lieut.-Colonel	H. B. Jeffreys, A. S. Pratt, R. E. Boothby, W. H. O'Neill.
Adjutant	— Captain	S. W. W. Blacker, C. E. Lawrie, A. G. Glanville, W. P. L. Davies.
74th Battery	— Major	R. G. McQ. McLeod.
77th do.	— ,,	E. M. Perceval, G. H. Geddes.
79th do.	— ,,	E. H. Armitage, R. F. McCrea.
Ammunition Column	— ,,	R. F. McCrea, G. Lewis, W. P. L. Davies, Sir J. Keane.

VII Brigade.

Brigade Commander	— Lieut.-Colonel	P. C. Newbigging, C. T. Blewitt, Hon. A. Sidney.
Adjutant	— Captain	E. C. Cameron, A. F. R. Thomson, W. Evans.
4th Battery	— Major	A. E. A. Butcher, H. Rouse.
38th do.	— ,,	H. E. Oldfield, H. G. Burrowes.
78th do.	— ,,	D. C. Carter, A. M. Kennard.
Ammunition Column	— ,,	W. H. Connolly, H. G. Burrowes.

VIII Brigade.

Brigade Commander	— Lieut.-Colonel	J. S. S. Barker, C. M. Barlow, W. Hanna, G. M. Yonge-Bateman.
Adjutant	— Captain	E. J. Duffus.
37th Battery	— Major	R. A. K. Montgomery, C. M. Ross-Johnson.
61st do.	— ,,	A. H. Gordon, E. J. Duffus.
65th do.	— ,,	W. Tylden, G. M. Yonge-Bateman.
Ammunition Column	— Captain	G. F. MacMunn.

IX Brigade.

Brigade Commander	— Lieut.-Colonel	A. J. Montgomery, W. A. Smith, C. V. B. Kuper, A. d'A. King.
Adjutant	— Captain	C. N. B. Ballard.
19th Battery	— Major	H. A. D. Curtis, R. D. Gubbins.
20th do.	— ,,	C. H. Blount, A. d'A. King.
28th do.	— ,,	A. Stokes, L. H. Ducrôt, J. W. M. Newton, A. B. Helyar, P. W. Game.
Ammunition Column	— ,,	J. R. Foster, K. Combe.

X Brigade.

Brigade Commander	— Lieut.-Colonel	J. McDonnell, F. Waldron.
Adjutant	— Captain	W. H. Onslow, C. F. Stevens, R. S. Hardman.
76th Battery	— Major	R. A. G. Harrison, H. M. Campbell, E. H. Paterson.
81st Battery	— ,,	H. A. Chapman, G. G. Simpson, E. S. Cleeve.
82nd do.	— ,,	A. S. Pratt, W. H. Connolly.
Ammunition Column	— ,,	A. Bell-Irving, R. S. Hardman.

XI Brigade.

Brigade Commander	— Lieut.-Colonel	E. M. Flint, A. Bell-Irving.
Adjutant	— Captain	A. T. Butler, H. E. Stockdale, E. P. Smith.
∧ 83rd Battery	— Major	W H. Darby, H. G. Smith, A. D. Young.
84th do.	— ,,	E. Guinness, F. H. Ward.
85th do.	— ,,	W. H. Williams, A. H. Hussey.
Ammunition Column	— ,,	G. G. Simpson, W. C. Staveley.

XII Brigade.

Brigade Commander	— Lieut.-Colonel	F. Waldron, J. S. S. Barker.
Adjutant	— Captain	P. T. C. Herbert, R. E. A. LeMottée.
43rd Battery	— Major	G. R. T. Rundle, R. Casement.
86th do.	— ,,	C. D. Guinness.
87th do.	— ,,	A. M. Balfour, C. J. U. Morris.
Ammunition Column	— ,,	S. E. G. Lawless.

XIII Brigade.

Brigade Commander	— Lieut.-Colonel	L. J. A. Chapman, J. W. Dunlop.
Adjutant	— Captain	A. H. S. Goff.
2nd Battery	— Major	P. H. Slee.
8th do.	— ,,	H. Chance.
44th do.	— ,,	B. F. Drake.
Ammunition Column	— Captain	T. S. Hichens.

XIV Brigade.

Brigade Commander	— Lieut.-Colonel	S. Watson, E. H. Pickwoad.
Adjutant	— Captain	E. S. Nairne.
39th Battery.	— ,,	J. Hanwell, S. E. G. Lawless.
68th do.	— Major	W. G. Massey.
88th do.	— ,,	G. W. Biddulph, W. H. Olivier.
Ammunition Column	— Captain	R. St.G. Gorton, E. V. D. Riddell.

XV Brigade.

Brigade Commander	— Lieut.-Colonel	H. H. Pengree.
Adjutant	— Captain	E. C. Sandars.
5th Battery	— Major	S. W. Lane.
9th do.	— ,,	A. S. Wedderburn, A. T. Butler.
17th do.	— ,,	T. K. E. Johnston.
Ammunition Column	— Captain	G. A. Skipwith, T. B. Wood, W. C. Symon.

Ammunition Park.

Commander	— Colonel	H. T. Lugard, Captain A. S. Buckle.
Adjutant	— Captain	H. G. Smith.
No. 1 Section	— Major	H. M. Campbell.
No. 2 do.	— ,,	C. T. Blewitt.
No. 3 do.	— ,,	H. G. Burrowes.

Volunteer Batteries.

Elswick	— Major	H. Scott, Captain C. J. U. Morris.
C.I.V.	— ,,	G. McMicking, Captain C. E. D. Budworth.

COLONIAL FIELD ARTILLERY.

Australian.

N.S.W. Battery	— Colonel	S. C. U. Smith.

Canadian.

Brigade Commander	— Lieut.-Colonel	C. W. Drury.
Adjutant	— Captain	H. C. Thacker.
C Battery	— Major	J. A. G. Hudon.
D do.	— ,,	W. G. Hurdman.
E do.	— ,,	G. H. Ogilvie.

Rhodesian.

Brigade Commander	— Lieut.-Colonel	G. Wright.
Adjutant	— Captain	E. B. Macnaghten.
1st Battery	— Major	A. ff. Powell.
2nd do.	— ,,	A. Paris, R.M.A.
3rd do.	— Captain	G. E. Giles.

South African.

P.A.O. Cape Artillery

Cape Garrison Artillery

Artillery Troop Cape Mounted Rifles	— Major	H. T. Lukin.
Natal Field Artillery	— Major	D. W. G. Taylor.

Diamond Fields Artillery

Local Ammunition Columns.

No. 1.	Bloemfontein.	(XV Brigade, R.F.A.)	—Captain W. C. Symon, A. S. P. McGhee.
No. 2.	Pretoria.	(III do. do.) —	do. C. StM. Ingham.
No. 3.	Middelburg.	(II do. do.) —	do. H. E. Carey.
No. 4.	Harrismith.	(VI do. do.) —	do. Sir J. Keane.
No. 5.	Belfast.	(IV do. do.) —	do. G. J. C. Stapylton.
No. 6.	Kroonstad.	(X do. do.) —	do. R. S. Hardman, H. L. Griffin.
No. 7.	Newcastle.	(I do. do.) —	do. H. W. A. Christie.
No. 8.	Krugersdorp.		—Lieut. & Qr.-Mr. W. Bass.
No. 9.	Mafeking.		—Major A. ff. Powell.

R.G.A.

(With designation at the beginning of the War.)

Mountain Artillery.

No. 4 Battery	— Major	H. C. C. D. Simpson, G. F. W. St.John.
No. 10 do.	— ,,	G. E. Bryant, T. R. C. Hudson, P. de S. Burney.

Siege Train.

Commander	— Lieut.-Colonel	T. Perrott, E. F. Johnson.
Adjutant	— Captain	A. C. Currie, C. R. Buckle, W. J. Napier, G. R. H. Nugent.
91st Coy. (15/S.)		
Right Half	— Major	J. R. H. Allen, F. J. Graeme.
Left Half	— Captain	M. B. Roberts, H. S. de Brett.
92nd do. (15/W.)	— Major	E. G. Nicolls.

Cape Peninsula.

Commander	— Colonel	C. M. Western.
Adjutant	— Captain	R. H. F. McCulloch.

Cape Town.

Commander	— Lieut.-Colonel	E. J. K. Priestley.
Adjutant	— Captain	E. A. Saunders, A. E. C. Myers.
Instr. Range-Finding	— Captain	C. G. E. Stevens-Nash.
68th Coy. (14/W.)	— Major	H. de T. Phillips.
31st do. (23/W.)	— ,,	G. D. Chamier, A. R. Stuart.

Natal.

Commander	— Lieut.-Colonel	P. Saltmarshe, F. A. Curteis. H. C. C. D. Simpson.
Adjutant	— Captain	A. J. Budd, J. B. Parry.

Heavy Batteries.

56th	Company —	(36/S.)	Major	R. C. Foster, T. F. Bushe.
57th	do.	— (16/S.)	do.	C. E. Callwell.
62nd	do.	— (2/S.)	do.	T. W. Powles.
63rd	do.	— (22/S.)	do.	F. A. L. Powell.
66th	do.	— (5/E.)	do.	N. B. Inglefield.
97th	do.	— (14/S.)	do.	W. L. Brooksmith, H. P. Hickman.
98th	do.	— (17/W.)	do.	M. B. G. Jackson, W. H. Darby, O. C. Williamson.
99th	do.	— (10/W.)	do.	F. E. Kent.
100th	do.	— (2/W.)	do.	F. A. Curteis.
101st	do.	— (6/W.)	do.	G. J. F. Talbot, E. F. Nelson.
102nd	do.	— (10/E.)	do.	C. E. Jervois.
103rd	do.	— (6/E.)	do.	A. B. Shute.

POM-POM SECTIONS.

Inspector - Major P. J. R. Crampton.
Depot - W. H. Robinson.

A. 1 Hon. N. A. Hood
2 C. Stirling, W. L. Foster, E. F. Calthrop
B. 1 F. R. Sedgwick, H. A. D. Simpson-Baikie
2 E. H. Stevenson, G. R. H. Nugent, H. E. Street
C — J. G. Rotton, R. S. de Winton
D — F. H. F. R. McMeekan, E. W. M. Powell, W. H. Moore, A. M. Fox
E — C. H. Ziegler, C. St. M. Ingham, H. R. W. M. Smith
F — W. H. Robinson, G. Baillie
G. 1 A. E. C. Myers, G. J. E. Stapylton
2 G. D. Wheeler
3 E. M. Connolly, F. L. C. Livingstone-Learmonth, B. Vincent
H. 1 H. B. Dodgson, F. A. Wilson
2 G. L. H. Howell
J — H. L. Griffin, T. R. C. Hudson
K — M. J. S. Dennis, W. A. S. Gemmell
L — J. C. Burnett
M. 1 W. F. T. Corrie
2 P. R. Wheatley, G. H. Rickard
N. 1 J. W. Kempson, A. A. Montgomery, E. C. Harrington
2 W. C. Symon, H. R. Peck
O. 1 H. H. Harvest, F. R. Patch, P. L. Holbrooke
2 C. E. Hill, S. F. Gosling
P. 1 A. Ellershaw, F. C. Poole
2 I. de L. Pollard-Lowsley, C. W. Scott
Q. 1 J. G. Rotton
2 P. H. Wilson
R. 1 G. W. Brierley, G. V. Davidson
2 J. F. A. Higgins

Pom-Pom Sections—*Continued.*

S.	1	H. Rouse, H. de B. Miller
	2	C. B. Simonds
T.	1	S. T. Hardwick
	2	C. D. Hope
U.	1	C. E. Hill
	2	H.E. Pilkington (N.Z.A.), A. R. G. Begbie, J. A. P. Robinson
¹ V	—	C. F. Brace, (A.A.)
¹ W	—	C. W. R. Dalyell, E. P. Smith, W. J. Maxwell-Scott (R. Scots)
X.	1	H. J. P. Jeffcoat, E. H. Phillips
	2	R. C. Littledale, W. C. Symon
Y	—	H. J. Brock, A. E. B. Fair, H. Allcard
Z	1	F. H. G. Stanton
	2	A. M. R. Mallock
A/A	—	E. C. W. D. Walthall, G. L. H. Howell, H. E. Carey
B/B	—	L. M. Dyson
C/C	—	S. R. Normand
D/D	—	E. B. Macnaghten
E/E	—	H. de B. Miller
F/F	—	R. O. Marton
G/G	—	R. T. Hill
H/H	—	S. F. Gosling, R. Geoghegan
J/J.	1	A. W. Disney-Roebuck, G. A. Hare
	2	R. E. Ramsden
K/K	—	L. M. Dyson
L/L	—	C. A. Ker, W. G. H. Salmond

GALLOPING MAXIMS

A	—	E. H. Stevenson
B	—	G. W. Brierley
C	—	S. T. Hardwick
D	—	H. A. D. Simpson-Baikie

---

¹ Armoured Train.

APPENDIX K.

ARMY COMMANDS AND STAFF APPOINTMENTS.

Commander-in-Chief—Field-Marshal LORD ROBERTS.

Divisional Commander—Lieut.-General SIR H. M. L. RUNDLE.

Column Commanders.

Colonel	J. S. S. Barker.	Lieut.-Colonel	Sir J. H. Jervis-White-Jervis.
Lieut.-Colonel	T. S. Baldock.		
,,	A. Bell-Irving.	,,	J. L. Keir.
Colonel	G. E. Benson.	Colonel	Sir C. S. B. Parsons.
Lieut.-Colonel	C. E. Callwell.	,,	A. N. Rochfort.
,,	J. W. Dunlop.	Lieut.-Colonel	A. B. Scott.
,,	J. W. G. Dawkins	Major	S. B. von Donop.
Major	J. P. DuCane.	Lieut.-Colonel	F. D. V. Wing.
,,	F. W. Heath.	,,	W. H. Williams.
Colonel	H. B. Jeffreys.	,,	W. L. White.

Local Commanders.

Major	Hon. H. W. Addington.	Major	W. H. Mills.
,,	G. H. Balguy.	Brig.-General	J. W. Murray.
,,	G. D. Chamier.	Major-General	G. T. Pretyman.
Colonel	L. J. A. Chapman.	Colonel	Sir C. S. B. Parsons.
,,	W. L. Davidson.	Lieut.-Colonel	Hon. A. Sidney.
,,	C. M. H. Downing.	,,	H. C. C. D. Simpson.
,,	F. H. Hall.	Major	F. G. Stone.
,,	R. A. G. Harrison.	,,	A. R. Stuart.
Major	H. S. Jeudwine.	,,	F. W. G. Tothill.
,,	R. M. B. F. Kelly.	Lieut.-Colonel	G. Wright.
Major-General	W. G. Knox.	Major	C. C. Wiseman-Clarke.
Lieut.-Colonel	E. A. Lambart.		

A.A.G. & D.A.A.G.

Colonel	J. Adye.	Captain	F. L. C. Livingstone-Learmonth.
Captain	J. G. Baldwin.		
,,	C. R. Buckle.	Lieut.-Colonel	A. W. Money.
Lieut.-Colonel	H. J. DuCane.	Major	G. F. Milne.
Major	W. E. Fairholme.	,,	R. A. K. Montgomery.
Captain	W. T. Furse.	,,	W. R. N. Madocks.
,,	G. T. Forestier-Walker.	,,	G. F. MacMunn.
Colonel	J. M. Grierson.	,,	R. H. Massie.
Lieut.-Colonel	E. J. Granet.	Captain	A. A. McHardy.
,,	A. H. Gordon.	Major	J. W. M. Newton.
,,	C. deC. Hamilton.	Captain	S. W. Robinson.
Major	G. F. Herbert.	,,	H. L. Reed.
,,	C. V. Hume.	Major	W. H. F. Taylor.
Captain	H. C. T. Hildyard.	Colonel	J. K. Trotter.
Major	N. B. Inglefield.	Major	F. J. A. Trench.
Lieut.-Colonel	J. T. Johnston.	Lieut.-Colonel	W. H. H. Waters.
Colonel	H. T. Lugard.		

Staff Officers.

Major	D. Arbuthnot.	Captain	A. A. Montgomery.
,,	A. M. Balfour.	,,	E. S. Nairne.
,,	A. E. A. Butcher.	,,	J. S. Ollivant.
,,	A. T. Butler.	,,	E. H. T. Parsons.
,,	G. D. Baker.	,,	H. R. Peck.
Captain	R. A. Bright.	Major	C. M. Ross-Johnson.
,,	W. E. Clark.	Captain	E. V. D. Riddell.
,,	E. F. Calthrop.	,,	G. H. Riach.
Lieutenant	L. W. laT. Cockcraft	,,	H. B. Roberts.
,,	A. N. Campbell.	,,	C. F. Rugge-Price.
Captain	A. H. N. Devenish.	,,	E. A. Saunders.
,,	A. D. Freeland.	Major	C. Stirling.
,,	F. W. Gosset.	,,	F. H. G. Stanton.
,,	R. StG. Gorton.	Captain	C. B. Simonds.
,,	E. W. Grove.	Major	W. A. M. Thompson.
Major	C. E. H. Heyman.	,,	J. A. Tyler.
Captain	V. R. Hine-Haycock.	Captain	E. F. Talbot-Ponsonby.
,,	K. J. Kincaid-Smith.	,,	H. H. Tudor.
,,	R. E. A. Le Mottée	,,	B. Vincent.
,,	C. B. Levita		

Personal Staff.

Captain	J. B. Aldridge.	Major	R. M. B. F. Kelly.
Lieutenant	H. W. Atlay.	Captain	A. D. Kirby.
Colonel	H. V. Cowan.	,,	W. E. Kemble.
Major	F. E. Cooper.	,,	H. T. Russell.
,,	D. J. M. Fasson.	Major	H. N. Schofield.
Captain	H. W. M. Harrington.	Lieutenant	F. M. C. Trench.
,,	R. H. Hare.		

In addition to the above a considerable number of artillery officers served in the Transport and Remount Departments, and as regimental officers with the Imperial Yeomanry and Colonial Corps.

APPENDIX L.

MEDALS & CLASPS.

Two medals were issued, generally known as the "Queen's"[1] and the "King's" medals. The former was granted for the whole of the war, the latter (in addition) to those who had served 18 months in South Africa and were there in 1902.

The following clasps were awarded :—

To the Queen's medal.

Belfast, Belmont, Cape Colony, Diamond Hill, Driefontein, Elandslaagte, Johannesburg, Defence of Kimberley. Relief of Kimberley, Defence of Ladysmith, Relief of Ladysmith, Laing's Nek, Defence of Mafeking, Relief of Mafeking, Modder River, Natal, Orange Free State, Paardeberg, Rhodesia, Talana, Transvaal, Tugela Heights, Wepener, Wittebergen.

To the King's medal.

South Africa 1901, South Africa 1902.

The clasps "South Africa 1901" and "1902" were also awarded in certain cases with the Queen's medal.

---

[1] Queen Victoria died in January, 1901.

CORRIGENDA.

Vol. I.

Page 2, line 26—*For* "Board of Control" *read* "Board of Ordnance".

,, 28, line 2 from bottom—*For* "Sir R. Whish" *read* "Sir W. S. Whish".

,, 58, the last two lines should read—"took the troop out to the Peninsula in 1809, but only reached the army on the 2nd August, too late to take part in the battle of".

,, 134, line 12—*For* "Jersey" *read* "Guernsey".

,, 263, last line—*For* "Lushai Campaign of 1868" *read* "Bhutan Campaign of 1864-5."

Vol. II.

Page 6, Table—In first column *enter* "10", opposite "Royal Military College, Canada"; and correct totals accordingly.

,, 10, last line—*Dele* "on the . . . . fought".

,, 25, line 26—*For* "Borden" *read* "Bordon".

,, 153, line 15—*For* "unarmed" *read* "unharmed".

,, 348, line 3 from bottom—*For* "countries" *read* "counties".

,, 379, ,, 5 ,, ,, —*For* "made" *read* "may".

,, 384, first line—*For* "elecrticity" *read* "electricity".

,, 402, line 13—*For* "part" *read* "parts".

,, 432, line 6—*Dele* "on".

,, 461, line 8—*After* "77" *add* "353".

LIST OF UNITS, ROYAL ARTILLERY, 1914.

(with Chapters in which mentioned).

R.H.A.

"B". XI.
"K". VII.
"M". XV, XVII.
"S". I.
"W". I.
"Z". XV.

R.F.A.

3. VII.
5. XV.
9. VII, XV.
10. VII.
11. X.
12. X, XI.
15. VI.
18. VII.
19. XIII, XXIII.
26. XV.
29. XV.
30. XV.
32. XVIII.
35. I, XV.
37. XVIII.
38. IX, XII.
43. X.
44. X.
45. XVII.
51. VII.
57. VII.
59. I.
62. X.
71. XV.
72. XV.
76. XII, XXIII.
80. VII.
86. XXIII, XXIV.
89. XXIII.
120. XVI.

R.G.A.—(Mountain).

1. I, VI, VII.
2. I, IX.
3. VI, VII, VIIII, IX.
4. VIII.
5. IX.

R.G.A.—(Mountain)—Continued.

6. VI, IX.
7. VI, VII, VIII.
8. VI, VII, VIII.
9. VI, VII, IX.

R.G.A.

1. IX.
7. IX.
16. VIII.
22. IX.
33. XIV.
54. VIII.
57. XIV, XVII.
64. X.
66. XVII.
70. XII.
77. X.
87. IX.
88. XIV.
90. IX, XIV.
91. IX.
93. XXIII.
94. XVIII, XXIII.
95. XXIII.
102. X.
107. VIII.
H.K.—Singapore XI.

Indian Mountain.

21 Kohat (F.F.) I, VI, VII, VIII.
22 Derajat do. I, VI, VII, VIII.
23 Peshawar do. I, VI, VII, VIII.
24 Hazara do. I, VI, VII, VIII, IX.
25 Bombay VI VII. IX, XIV.
26 Jacob's VII, IX.
27 VI, VII, IX.
28 VII, IX, XIX.
29 VII.
30 VII, VIII.
31 VII.
Frontier Garrison Arty. (F.F.). VI.

Reduced Units.

6/25. VIII.
53 R.G.A. I.
92 ,, IX, XV.
103 ,, I.

551

INDEX.

Note.—In this Index the names of all artillerymen are printed in italics, and, to avoid frequent changes, ranks are omitted in the case of officers. At the end will be found a list of artillery units mentioned in the text, but, in order to save space, those units which took part in the Afghan and South African Wars, and which are shown in the Appendices to Chapters V and XXXVII, have not been repeated. And the same course has been taken with regard to the names of artillery officers holding commands and staff appointments in those wars.

K2

INDEX. 555

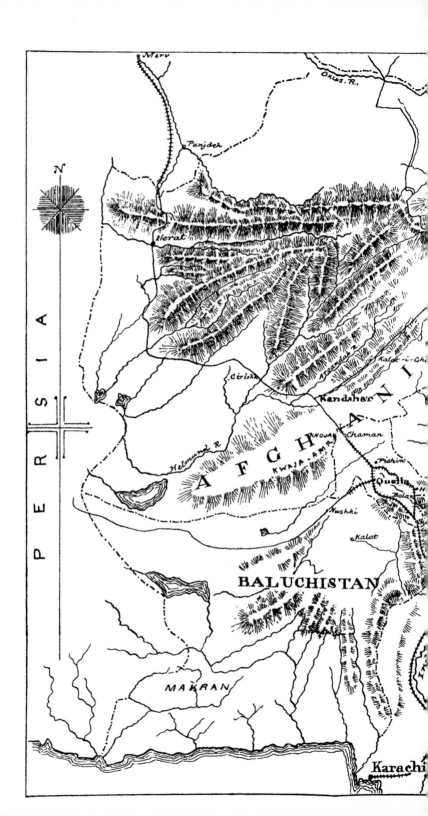

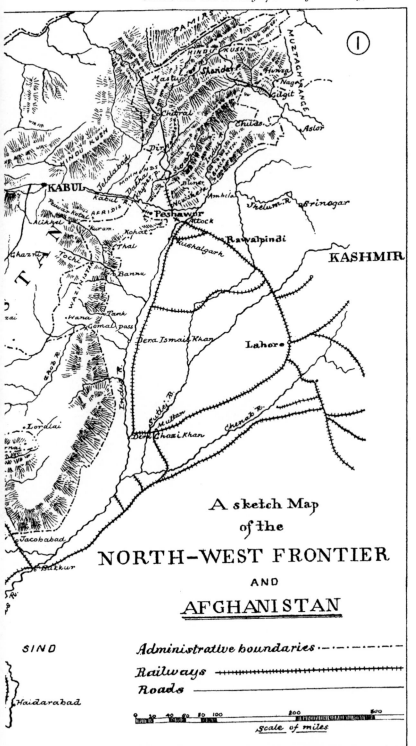

A sketch Map
of the
NORTH—WEST FRONTIER
AND
AFGHANISTAN

Administrative boundaries ·—·—·—·—·—·

Railways +++++++++++++++++++++++++++++

Roads ————————————————

scale of miles

Peiwar Kotal from British camp

Biwouac Dec. 1st 1878
9400

Line of Afghan retreat by Daftani pass

Camp Dec. 3rd 1878
8300

from Sapurgardan
and Ali Kheyl

Line of Afghan retreat

Shujurgardan

Afghan camp
8600

PEIWAR KOTAL

Spur from Sikaram

9400

one gun

Katerai Tur

Major Palmer's Route

English miles

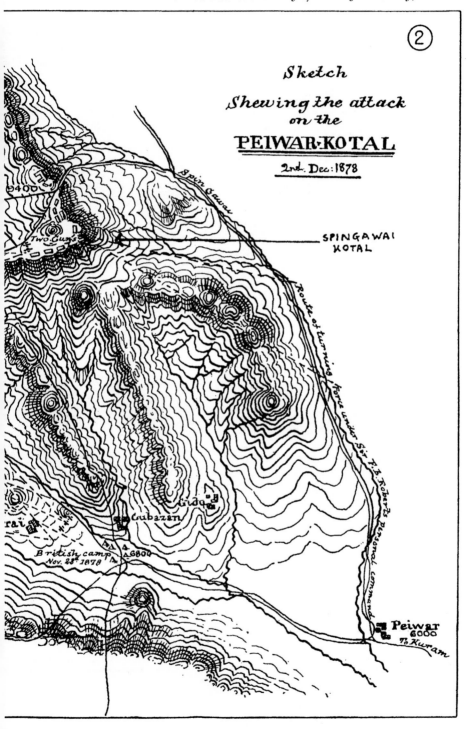

Sketch
Shewing the attack
on the
PEIWAR-KOTAL
2nd. Dec: 1878

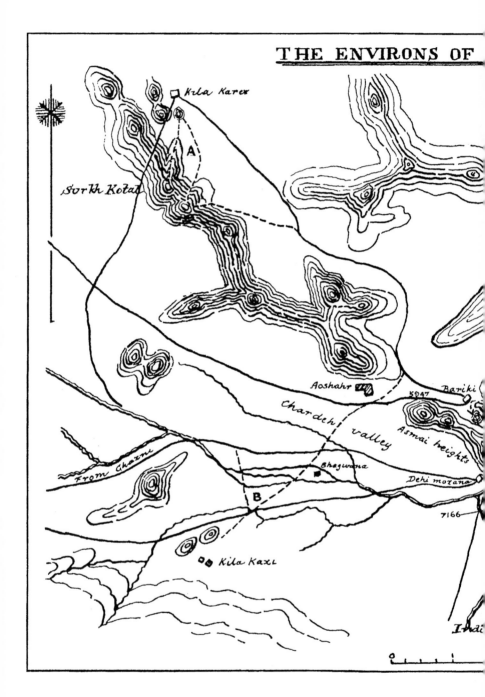

Kila Karex

A

Sorkh Kotal

Aoshahr

Bariki

5947

Chardeh valley

Asmai heights

From Ghazni

Bhagwana

Dehi mozana

7166

B

Kila Kazi

Indi

KABUL   DEC. 11 to 14, 1879.   ③

Wazirabad Lake

6007

Dimaru

Sherpur Cantonment

D

Kabul river

Logar river

KABUL

Siah Sang

Bala Hissar

C

Takti Shah

Beni Hissar

5780

E

Action of Dec 10ᵗʰ about A
   „   „   „ „ 11ᵗʰ   „   B
   „   „   „ „ 13ᵗʰ   „   C
   „   „   „ „ 15ᵗʰ   „   D

The field of CHARASIA
2 or 3 miles south of E

5 miles

AHMED KHEL.
19th April 1880.

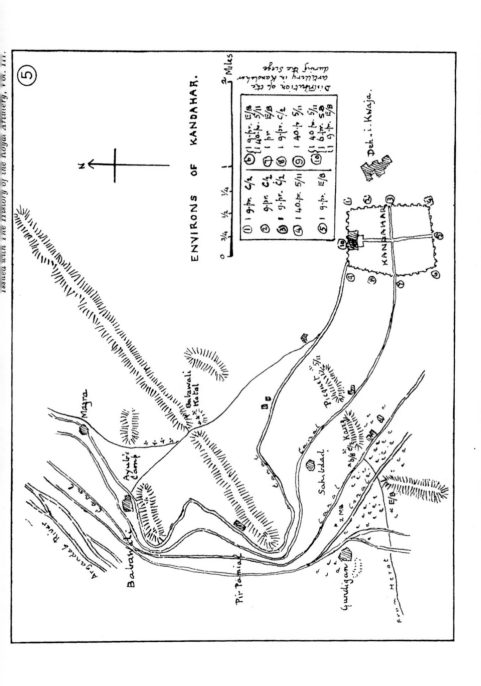

⑤

ENVIRONS OF KANDAHAR.

Distribution of the
artillery in Kandahar
during the siege.

①	9-pr.	C/2		⑥	9-pr.	E/B
②	9-pr.	C/2		⑦	40-pr.	E/B
③	9-pr.	C/2		⑧	9-pr.	C/2
④	40-pr.	5/11		⑨	40-pr.	5/11
⑤	9-pr.	E/B		⑩	40-pr.	5/11
					9-pr.	S.B
					9-pr.	E/B

N

0   ¾   ½   ¼   1   2 Miles

KANDAHAR.

Deh-i-Kwaja.

Argandab River

Magra

Ayub's Camp

Baba Wali

Baba Wali Kotal

Pir Paimal

Picquet 5/11

Sahibdad

Canal

Kotal

Karez

Gundigan

from Herat

⑥

Men stationed to roll down stones

Stone shoots

Nisa Gol

Enemy's sangar

Snow line

Snow line

Enemy's sangar

Large Sangar
Covering mom Road

Impracticable

Snow line

Enemy's Line of Sangars

Impracticable

Impracticable

Snow Slopes

7 pdr. guns

Road to Chitral

Shale Slopes

From Mastuj 4 miles

Reconnaissance sketch of
ENEMY'S POSITION
12th Apr. 1895

Showing position of British troops and
points of passage of the Gol during attack
on enemy's position on 13th April 1895
by Lieut: (now Major-General Sir) W. Bynron.

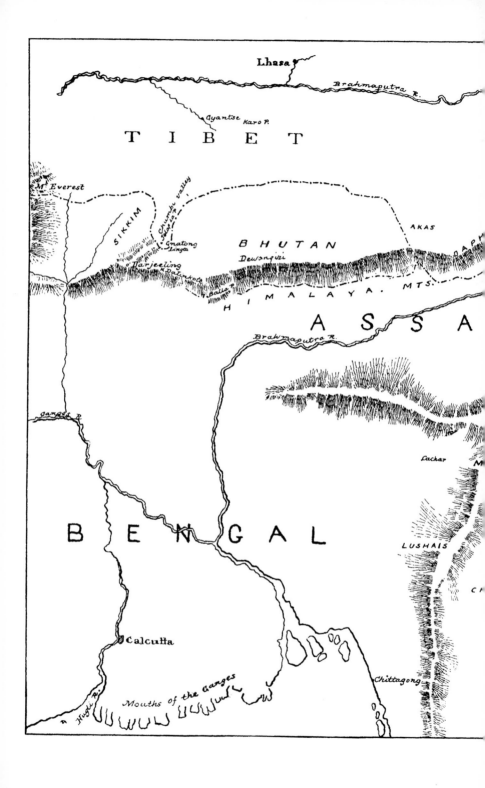

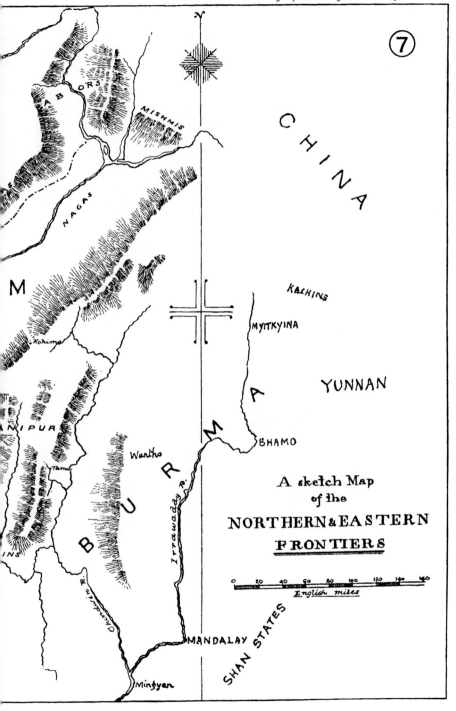

A sketch Map
of the
NORTHERN & EASTERN
FRONTIERS

English miles

Issued with The History of the Royal Artillery, Vol. III.

MAGDALA, 13th April 1868.

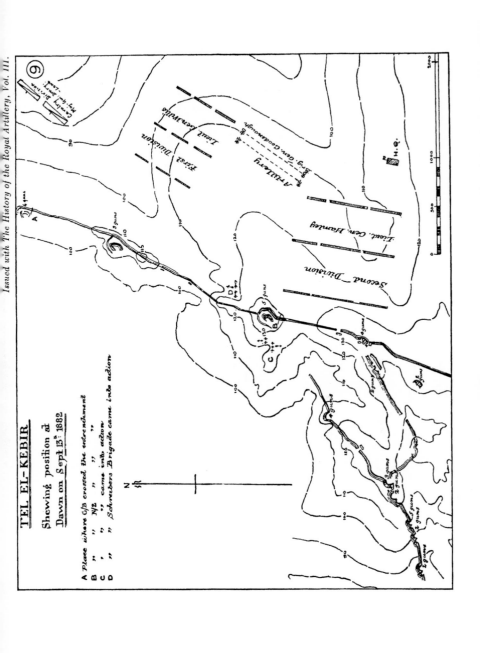

(9)

TEL EL-KEBIR

Shewing position at
Dawn on Sept 13th 1882

A Place where c/B crossed the entrenchment
B   ,,   ,,   N/2   ,,   ,,
C   ,,   ,,   ,, came into action
D   ,,   ,, Schreiber's Brigade came into action

THE NILE

KOROSKO TO KHARTOUM.

RED SEA

miles
0   50   100   200

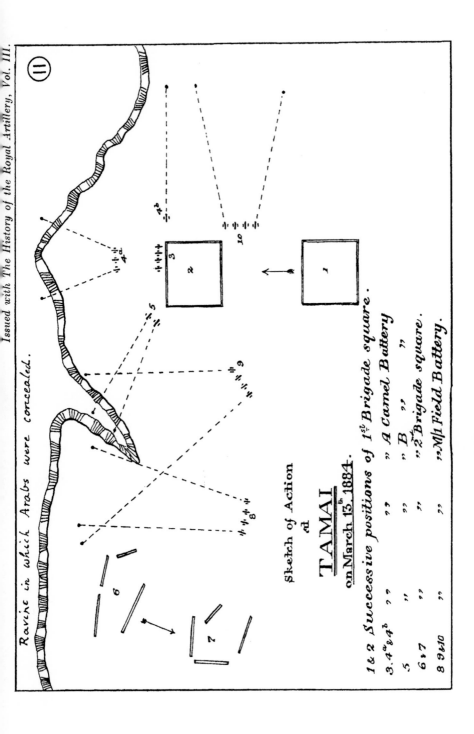

Ravine in which Arabs were concealed.

Ⅱ

Sketch of Action
of
TAMAI
on March 13th 1884.

1 & 2 Successive positions of 1st Brigade square.
3.4a & 4b ,, ,, ,, ,, A Camel Battery.
5 ,, ,, ,, ,, B ,, ,,
6 & 7 ,, ,, ,, ,, 2d Brigade square.
8 9 &10 ,, ,, ,, ,, M/1 Field Battery.

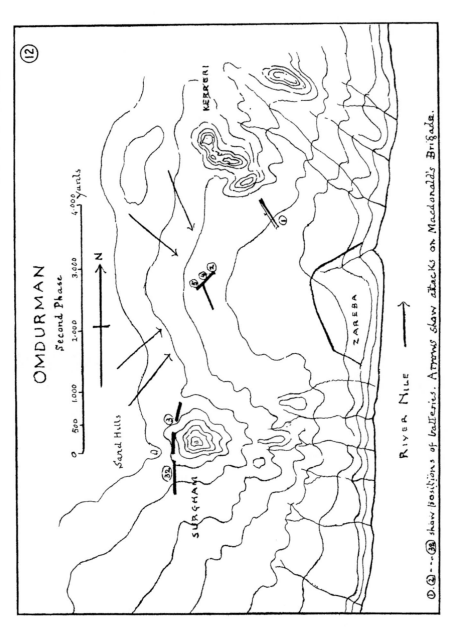

OMDURMAN

Second Phase

① ② - - - 32 show positions of batteries. Arrows show attacks on Macdonald's Brigade.

13

RHODESIA

PORTUGUESE

Limpopo or Crocodile R.

Olifants R.

SWAZI-LAND

Pietersburg

Hector Spruit
Komati Poort
Delagoa Bay
Barberton

Pieterrs River
Nylstroom
Warmbad

Pilgrim Rest
Lydenburg
Machadodorp
Nooitgedacht
Belfast
Burgher
Middelburg

Bakenlaagte
Lake Chrissie
Ermelo
Carolina

Bushy Veld

TRANSVAAL

Magaliesberg
Krugersdorp
The Rand

Rustenburg

PRETORIA
JOHANNESBURG
Elandsfontein
Heidelberg
Doorn R.
Klip R.

Greylingstad
Standerton
Vaal R.

Lichtenburg
Ottoshoop
Frederikstad
Potchefstroom
Hartebeestefontein
Klerksdorp

BECHUANALAND

To Buluwayo

Mafeking

Vryburg

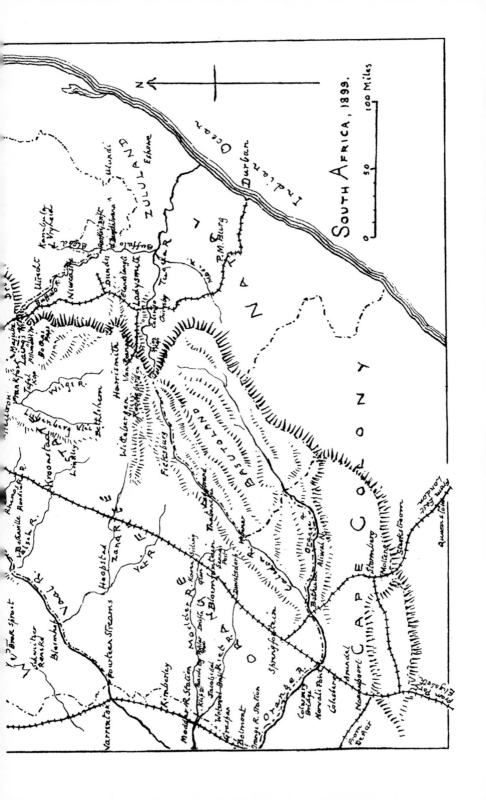

SOUTH AFRICA, 1899.

LOMBARDS KOP
30th. Oct. 1899

British ▭ ▬ ⫿⫿ ⫿⫿ Boers ▭ ⑴

2.000   1.000   0   1
Yards   Miles

Ladysmith from Gun Hill
16.5.06

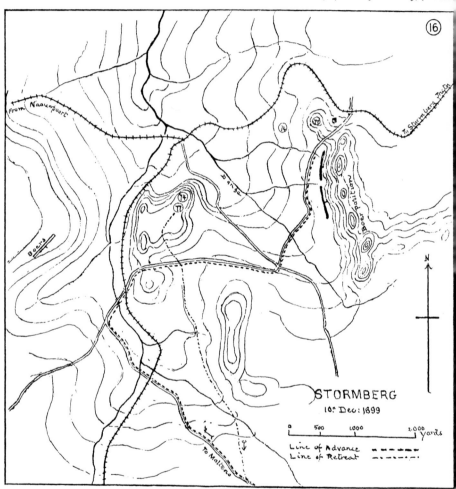

(16)

STORMBERG

10ᵗʰ Dec: 1899

| 0 | 500 | 1,000 | 2,000 yards |

Line of Advance — — — —
Line of Retreat — . — . — .

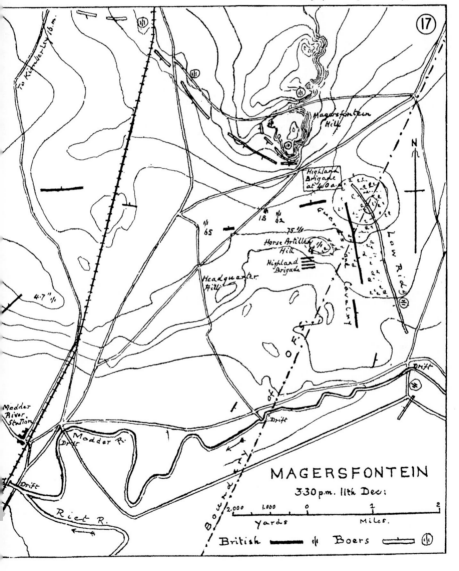

MAGERSFONTEIN

3.30 p.m. 11th Dec:

Yards   Miles.

British ━━━━━   Boers ◁━━━━▷

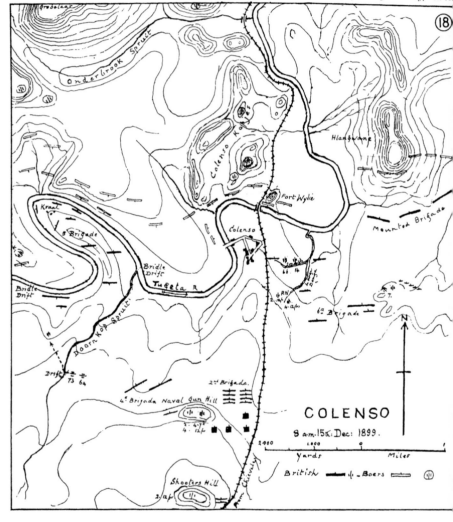

COLENSO

8 a.m. 15th. Dec: 1899.

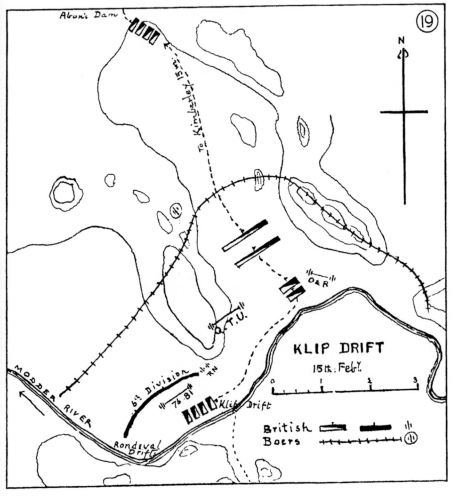

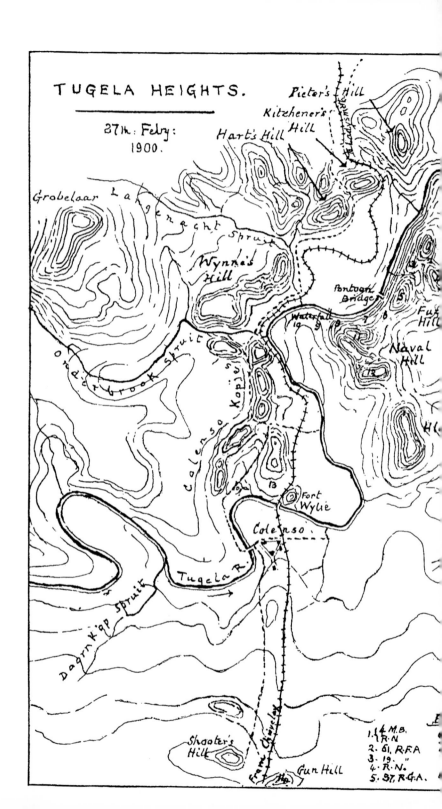

TUGELA HEIGHTS.

27th: Feby:
1900.

Pieter's Hill
Kitzhener's Hill
Hart's Hill

Grobelaar Langenecht Spruit

Wynne's Hill

Pontoon Bridge

Waterfall

Fuz Hill

Naval Hill

Onderbrook Spruit

Colenso Kopje

Fort Wylie

Colenso.

Tugela R.

Dagrn Kop Spruit

Shooter's Hill

From Chieveley

Gun Hill

1. 4 M.B.
   R.N
2. 61. R.F.A
3. 19. "
4. R. N.
5. 37, R.G.A.

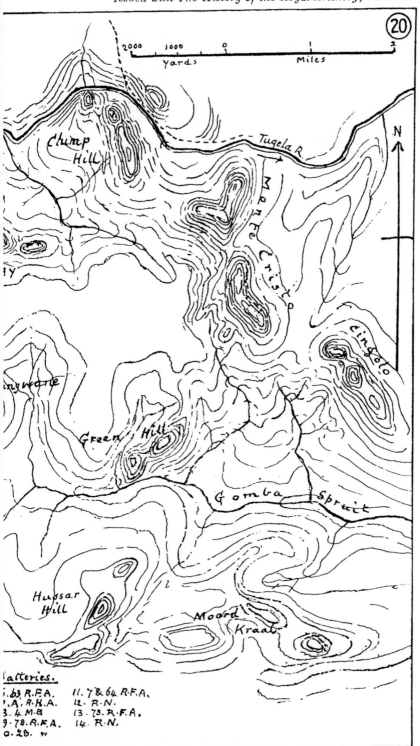

20

2000 1000 0 1 2
Yards   Miles

N

Chump Hill

Tugela R.

Monte Cristo

Cingolo

Green Hill

Gomba Spruit

Hussar Hill

Moord Kraal

Batteries.
1. 63. R.F.A.   11. 7 & 64 R.F.A.
7. A. R.H.A.   12. R.N.
3. 4. M.B   13. 73. R.F.A.
9. 78. R.F.A.   14. R.N.
0. 28. "

Grobelaar — Pangwani — Onderbrook Spruit — Wynne's Hill — Hart's Hill — Kitchener's Hill — Pieter's Hill — Site of Pontoon Bridge

Tugela R.

Langwacht Spruit

Onderbrook Spruit

Colenso Kopjes

Boer Position on 27th February 1900,

showing shelters thrown up in previous attacks,

from Hlangwane.

17.8.01.

㉑

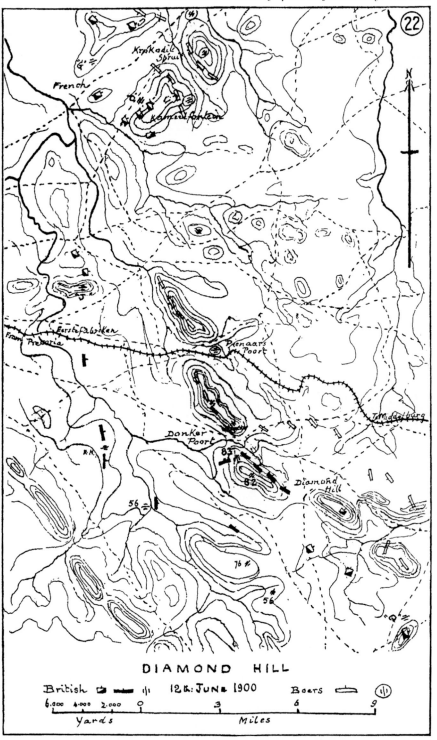

DIAMOND HILL

British 🏠 ▬▬ ⫙ 12ᵗʰ JUNE 1900 Boers ⬚ ⫙

Yards   Miles

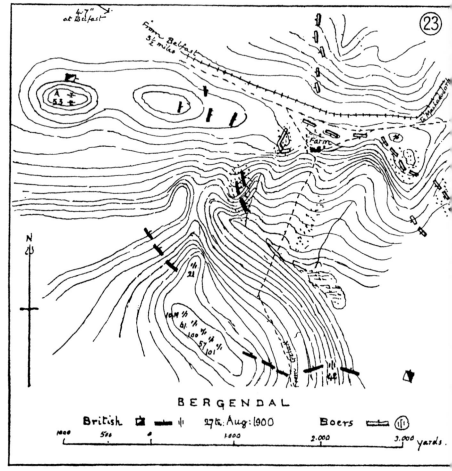

BERGENDAL

British   27th Aug. 1900   Boers